ELECTRICITY AND MAGNETISM

Gary Gladding Mats Selen Tim Stelzer
University of Illinois at Urbana-Champaign

w. h. freeman
Macmillan Learning
New York

System Development: Michel Herquet, Geoffrey Piroux, and Michael L. Scott
Book Design, Illustrations, and Animations: Michael L. Scott

Publisher: Katherine Parker
Acquisitions Editor: Alicia Brady
Marketing Manager: Maureen Rachford
Marketing Assistant: Cate McCaffery
Managing Editor: Lisa Kinne
Project Editor: Jodi Isman
Director of Design, Content Management: Diana Blume
Cover Designer: Cambraia F. Fernandes
Print and Binding: Mercury Print Productions

ISBN 13: 978-1-319-06650-5
ISBN 10: 1-319-06650-X

First Printing

Macmillan Learning
W.H. Freeman and Company
One New York Plaza
Suite 4500
New York, NY 10004-1562

www.flipitphysics.com

CONTENTS IN BRIEF

CONTENTS

9 ELECTRIC CURRENT / 99

10 KIRCHHOFF'S RULES / 111

11 *RC* CIRCUITS / 123

PART III MAGNETISM

12 MAGNETISM / 135

17 FARADAY'S LAW / 201

PART IV AC CIRCUITS

18 INDUCTION AND *RL* CIRCUITS / 215

19 *LC* AND *RLC* CIRCUITS / 229

20 AC CIRCUITS / 245

21 AC CIRCUITS: RESONANCE AND POWER / 259

PREFACE

Welcome to FlipItPhysics, a ground-breaking learning environment for the calculus-based physics course.

STUDENTS

The textbook you hold in your hands is a companion to the online materials that precede and follow each lecture. Its streamlined, readable narrative emphasizes essential concepts and problem-solving strategies.

The textbook is to be used in tandem with the FlipItPhysics system:

1) **PreLectures** are animated, narrated introductions to the core concepts. Each PreLecture is approximately 20 minutes in length and includes embedded concept questions that allow students to check their understanding along the way.

2) Completion of the PreLecture unlocks a series of multiple choice and free response questions that gauge comprehension of the material. Results provide important feedback that instructors can use to tailor lectures to meet their students' needs.

3) With more useful preparation before class, the **Lecture** is transformed, becoming a place where students and instructors can interact, build on common understanding, and begin to transform facts into true knowledge.

4) Finally, FlipItPhysics offers a series of print and online **Problems and Interactive Tutorials**, which provide problem-solving models and include sophisticated, answer-specific feedback.

INSTRUCTORS

FlipItPhysics allows instructors to take advantage of a better-prepared student audience. It provides valuable feedback and tools, which facilitate the incorporation of proven just-in-time teaching methods and peer-instruction elements. Instructors have access to a wealth of data showing students' interaction with the course material, from the number of attempts on a given problem to the amount of time spent watching a PreLecture.

A robust suite of instructor resources makes FlipItPhysics implementation straightforward and flexible:

Lecture PowerPoints will help make your transition to the FlipItPhysics system easy and efficient. Slides correspond directly with PreLecture and CheckPoint content, and include designated spots for student data, a detailed notes section, and embedded clicker questions that can be used for classroom response.

Art Images and **Tables** are available to instructors in JPEG or PPT format.

Worked Solutions for all **Standard Exercises** are available to adopters.

The functionality of FlipItPhysics is intuitive and practical. Course setup is accomplished in just a few short steps. The calendar allows for drag-and-drop adjustment of course assignments. The gradebook monitors individual and class performance, and exports scores to standard campus course management systems.

Extensive and intuitive system support is an integral part of the FlipItPhysics system. Users have a comprehensive array of support options, from informative "sidebar" guidance that appears next to major functions, to a help-menu with detailed text-based and multimedia based instructions. Technical support can be reached by phone.

1-800-936-6899

Monday-Thursday: 7:00 a.m. to 3:00 a.m. EST
Friday: 7:00 a.m. to 11:00 p.m. EST
Saturday: 11:30 a.m. to 8:00 p.m. EST
Sunday: 11:30 a.m. to 11:00 p.m. EST

Our Approach

The FlipItPhysics approach is grounded in physics education research. The system was developed for over 10 years at the University of Illinois at Urbana-Champaign, and has been tested by thousands of student users and instructors at more than 75 institutions. Published research results show that students learn more when using FlipItPhysics and have a more positive outlook on physics and the lecture:

Stelzer, T., Gladding, G., Mestre, J.P., Brookes, D.T. (2009). Comparing the efficacy of multimedia modules with traditional textbooks for learning introductory physics content. *American Journal of Physics*, 77(2), 184-190.

Stelzer, T., Brookes, D.T., Gladding, G., Mestre, J.P. (2010). Impact of multimedia learning modules on an introductory course on electricity and magnetism. *American Journal of Physics*, 77(7), 755-759.

Chen, Z., Stelzer, T., Gladding, G. (2010). Using multimedia modules to better prepare students for introductory physics lectures. *Physics Review Special Topics – Physics Education Research*, 6, 010108, 1-5.

Sadaghiani, Homeyra R. (2011). Using multimedia learning modules in a hybrid-online course in electricity and magnetism. *Physics Review Special Topics – Physics Education Research*, 7, 010102, 1-7.

ABOUT THE AUTHORS

GARY GLADDING

Professor Gary Gladding, a high energy experimentalist, joined the Department of Physics at Illinois as a research associate after receiving his Ph.D. from Harvard University in 1971. He became assistant professor in 1973 and has, since 1985, been a full professor. He has performed experiments at CERN, Fermilab, the Stanford Linear Accelerator Center, and the Cornell Electron Storage Ring. He served as Associate Head for Undergraduate Programs for 13 years. He was named a Fellow of the American Physical Society for his contributions to the improvement of large enrollment introductory physics courses.

Since 1996, Professor Gladding has led the faculty group responsible for the success of the massive curriculum revision that has transformed the introductory physics curriculum at Illinois. This effort has involved more than 50 faculty and improved physics instruction for more than 25,000 science and engineering undergraduate students. He has shifted his research focus over the last ten years to physics education research (PER) and currently leads the PER group. He is also heavily involved in preparing at-risk students for success in physics coursework through the development of Physics 100. Professor Gladding was also a key player in the creation and development of i>clicker™.

MATS SELEN

Professor Mats Selen received his bachelor's degree in physics from the University of Guelph (1982), an M.Sc. in physics from Guelph (1983), and an M.A. in physics from Princeton University (1985). He received his Ph.D. in physics from Princeton (1989). He was a research associate at the Laboratory for Elementary Particle Physics (LEPP) at Cornell University from 1989 to 1993. He joined the Department of Physics at Illinois in 1993 as an assistant professor and, since 2001, has been a full professor. He was named a Fellow of the American Physical Society in 2006 for his contributions to particle physics.

Since arriving at Illinois, he has been a prime mover behind the massive curriculum revision of the calculus-based introductory physics courses (Physics 211-214), and he was the first lecturer in the new sequence. He created an undergraduate "discovery" course where freshmen create their own physics demonstrations, and developed the Physics Van Outreach program, in which physicists visit elementary schools to share enthusiasm for science. Professor Selen played a key role in the development of i>clicker™.

TIMOTHY J. STELZER

Professor Timothy Stelzer received his bachelor's degree in physics from St. John's University (1988) and his Ph.D. in physics from the University of Wisconsin-Madison (1993). After working as a senior research assistant in the Center for Particle Theory at Durham University (UK), he joined the Department of Physics at the University of Illinois as a postdoctoral research associate in 1995. Active in theoretical high-energy physics, he is currently a research associate professor at the university.

Professor Stelzer has been heavily involved with the Physics Education Group at Illinois, where he has led the development and implementation of tools for assessing the effectiveness of educational innovations in the introductory courses and expanding the use of Web technology in physics pedagogy. He was instrumental in the development of the i>clicker™ and is a regular on the University's "Incomplete List of Teachers Ranked as Excellent by Their Students." He was named University of Illinois Distinguished Teacher-Scholar in 2009.

ABOUT THE DEVELOPERS

MICHEL HERQUET

Michel received a M.Sc. in Theoretical Physics from the University of Mons in 2003 and a Ph.D. in Particle Physics from the University of Louvain in 2008. He also worked as a postdoctoral researcher at the Nikhef Institute in Amsterdam between 2008 and 2010. His research work was focused on phenomenology at CERN experiments and Monte Carlo simulations, in particular in the context of the hunt for the Higgs boson. He is one of the authors of MadGraph 4 and 5, two highly successful simulation packages totalling over 2000 citations. In 2010, he joined McKinsey & Company, a global management consulting firm, as a junior associate consultant and worked on projects in strategy and marketing/sales for clients in various industries. One year later, he helped in founding Novapta Consulting. He is also invited lecturer at the Louvain University, teaching Physics both at elementary and advanced levels.

GEOFFROY PIROUX

Geoffroy obtained a M.Sc. in Physics from the University of Namur in 1999 and a Ph.D. in Mathematical Physics from the University of Louvain in 2006. In parallel to his research, he studied Philosophy and obtained a B.A. degree in 2003.

His main research work focused on conformal field theories and their links with two dimensional critical dynamic models such as the sandpile model. In 2006, he joined Sopra Banking Software and worked in the product department as solution expert in the loans, treasury and risk management domains. He was also the manager of the compliance team that ensures the compatibility of the solution regarding regulatory, legal and third party products evolutions. Geoffroy is still involved in the academic field as invited lecturer at the University of Louvain where he teaches Physics for M.S. Business Engineering students.

MICHAEL L. SCOTT

Michael Scott received his bachelor's degree in physics and mathematics from the University of Indianapolis (2000) and both an M.S. in physics (2002) and a Ph.D. in physics (2008) from the University of Illinois at Urbana-Champaign. Dr. Scott's work at Illinois was in the field of physics education research (PER), with a focus on the effectiveness of multiple-choice exams in large introductory physics courses and on how explicit reflection can enhance physics learning. Michael received several teaching awards as a teaching assistant at Illinois, including the Scott Anderson Award (2002) and the AAPT Outstanding Teaching Assistant Award (2003) given by the Deparment of Physics. He also appeared numerous times on the University's "Incomplete List of Teachers Ranked as Excellent by Their Students."

As a member of the smartPhysics team, Michael leads the development of the creative media, which includes animating the PreLectures and Problem-Solving Tutorials in Adobe Flash®. His responsibilities also include website design, textbook design, and illustrations.

PART I: ELECTRICTY

UNIT

1

COULOMB'S LAW

1.1 Overview

Electricity and magnetism are the central topics to be developed in this course. The one new big idea we will develop is a modified description of the world to include the presence of *fields*. In particular, to account for new kinds of phenomena, we will need to introduce the concepts of electric and magnetic fields.

In this first unit, we will introduce the concept of electric charge, the ultimate source of these electric and magnetic fields. We'll discuss some basic features of electric charge and then introduce Coulomb's law, the quantitative description of the forces between charges. We'll then discuss the principle of superposition, which allows us to determine the forces when more than two charges are present. Finally, we will apply this principle of superposition to a couple of examples to see how it works.

1.2 Electric Charge

We've just said that we will need to change our basic framework to include electric and magnetic fields as fundamental entities, and these fields are determined by the distribution and motion of electric charge. The first question then is what is this electric charge anyway?

This question is not an easy one to answer. During the mechanics course, we talked about a box sliding down an inclined plane, but we never asked "what is a box?" What a box is seems obvious, and the reason a box slides down the plane is that the box has mass. What is this mass anyway? This question didn't seem too important during the mechanics course because you have an intuitive sense of mass. However, your intuitive sense of mass probably comes from your understanding of weight, which we understand in terms of the gravitational force. In other words, we know about mass because it is the thing that is responsible for gravitational forces. Well, **electric charge** is the thing that is responsible for electric forces. That's really about as good a description as we can make.

Accepting that electric charge is the thing that is responsible for electric forces, we can learn more about charge by investigating its properties. The first important property of electric charge is that, unlike mass, there are two kinds of electric charge: positive and negative. Because there is only one kind of mass, the gravitational force exerted by one object on another object is always attractive. However, the existence of both positive and negative charges results in both attractive and repulsive electrical forces being exerted between charges. In particular, we say unlike charges attract and like charges repel.

Figure 1.1 shows the magnitudes and directions of these forces. It is important to note that the forces on each pair of charges are in opposite directions and have equal magnitude, in accordance with Newton's third law.

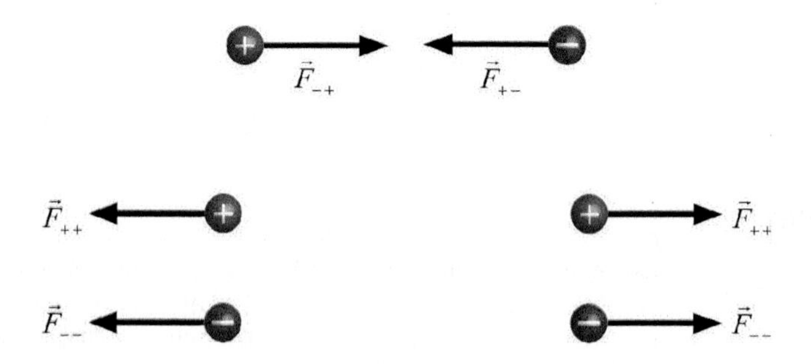

FIGURE 1.1 Directions and magnitudes of forces exerted by pairs of charges on each other. (The charges have been colored for clarification.)

1.3 Distributions of Electric Charge in Matter

This electric force is ultimately responsible for the stability of atoms. Atoms have positive charges (**protons**)and neutral charges (**neutrons**) in the nucleus and negative charges (**electrons**) far outside. The attractive forces between protons and electrons hold the atoms together. You might wonder how a nucleus stays together with all those positive charges so close together. The answer is that an attractive force exists (called the strong force) between these protons and neutrons at small distances that is much stronger than the electric force. We will not discuss this strong force, though, during this course. The magnitude of the charge of the proton is exactly equal to the magnitude of the charge on the electron. Since there are equal numbers of electrons and protons in an atom, atoms are electrically neutral.

In dealing with macroscopic objects that are charged, we need to discuss how charge is distributed in these objects. In fact, how the charge is distributed depends on the internal structure of the particular material. During this course, we will adopt a very simple model in which there are only two kinds of materials, **conductors** and **insulators**. We will assume that the charge carriers (the electrons) are totally free to move in conductors, such as metals. We will discuss what actually determines the motions of these electrons later. We will also assume that charges are fixed and cannot move in insulators, such as plastics.

1.4 Coulomb's Law

We now discuss the quantitative statement of the force law between two charges. The important features of this law were discovered by Charles Augustin de Coulomb in 1785, using the torsion balance to precisely measure the forces.

Consider two charges, q_1 and q_2, which are separated by a distance $\vec{r}_{12}$, as shown in Figure 1.2.

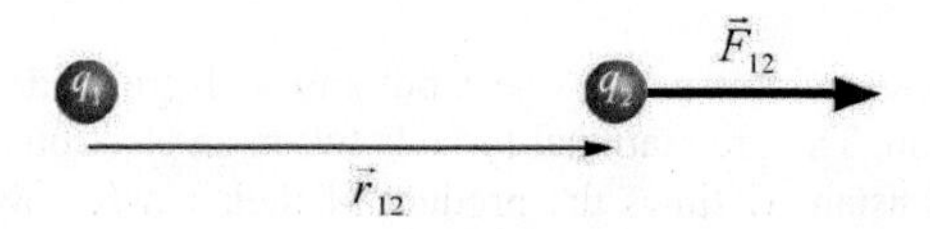

FIGURE 1.2 $\vec{F}_{12}$ is the force that charge q_1 exerts on charge q_2.

The force that q_1 exerts on q_2 is directed along the line connecting the charges and is directly proportional to the product of the charges and inversely proportional to the square of the distance between them. In particular, we write **Coulomb's law** as

$$\vec{F}_{12} = k\frac{q_1 q_2}{r_{12}^2}\hat{r}_{12}$$

where $\hat{r}_{12}$ is a unit vector in the direction of $\vec{r}_{12}$, and k is a constant. Note that we have adopted a subscript convention for forces in which the first index refers to the charge that exerts the force and the second index refers to the charge that experiences the force, i.e.,

$\vec{F}_{12}$ is the force "by" q_1 "on" q_2. We will adopt the use of SI units during this course in which charge is measured in coulombs (C), distance in meters (m), and force in newtons (N). The charge on an electron, usually denoted as *e*, is equal to 1.6×10^{-19} C, and the constant *k* has the value of $8.99\times10^9\ \text{N}\cdot\text{m}^2/\text{C}^2$.

Note that we've drawn $\vec{F}_{12}$ as a repulsive force. Therefore, we know that the charges must have the same sign. If we treat q_1 and q_2 as signed numbers, then the vector form of Coulomb's law (as shown above) determines the correct direction automatically. In other words, if q_1 and q_2 have the same sign, the product $q_1 q_2$ is positive, and the force is repulsive (i.e., along $\vec{r}_{12}$). If q_1 and q_2 have opposite signs, then the product $q_1 q_2$ is negative and the force is attractive (i.e., along $-\vec{r}_{12}$).

We can obtain an equation for $\vec{F}_{21}$, the force q_2 exerts on q_1, by simply interchanging the subscripts 1 and 2 in the previous equation.

$$\vec{F}_{21} = k\frac{q_2 q_1}{r_{21}^2}\hat{r}_{21}$$

This situation is shown in Figure 1.3. Note that these equations directly demonstrate Newton's third law. Since $\vec{r}_{12}$ is equal in magnitude but opposite in direction to $\vec{r}_{21}$, the force by charge q_1 on charge q_2 is equal and opposite to the force by charge q_2 on charge q_1.

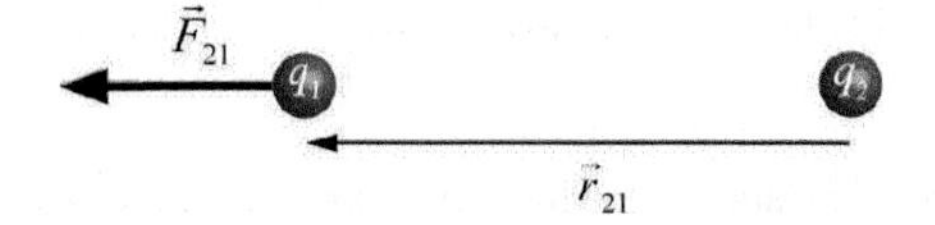

FIGURE 1.3 $\vec{F}_{21}$ is the force that charge q_2 exerts on charge q_1.

For example, we can calculate both the Coulomb and gravitational forces between an electron and a proton. The gravitational force between an electron and a proton is equal to the gravitational constant *G* times the product of their masses divided by the square of their distance of separation ($F_G = Gm_e m_p / r^2$). The Coulomb force between an electron and a proton is equal to the electrical constant *k* times the product of their charges divided by the square of their distance of separation ($F_C = ke^2 / r^2$).

Consequently, the ratio of the Coulomb force to the gravitational force does not depend on the distance of separation and is just equal to the ratio of the constants (k / G) times the ratio of the product of the charges to the product of the masses.

$$\frac{F_C}{F_G} = \frac{k}{G}\frac{e^2}{m_e m_p}$$

Putting in the numbers, we see that the Coulomb force between an electron and a proton is much stronger (almost 40 orders of magnitude stronger!) than the gravitational force between them. This calculation should give you confidence that it is the Coulomb force, not the gravitational force, that holds atoms together.

1.5 Superposition

We now know how to determine the force that is exerted on charge q_1 by another charge q_2. For example, if a negative charge is placed directly to the right of a positive charge, the force on the positive charge will be to the right (toward the negative charge). If a negative charge is placed directly below a positive charge, the force on the positive charge will be down.

What is the force on a positive charge if there is a negative charge directly to the right *and* another negative charge directly below it?

The answer is that the net force on the positive charge is the *vector sum* of the forces exerted on it due to the other two charges. This situation is shown in Figure 1.4. The net force on charge q_1 due to charges q_2 and q_3 is just equal to the vector sum of the forces that charges q_2 and q_3 would exert on charge q_1, individually (i.e., $\vec{F}_{21} + \vec{F}_{31}$).

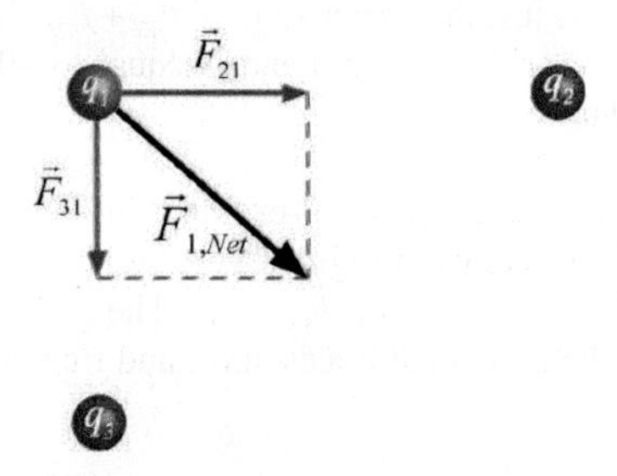

FIGURE 1.4 $\vec{F}_{1,Net}$ is the total force on q_1 due to q_2 and q_3. It is obtained by taking the vector sum of the individual forces $\vec{F}_{21}$ and $\vec{F}_{31}$.

It's as simple as that. The total electric force on a charge is just the vector sum of the electric forces exerted on it by all other charges. We'll next work through a couple of examples that illustrate the use of the principle of superposition with Coulomb's law.

1.6 Examples

We'll start with a one-dimensional problem—a negative charge q_2 is situated halfway between two positive charges q_1 and q_3, as shown in Figure 1.5. What is the force on q_3?

To calculate this net force, we simply calculate the forces on q_3 due to q_1 and q_2 separately and then add these forces together. In each case, the force exerted by one charge on another charge can be calculated from Coulomb's law.

$\vec{F}_{13}$ is the force between the two positive charges, and is therefore repulsive and points to the right as shown. Using the values shown for the distances and the charges, we obtain the magnitude of F_{13} to be equal to +0.018 N.

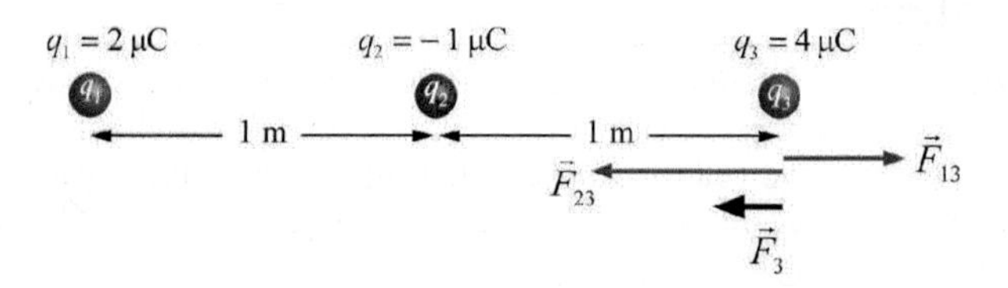

FIGURE 1.5 $\vec{F}_3$ is the total force exerted on charge q_3; it is equal to the vector sum of $\vec{F}_{13}$, the force exerted by charge q_1 on charge q_3, and $\vec{F}_{23}$, the force exerted by charge q_2 on charge q_3.

$\vec{F}_{23}$ is the force exerted by a negative charge on a positive charge, and is therefore attractive and points to the left as shown. Using the values shown for the distances and the charges, we obtain the magnitude of F_{23} to be equal to –0.036 N.

The total force $\vec{F}_3$ on q_3 is just the vector sum of $\vec{F}_{13} + \vec{F}_{23}$. In this one-dimensional case, the vector sum is also the arithmetic sum and is equal to –0.018 N. The resultant force points to the left as shown.

We'll now transform this one-dimensional example into a two-dimensional example by rotating q_2 through 90°, as shown in Figure 1.6. Now, we need to calculate the net force $\vec{F}_3$ on q_3 by making the vector sum of $\vec{F}_{13} + \vec{F}_{23}$. The general procedure for making the vector sum is to resolve the individual forces into x and y components and then simply add the components.

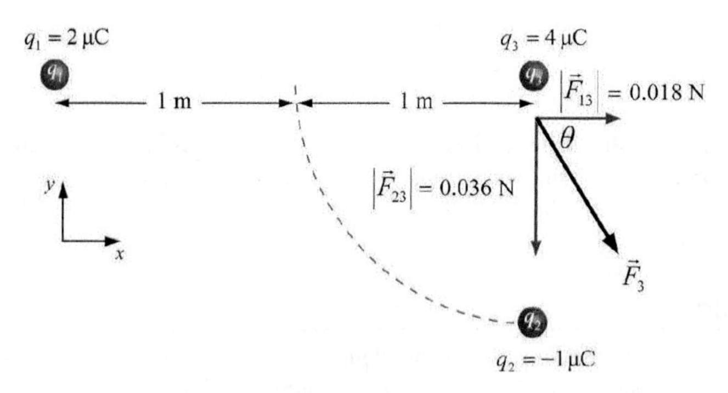

FIGURE 1.6 $\vec{F}_3$ is the total force exerted on charge q_3; it is equal to the vector sum of $\vec{F}_{13}$, the force exerted by charge q_1 on charge q_3, and $\vec{F}_{23}$, the force exerted by charge q_2 on charge q_3.

In this case, resolving the forces is trivial because $\vec{F}_{13}$ is in the $+x$ direction and $\vec{F}_{23}$ is in the $-y$ direction. Consequently, the x-component of $\vec{F}_3$ is just F_{13} (+0.018 N) and the y-component of $\vec{F}_3$ is just the negative of F_{23} (–0.036 N).

We can find the magnitude of $\vec{F}_3$ by just taking the square root of the sum of the squares of the x and y components. We obtain $F_3 = 0.040\,\text{N}$ from this procedure. To find the direction of $\vec{F}_3$, we can use trigonometry, namely $\tan\theta = F_y / F_x = -2$, which yields $\theta = -63°$.

1.7 Summary

We began by introducing a central new idea that will be developed during this course, namely the need to introduce a new kind of entity into the world, electric and magnetic fields, in order to describe a host of new phenomena. These fields are determined by the distributions and motions of another new kind of thing: electric charge. There are two types of electric charge, positive and negative, which result in both attractive and repulsive electric forces.

The equation describing the force that is exerted between two charges, q_1 and q_2, is called Coulomb's law.

$$\vec{F} = k\frac{q_1 q_2}{r^2}\hat{r}$$

The direction of the force is always along the line connecting the charges, and the magnitude is directly proportional to the product of the charges and inversely proportional to the square of the distance between them.

Finally, we determined the total force on a given charge due to an arbitrary collection of other charges by applying a principle of superposition, namely that this total force is just the vector sum of all the forces exerted on it by the other individual charges.

MAIN POINTS

Coulomb's Law

The direction of the force between two charges is always along the line connecting the charges, and the magnitude is directly proportional to the product of the charges and inversely proportional to the square of the distance between them.

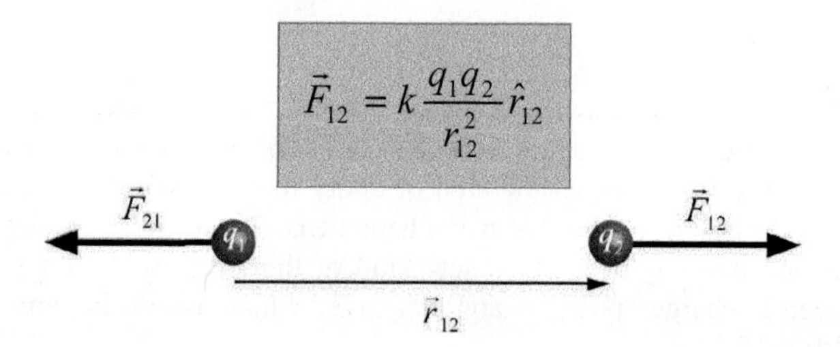

Superposition Principle

The total force on a given charge due to an arbitrary collection of other charges is equal to the vector sum of all the forces exerted on it by the other individual charges.

PROBLEMS

1. Point Charges in 1-D: A point charge $q_1 = -2.3\ \mu\text{C}$ is located at the origin of a coordinate system. Another point charge $q_2 = 5.5\ \mu\text{C}$ is located along the x-axis at a distance $x_2 = 9.5\ \text{cm}$ from q_1. (a) What is $F_{12,x}$, the value of the x-component of the force that q_1 exerts on q_2? (b) Charge q_2 is now displaced a distance $y_2 = 2.2$ cm in the positive y-direction. What is the new value for the x-component of the force that q_1 exerts on q_2? (c) A third point charge q_3 is now positioned halfway between q_1 and q_2. The net force on q_2 now has a magnitude of $F_{2,Net} = 4\ \text{N}$ and points away from q_1 and q_3. What is the value (sign and magnitude) of the charge q_3? (d) How would you change q_1 (keeping q_2 and q_3 fixed) in order to make the net force on q_2 equal to zero?

(i) Increase its magnitude and change its sign.
(ii) Decrease its magnitude and change its sign.
(iii) Increase its magnitude and keep its sign the same.
(iv) Decrease its magnitude and keep its sign the same.
(v) There is no change you can make to q_1 that will result in the net force on q_2 being equal to zero.

(e) How would you change q_3 (keeping q_1 and q_2 fixed) in order to make the net force on q_2 equal to zero?

(i) Increase its magnitude and change its sign.
(ii) Decrease its magnitude and change its sign.
(iii) Increase its magnitude and keep its sign the same.
(iv) Decrease its magnitude and keep its sign the same.
(v) There is no change you can make to q_3 that will result in the net force on q_2 being equal to zero.

2. Point Charges in 2-D: Three charges ($q_1 = 7.1\ \mu\text{C}$, $q_2 = -3.3\ \mu\text{C}$, and $q_3 = 1.2\ \mu\text{C}$) are located at the vertices of an equilateral triangle with side $d = 7.1$ cm as shown. (a) What is $F_{3,x}$, the value of the x-component of the net force on q_3? (b) What is $F_{3,y}$, the value of the y-component of the net force on q_3? (c) A charge $q_4 = 1.2\ \mu\text{C}$ is now added as shown. What is $F_{2,x}$, the x-component of the new net force on q_2? (d) What is $F_{2,y}$, the y-component of the new net force on q_2? (e) What is $F_{1,x}$, the x-component of the new net force on q_1? (f) How would you change q_1 (keeping q_2, q_3 and q_4 fixed) in order to make the net force on q_2 equal to zero?

(i) Increase its magnitude and change its sign.
(ii) Decrease its magnitude and change its sign.
(iii) Increase its magnitude and keep its sign the same.
(iv) Decrease its magnitude and keep its sign the same.
(v) There is no change you can make to q_1 that will result in the net force on q_2 being equal to zero.

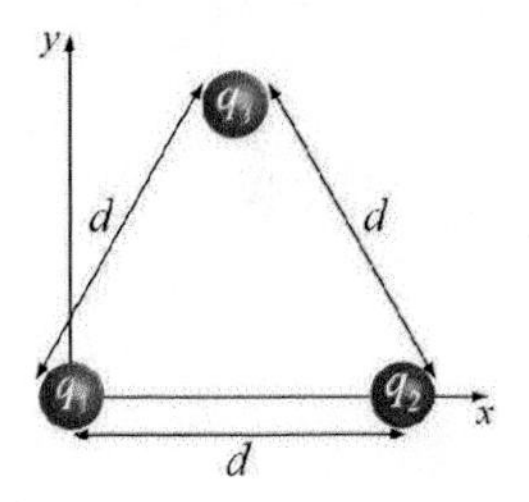

FIGURE 1.7 Problem 2 parts (a)-(b)

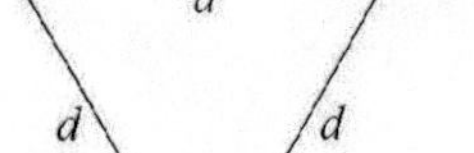

FIGURE 1.8 Problem 2 parts (c)-(f)

3. Three Charges (INTERACTIVE EXAMPLE): Three charges are fixed in place as shown. The squares in the grid have sides of length $s = 0.24$ m. The magnitude of q is 3 μC ($1\ \mu C = 10^{-6}$ C), while the magnitude of Q is 4.5 μC. What is the magnitude of the net force on q due to the other two charges?

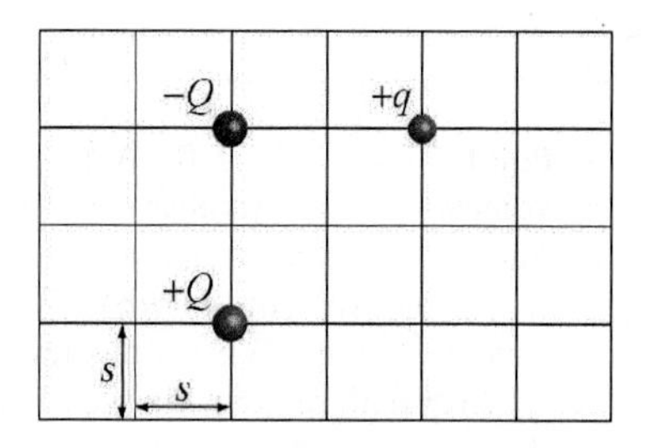

FIGURE 1.9 Problem 3

UNIT 2

ELECTRIC FIELDS

2.1 Overview

In this unit, we will introduce an important new concept, the electric field. Coulomb's law and the principle of superposition ensure that the net electric force on a charge will be proportional to the magnitude of that charge. Therefore, we will define the net electric field at a given point in space as the force per unit charge at that point.

We will explore some consequences of this definition by examining several examples. First, we'll determine the electric fields produced by discrete point charges—initially a single charge and then an electric dipole. We'll then discuss the more general cases of determining electric fields produced by continuous charge distributions. The specific example we will discuss is an infinite line of charge. Finally, we'll compare the electric field generated by a point charge (spherical symmetry) with the electric field generated by an infinite line of charge (cylindrical symmetry).

2.2 Electric Fields

During the last unit, we learned that we can use Coulomb's law to calculate the force on one charge due to another single charge. Furthermore, we learned that Coulomb's law obeys a superposition principle that allows us to calculate the total force on one charge due to a collection of charges by taking the vector sum of the forces on that charge due to each of the other charges individually.

In the example shown in Figure 2.1, the positive charge q_1 exerts force $\vec{F}_{12}$ on the positive charge q_2. We know that we can calculate this force from Coulomb's law. The magnitude of the force is proportional to the product of the charges ($q_1 q_2$) and is inversely proportional to the square of the distance between them (r^2). Since these are both positive charges, the force is repulsive and $\vec{F}_{12}$ points away from q_1, as shown.

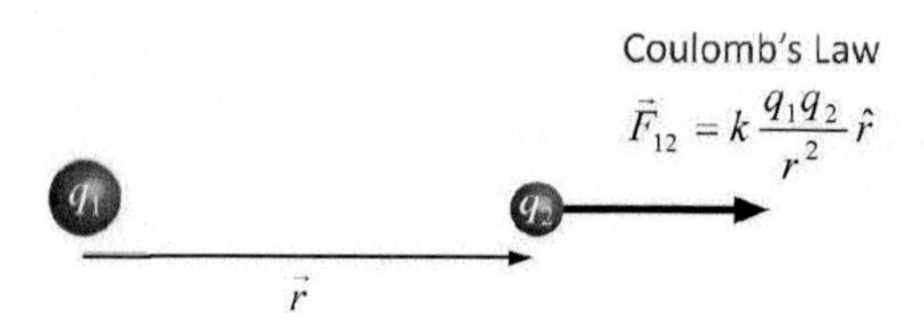

FIGURE 2.1 The force charge q_1 exerts on charge q_2, $\vec{F}_{12}$, is given by Coulomb's law and is proportional to the product of the charges and inversely proportional to the distance between them.

Suppose we now double the magnitude of the charge q_2, while keeping everything else the same. How will the total force on q_2 change? Since F_{12} is proportional to q_2, the magnitude of F_{12} will simply double and its direction will remain the same.

Something is staying constant here and that something is the *force on* q_2 *divided by the magnitude of* q_2. This constant, the force per unit charge, is associated with the *position* of q_2. Specifically, the force per unit charge at the position of q_2 is totally determined by the magnitude and location of q_1. We will call this quantity $\vec{E}_1$, the **electric field** produced by the charge q_1.

$$\vec{E}_1 \equiv \frac{\vec{F}_{12}}{q_2}$$

It is important to note, however, that the electric field is a much more significant quantity than you might guess from this simple definition. Electric fields are in fact quite real and not just mathematical constructs. Energy is stored in electric fields and information about changes in the charge distribution is transmitted throughout space by these fields at a finite speed—the speed of light.

We will now consider some examples that illustrate the determination of the electric fields that are produced by both discrete and continuous static charge distributions.

2.3 Electric Field from a Point Charge

We'll start with the simplest example: the electric field produced by a single point charge Q. The definition of the electric field suggests we can calculate the electric field produced by this charge at some point P by determining the force this charge would exert on a unit positive charge located at point P, as shown in Figure 2.2.

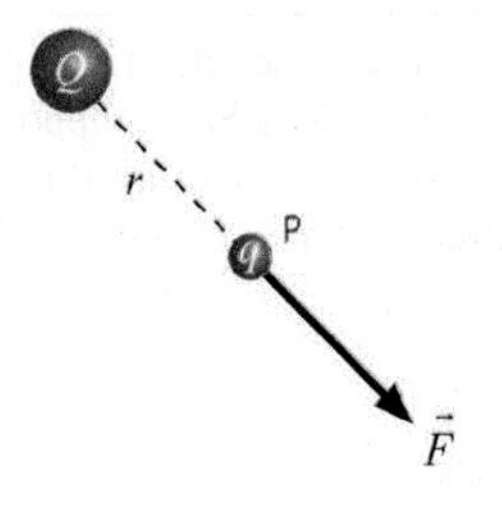

FIGURE 2.2 The electric field produced by the point charge Q at point P can be determined by calculating the force $\vec{F}$ exerted by Q on a unit positive charge q located at point P.

The first thing to notice is that the direction of this force will always be radial and pointing outward, since like charges repel. Consequently, the electric field at any point will also always point radially outward from Q. The magnitude of the field can be calculated directly from Coulomb's law.

$$\vec{E} = \frac{\vec{F}}{q} = k\frac{Q}{r^2}\hat{r}$$

The magnitude of the electric field due to charge Q is directly proportional to Q, and inversely proportional to the square of the distance from Q.

Figure 2.3 shows a plot of E, the magnitude of the electric field produced by Q as a function of r, the distance from Q. That is, for every value of r on the horizontal axis, the corresponding value of E is plotted vertically above it. We see that as r increases, E decreases as $1/r^2$. For example, if we increase r by a factor of two, E decreases by a factor of four, since the electric field of a point charge is inversely proportional to the square of the distance from the charge.

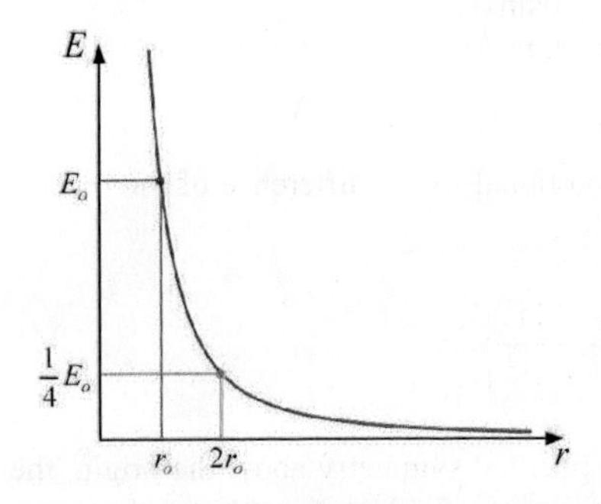

FIGURE 2.3 A plot of the magnitude of the electric field, E, produced by a single point charge as a function of r, the distance from the charge, illustrating that E decreases as $1/r^2$.

It is interesting to note that this solution is consistent with the spherical symmetry of the problem. In particular, a single point charge must produce a spherically symmetric field.

That is, this charge distribution (a single point charge) looks identical from all points that are the same distance from the charge. The magnitude of the field can then only depend on r, the distance to the charge. We'll now take a brief look at another example in which the charge distribution does not display this spherical symmetry.

2.4 Electric Field from an Electric Dipole

We'll now discuss another example using discrete charges, the electric dipole. The **electric dipole** consists of two charges that have equal magnitudes (Q) and opposite signs. We will encounter electric dipoles several times during this course because dipoles are of great practical interest, ranging from the study of molecular systems to antennas.

To calculate the electric field at point P produced by an electric dipole, we just need to calculate the fields produced by each charge separately and then take the vector sum. We will delay doing this general calculation until we have some more powerful tools—specifically, an understanding of the electric potential. For now, we restrict ourselves to calculating the electric field along the axis connecting the two charges. In particular, we will determine the electric field at points A and B that are equidistant from the origin, as shown in Figure 2.4.

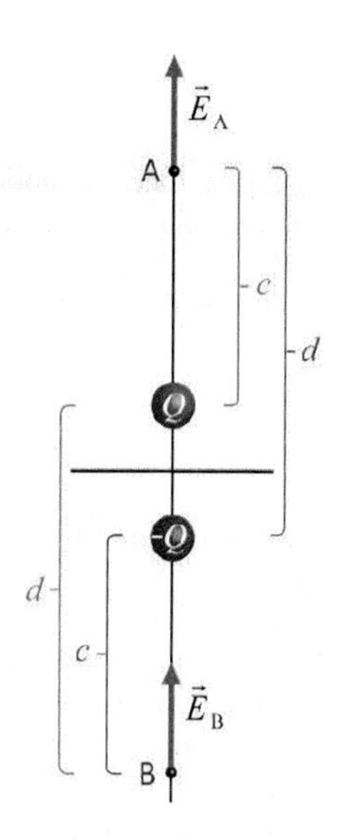

FIGURE 2.4 An electric dipole.

At point A, the electric field from the positive charge points up and the field from the negative charge points down. The magnitude of the field from the positive charge is larger than the magnitude of the field from the negative charge since point A is closer to the positive charge than it is to the negative charge. Therefore, the electric field $\vec{E}_A$ at point A points away from the dipole.

At point B, the situation is reversed; the magnitude of the field from the negative charge is now larger than the magnitude of the field from the positive charge since point B is closer to the negative charge than it is to the positive charge. Consequently, the electric field $\vec{E}_B$ at point B points *toward* the dipole.

The magnitude of the field at either point is proportional to the difference of the inverse squares of the distances to each charge. In particular,

$$E = kQ\left\{\frac{1}{c^2} - \frac{1}{d^2}\right\}$$

Note that the electric dipole field does not have spherical symmetry about the origin, the midpoint of the dipole. Even though the magnitudes of the fields at A and B are the same, the directions of the fields are different; the field at A points *away* from the origin while the field at B points *toward* the origin. This difference is simply due to the fact that A is closer to the positive charge while B is closer to the negative charge.

2.5 Electric Field from an Infinite Line of Charge

Up until now, we have only considered charge distributions that involve point charges. We'll now consider an example in which the charge distribution is continuous. In particular, we will calculate the electric field produced by an infinite line of charge.

The big idea here is exactly the same as for discrete charge distributions—to calculate the net electric field, we calculate the fields produced by individual charges and then take the vector sum of these fields. The only thing that changes is that the sum over a finite number of charges becomes an integral over the continuous charge distribution. We'll now calculate the electric field at the point P on the y-axis, directly above the line of charge, as shown in Figure 2.5.

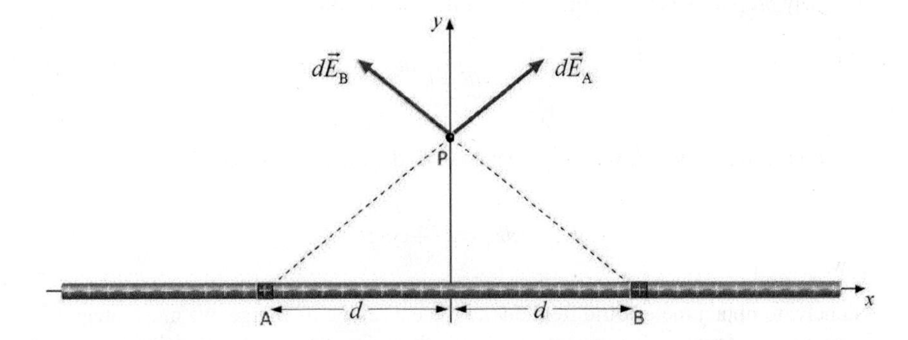

FIGURE 2.5 The electric field $\vec{E}$ produced at point P due to two segments, A and B, of an infinite line of charge.

First, consider the electric field due to a small segment, A, of the line of charge to the left of P. The magnitude of this field is given by Coulomb's law, and the direction of the field points directly away from the segment. Next, consider the electric field due to a small segment, B, of the line of charge to the right of P. The magnitude of the electric field from segment B is the same as the magnitude of the electric field from segment A since A and B are equidistant from point P. However, the electric field from B has a negative x-component that exactly cancels the positive x-component of the electric field from A.

As we integrate from $-\infty$ to $+\infty$, we see that for every segment of the line to the left of P there is another segment of the line to the right of P whose x-component of the field exactly cancels the x-component from the initial segment, just as we saw for A and B. Consequently, the total field at point P must point in the y direction.

To calculate the magnitude of the net field at P, we begin by writing an expression for the y-component of the electric field due to a small segment of the charged line that we can treat as a point source, as illustrated in Figure 2.6. The magnitude, dE, of the electric field

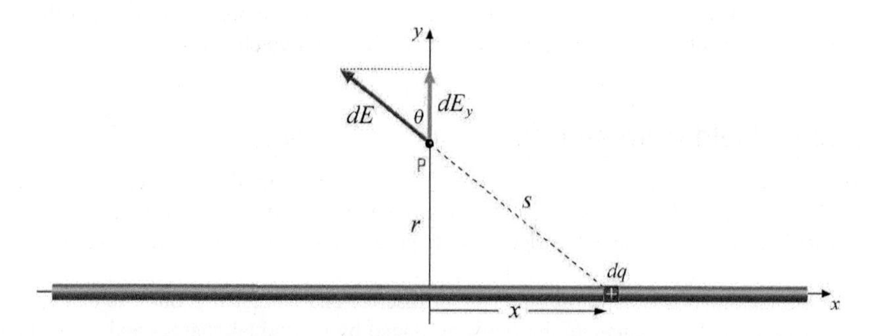

FIGURE 2.6 The electric field $d\vec{E}$ produced at point P due to an element dq of an infinite line of charge.

due to this segment is proportional to the amount of charge in the segment, dq, and is inversely proportional to s^2, the square of its distance from P.

$$dE = k\frac{dq}{s^2}$$

To find the total field, we will only need to consider the y-component:

$$dE_y = k\frac{dq}{s^2}\cos\theta$$

Namely, to obtain the electric field due to the entire line of charge, we must integrate the contributions from each segment as x goes from $-\infty$ to $+\infty$. To actually perform this integral, we will need to rewrite this expression so that we have only one variable inside the integral. We start by replacing dq by $\lambda\, dx$, where λ is the **linear charge density**, i.e., the amount of charge, as measured in Coulombs, in each meter of the line.

$$dE_y = k\frac{\lambda\, dx}{s^2}\cos\theta$$

Because both the distance s and the angle θ change with x, it is convenient to rewrite this integral such that only one variable changes. In this case, we will choose θ to be the variable of integration. We eliminate s and x from this equation, using the following expressions:

$$s = \frac{r}{\cos\theta} \qquad x = r\tan\theta \qquad dx = \frac{r}{\cos^2\theta}d\theta$$

The equation for the net electric field then becomes

$$E = \int dE_y = \int_{-\pi/2}^{+\pi/2} k\frac{\lambda}{r}\cos\theta \, d\theta = 2k\frac{\lambda}{r}$$

Thus, we see that the net electric field at point P is directly proportional to the charge density of the line and inversely proportional to the distance of point P from the line.

It's interesting to compare this result for the infinite line of charge to our earlier result for the point charge. The point charge has spherical symmetry about the point while the infinite line of charge has cylindrical symmetry about the line.

The electric field produced by a point charge can only depend upon r, the distance from the point. This distance is represented by a vector that can have components in all three dimensions (x, y, and z). The field produced by an infinite line of charge can only depend on r, the distance from the line. This distance is represented by a vector that has components in only two dimensions (x, y). This difference in the dimensions of r (i.e., the number of components of r) is reflected in the spatial dependence of the fields. The electric field of a point charge decreases as $1/r^2$, while the electric field of a line of charge decreases as $1/r$.

2.6 Summary

We began by defining the electric field at any point in space as the total Coulomb force per unit positive charge at that point. This shift from discussing forces between existing charges to fields that exist at points in space is a major reorientation, and will define our focus for the remainder of the course.

We then applied this definition to calculate the electric fields produced by specific charge distributions. We first considered the $\vec{E}$ field produced by a single point charge. We found that this field has spherical symmetry about the point charge; the direction of the field is radial and the magnitude of the field falls off as $1/r^2$. We then added another charge of opposite sign, creating an electric dipole. This addition destroys the spherical symmetry and thereby complicates the general calculation. The procedure for the calculation is the same, however. Calculate the field due to each charge separately and then take the vector sum of these fields to determine the total field. Finally, we did an example involving a continuous charge distribution, namely an infinite line of charge. We found that this $\vec{E}$ field has cylindrical symmetry about the line of charge; the direction of the field is perpendicular to the line and the magnitude of the field falls off as $1/r$.

MAIN POINTS

Definition of Electric Field

The electric field at any point in space is defined as the total Coulomb force exerted on a unit positive charge if it were fixed at that point.

$$\vec{E} \equiv \frac{\vec{F}}{q}$$

Electric Field Produced by a Distribution of Discrete Charges

The electric field produced by a distribution of discrete charges is equal to the vector sum of the fields produced by each charge individually.

Point Charge

Dipole

-Q

$$\vec{E}_{Net} \equiv \frac{\vec{F}_{Net}}{q} = \frac{1}{q}\sum_i \vec{F}_{iq} = k\sum_i \frac{Q_i}{r_{iq}^2}\hat{r}_{iq}$$

Electric Field Produced by a Continuous Charge Distribution

The electric field produced by a continuous charge distribution is equal to the integral of the field produced by each element of charge in the distribution.

Infinite Line of Charge

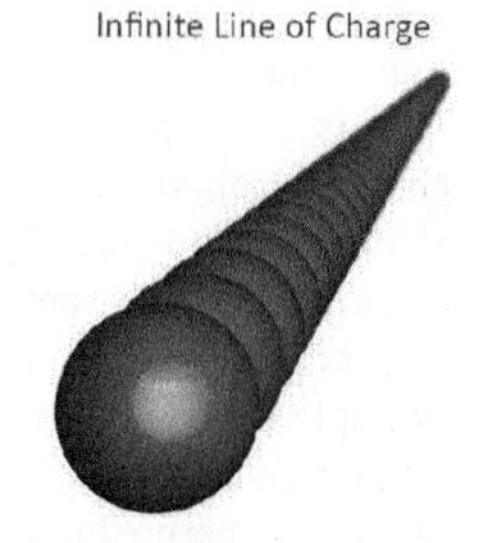

$$\vec{E}_{Net} \equiv \frac{\vec{F}_{Net}}{q} = \frac{1}{q}\int d\vec{F} = k\int \frac{dQ}{r^2}\hat{r}$$

PROBLEMS

1. Electric Field from Point Charges: Two point charges ($q_1 = -2.2\ \mu\text{C}$ and $q_2 = 6.3\ \mu\text{C}$) are fixed along the x-axis, separated by a distance $d = 9.3$ cm. Point P is located at $(x, y) = (d, d)$.

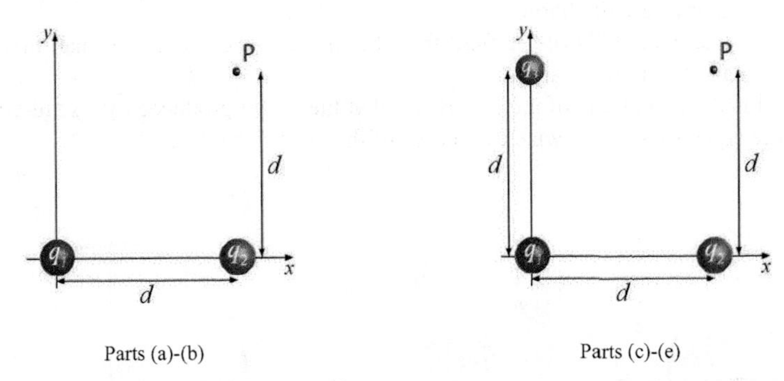

FIGURE 2.7 Problem 1

(a) What is $E_x(\text{P})$, the value of the x-component of the electric field produced by q_1 and q_2 at point P? (b) What is $E_y(\text{P})$, the value of the y-component of the electric field produced by q_1 and q_2 at point P? (c) A third point charge $q_3 = 2.9\ \mu\text{C}$ is now positioned along the y-axis at a distance $d = 9.3$ cm from q_1 as shown. What is $E_x(\text{P})$, the x-component of the field produced by all three charges at point P? (d) Suppose all charges are now doubled (i.e., $q_1 = -4.4\ \mu\text{C}$, $q_2 = 12.6\ \mu\text{C}$, $q_3 = 5.8\ \mu\text{C}$), how will the electric field at P change?

- (i) Its magnitude will increase by less than a factor of two and its direction will remain the same.
- (ii) Its magnitude will increase by less than a factor of two and its direction will change.
- (iii) Its magnitude will double and its direction will remain the same.
- (iv) Its magnitude will double and its direction will change.

(e) How would you change q_1 (keeping q_2 and q_3 fixed) in order to make the electric field at point P equal to zero?

- (i) Increase its magnitude and change its sign.
- (ii) Decrease its magnitude and change its sign.
- (iii) Increase its magnitude and keep its sign the same.
- (iv) Decrease its magnitude and keep its sign the same.
- (v) There is no change you can make to q_1 that will result in the electric field at point P being equal to zero.

2. Electric Field from Arc of Charge: A total charge $Q = 0.1\ \mu\text{C}$ is distributed uniformly over a quarter circle arc of radius $a = 9.4$ cm as shown. (a) What is λ the linear charge density along the arc? (b) What is E_x, the value of the x-component of the electric field at the origin $(x, y) = (0,0)$? (c) What is E_y, the value of the y-component of the electric field at the origin $(x, y) = (0,0)$? (d) How does the magnitude of the electric field at the origin for the quarter-circle arc you have just calculated compared to the electric field at the origin produced by a point charge $Q = 0.1\ \mu C$ located a distance $a = 9.4$ cm from the origin along a 45° line, as shown in the figure?

(i) The magnitude of the field from the point charge is less than that from the quarter-arc of charge.

(ii) The magnitude of the field from the point charge is equal to that from the quarter-arc of charge.

(iii) The magnitude of the field from the point charge is greater than that from the quarter-arc of charge.

(e) What is the magnitude of the electric field at the origin produced by a semi-circular arc of charge equal to $0.2\ \mu C$, twice the charge of the quarter-circle arc?

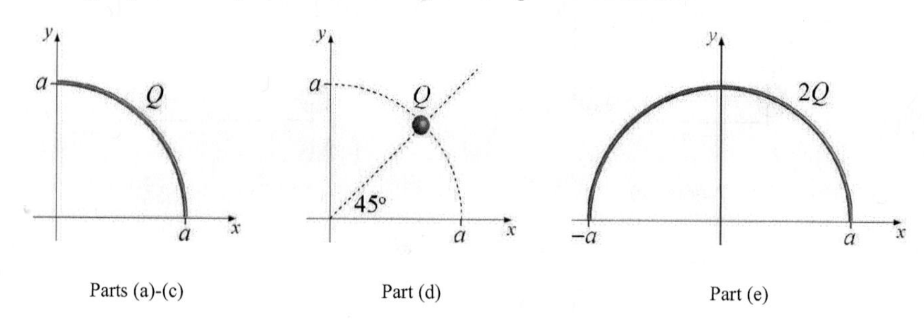

FIGURE 2.8 Problem 2

3. Zero: Two charges are fixed in place on the x-axis. The first charge is located at $x = 0$ and has a charge of $q = -2\ \mu C$. The second charge is located at $x = d$, where $d = +12$ cm, and has a charge of $Q = +4\ \mu C$. Find a point $x = x_o$ along the x-axis at which the electric field is zero.

UNIT

3

ELECTRIC FLUX AND FIELD LINES

3.1 Overview

We will continue our development of the concept of electric fields by constructing a geometric representation of an electric field using electric field lines. These field lines exist throughout space. The direction of the lines indicates the direction of the electric field and the density of the lines indicates the magnitude of the electric field.

We will introduce a new mathematical quantity, called the electric flux, which quantifies the number of field lines passing through a given surface. We will conclude with the development of a most important relationship, namely one of Maxwell's four equations that describe classical electromagnetic theory, Gauss' law.

3.2 Electric Flux

We will now construct a geometric representation of the electric field produced by a point charge (Figure 3.1). The electric field produced by a point charge is spherically symmetric; its direction is radial from the point charge. Therefore, we'll start by drawing radial straight lines that emanate from the point charge and are distributed in a spherically symmetric way. These lines provide an unambiguous representation of the direction of the electric field at any point in space. The amazing thing is these lines also provide information about the magnitude of the electric field at any point in space.

Consider a sphere that has a radius R and is centered on the point charge. The radial lines are distributed uniformly over the surface of that sphere. As we increase the radius of the sphere, the number of lines passing through the surface stays fixed, but the surface area increases as R^2. Therefore, the density of these lines passing through the surface decreases as $1/R^2$, exactly the same decrease that the magnitude of the electric field experiences. Consequently, we can represent the magnitude of the electric field at any point P as being proportional to the density of lines passing through the surface of the sphere that contains point P. If we make the number of lines drawn proportional to the electric charge, these lines provide a complete picture of the electric field at every location in space. For example, if we increase the magnitude of the charge by a factor of two, we'll have twice as many lines emanating from the charge. The density of these lines at any fixed distance from the charge will therefore also double, indicating that the electric field at that distance has also doubled. We call these lines the **electric field lines** and we will use them to represent the electric field produced by arbitrary charge distributions.

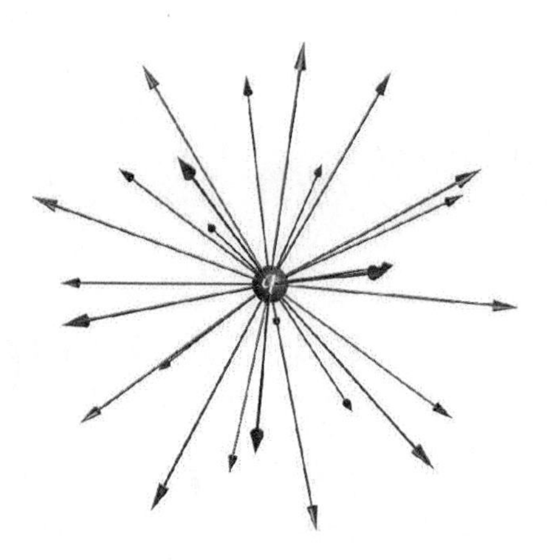

FIGURE 3.1 Radial electric field lines drawn from a point charge. The direction of the lines indicate the direction of the field.

It is common to rewrite Coulomb's law in a way that reinforces this geometric interpretation by introducing a new constant, ε_o, such that:

$$\vec{E} = \frac{1}{4\pi\varepsilon_o}\frac{q}{r^2}\hat{r} \equiv \frac{\Phi}{A}\hat{r}$$

where $A = 4\pi r^2$ is the surface area of a sphere of radius r centered on the charge, $\varepsilon_o = 1/(4\pi k) = 8.85\times10^{-12}\ \text{C}^2/(\text{N}\cdot\text{m}^2)$, and $\Phi = Q/\varepsilon_o$ is the electric flux, which we will discuss in more detail below.

This geometric representation for electric fields produced by a point charge will prove to be very powerful, being applicable to more general physical situations as well. The number of lines represents the magnitude of the charge that produces an electric field throughout space. The direction of the electric field is given by the direction of the lines and the magnitude of the field is proportional to the density of those lines (the number of lines per unit area).

We now want to use this geometric representation of an electric field from a point charge to further develop the concept of the **electric flux,** Φ. The electric flux is a quantitative expression of the number of lines passing through a surface. Because we represented the electric field of the point charge as the number of radial lines per unit area, it seems like the flux through a surface should be related to the product of the electric field and the area of that surface. Certainly for a sphere centered on the point charge Q, the flux is just this product $\Phi = EA = Q/\varepsilon_o$, as introduced in the expression for Coulomb's law above.

This expression for Φ is correct in this case because the electric field at the surface is always perpendicular to the surface and its magnitude is the same at all points on the surface. We need to extend this definition for electric flux so that it covers cases in which these conditions are not met.

Consider the flux through two different surfaces in the presence of a uniform electric field (Figure 3.2). Surface 1 is a rectangle, has a surface area A_1, and is perpendicular to the field lines. Surface 2 is also a rectangle but has a larger surface area A_2 and is tilted at an angle α to the field lines. Clearly the same number of lines that pass through A_1 also pass through A_2. Therefore, the flux through A_1 *must be the same* as the flux through A_2. We note that A_1 is in fact the projection of A_2 perpendicular to the lines. That is, A_1 is equal to A_2 multiplied by cos(α). By representing a surface area A as a vector whose direction is perpendicular to the surface and whose magnitude is equal to the area, we can define the flux for either A_1 or A_2 as $\Phi = \vec{E} \cdot \vec{A}$.

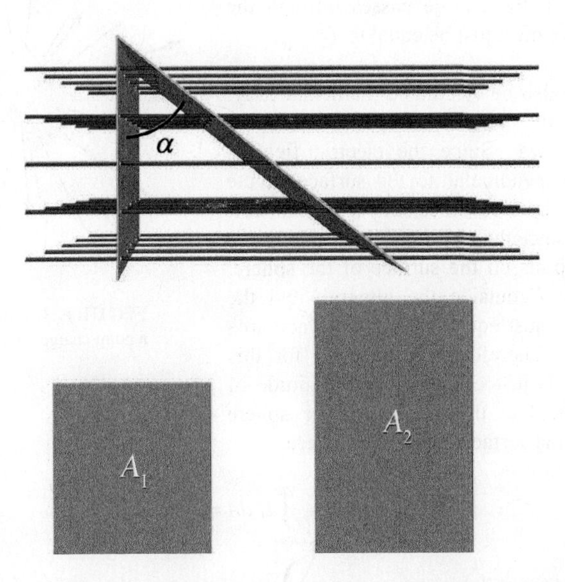

FIGURE 3.2 Flux lines intersecting two rectangles inclined at an angle a. The flux through each surface is identical and equal to $\vec{E} \cdot \vec{A}$.

This expression for the electric flux can be used for cases in which the electric field is constant at the surface and the surface is a plane. To generalize the expression so that it applies to cases where the electric field is not constant, we simply apply calculus. Namely, if we consider a small enough area on the surface (let's call it $d\vec{A}$), the electric field can be considered constant (let's call it $\vec{E}$), and we can define the flux through that small surface (let's call it $d\Phi$) as

$$d\Phi = \vec{E} \cdot d\vec{A}$$

To find the flux through the complete surface, we just need to add up the $d\Phi$ contributions from all of the small areas $d\vec{A}$. This summation is represented as the integral:

$$\Phi \equiv \int_{surface} \vec{E} \cdot d\vec{A}$$

We've now arrived at the mathematical expression for the electric flux, the quantity that corresponds to the number of field lines that pass through a surface. We'll now work through some examples that clarify the concept of electric flux.

3.3 Examples

Consider a sphere that has a radius R and is centered on a single point charge Q (Figure 3.3). The electric flux Φ is a measure of the number of lines passing through the surface. Because every line from the charge passes through the sphere, the flux must just be equal to Q/ε_o.

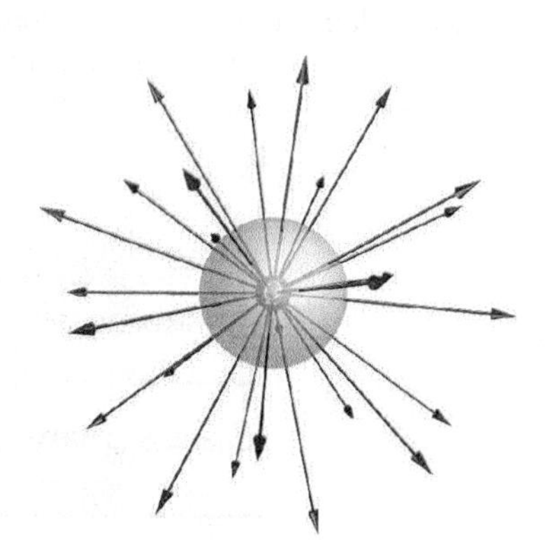

FIGURE 3.3 Radial lines from a point charge intersecting a sphere.

This flux can also be calculated mathematically. In particular, the flux is always equal to the integral of $\vec{E} \cdot d\vec{A}$. Since the electric field is everywhere perpendicular to the surface of the sphere, $\vec{E} \cdot d\vec{A}$ here becomes simply E times dA. Furthermore, since the electric field has the same value at all points on the surface of the sphere, we can move E outside the integral, and the integral is then just equal to A, the surface area of the sphere. Therefore, we see that for this case, the flux is just equal to the magnitude of the electric field at the surface of the sphere multiplied by the surface area of the sphere.

$$\Phi \equiv \int_{surface} \vec{E} \cdot d\vec{A} = \int_{surface} E\, dA = E \int_{surface} dA = EA$$

Now let's take one hemisphere of the sphere in Figure 3.3 and expand it to a radius of $2R$. Then, let's connect the two hemispheres with a circular plane at the equator, as shown in Figure 3.4. Note that all of the lines still pass through our new object, which is now composed of the three surfaces. Therefore, the total electric flux through this new object must be the same as the flux through the initial sphere of radius R. In fact, you can see that exactly half of the lines pass through the small hemisphere, exactly half of the lines pass through the large hemisphere, and no lines pass through the plane. When we calculate the flux, we verify this observation. Namely, the surface area of the hemisphere of radius $2R$ is four times the surface area of the hemisphere of radius R, but the electric field at the surface of the large hemisphere is one-fourth of the electric field at the surface of the small hemisphere. Consequently, the flux through each hemisphere ($\Phi = EA$) is the same and just equal to half of the charge Q divided by ε_o.

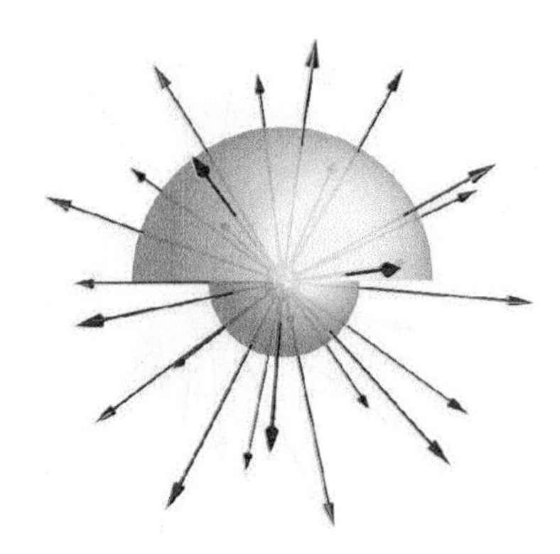

FIGURE 3.4 Radial lines from a point charge intersecting two hemispheres with different radii.

What about the plane connecting the hemispheres? We see that there are no lines passing through the surface; therefore we expect the electric flux to be zero. In fact, the flux is zero because the direction of the electric field (shown by the direction of the lines) is parallel to the surface. $\vec{E} \cdot d\vec{A}$ is equal to zero everywhere on this surface.

3.4 Gauss' Law

Consider a spherical shell that has a radius R and is centered on a single point charge Q. In our pictorial view of the electric field, the number of field lines leaving this charge is proportional to Q and the flux through any surface is proportional to the number of lines passing through the surface. To simplify the language in this section, let's just take the number of lines leaving the charge to be Q/ε_o, which results in the flux through the spherical shell being simply a count of the number of lines leaving the charge (Q/ε_o).

Now, if we move the charge around but keep it inside the shell, the number of lines leaving the shell (and therefore the net flux out of the shell) does not change (Figure 3.5(a)). We chose a spherical surface for simplicity, but this result holds for any surface that encloses the charge. Once we move the charge outside the shell, however, we have a qualitatively different situation. In this case, every line that enters the shell also leaves the shell; hence the net flux out of the shell is zero (Figure 3.5(b)).

We'll now discuss the more general case of a collection of N point charges. Consider a spherical shell that encloses a subset of these charges. What happens to the electric flux out of the shell in this case? Each charge Q_i inside the shell generates Q_i/ε_o lines, which all leave the shell, and therefore contributes the quantity Q_i/ε_o to the total flux.

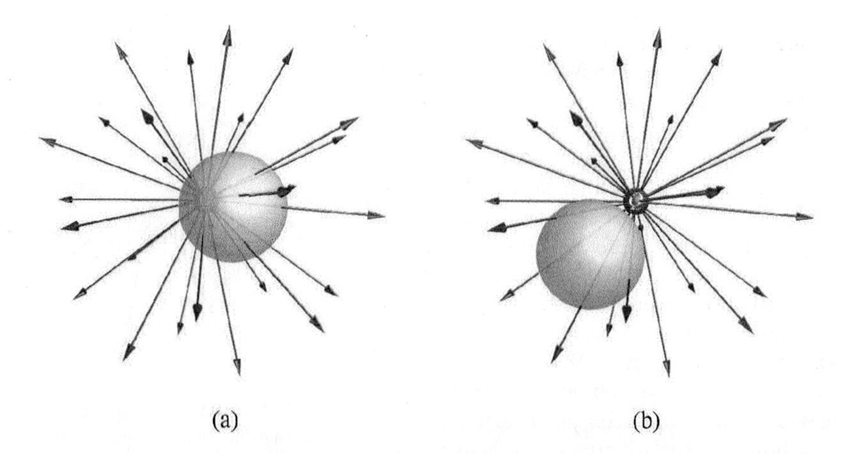

FIGURE 3.5 (a) *Charge enclosed by sphere*: All flux lines pass through sphere: $\Phi = Q/\varepsilon_o$. (b) *Charge not enclosed by sphere:* As many flux lines leave the sphere as enter the sphere: $\Phi = 0$.

Any charge outside the spherical shell will have a contribution of zero to the net flux because every line from such a charge that enters the shell will also leave it.

We now have our main result: *the total electric flux out of a closed surface is just proportional to the total charge enclosed by that surface.*

Although this concept may seem somewhat trivial, it is in fact one of the pillars of electromagnetism. Combining the above result with our integral equation for calculating the electric flux due to an electric field, we obtain what is known as **Gauss' law**:

$$\oint_{surface} \vec{E} \cdot d\vec{A} = \frac{Q_{enclosed}}{\varepsilon_o}$$

3.5 An Infinite Line of Charge and Gauss' Law

We'll now verify that Gauss' law works for our previous example of an infinite line of charge. Recall that we determined that the direction of the electric field produced by an infinite line of charge is perpendicular to the line, while the magnitude of the electric field is directly proportional to the linear charge density and inversely proportional to the distance from the line.

We'll now calculate the electric flux produced by the line of charge that passes through a cylindrical shell that has a length L, a radius R, and is centered on the line (Figure 3.6). The flux through the two ends of the cylindrical shell must be zero since the electric field is parallel to those surfaces, giving $\vec{E} \cdot d\vec{A} = 0$ everywhere—just like the case of the circular plane that connected the two hemispheres (Figure 3.4). Therefore, the total electric flux through the cylindrical shell is just the flux through the barrel of the shell.

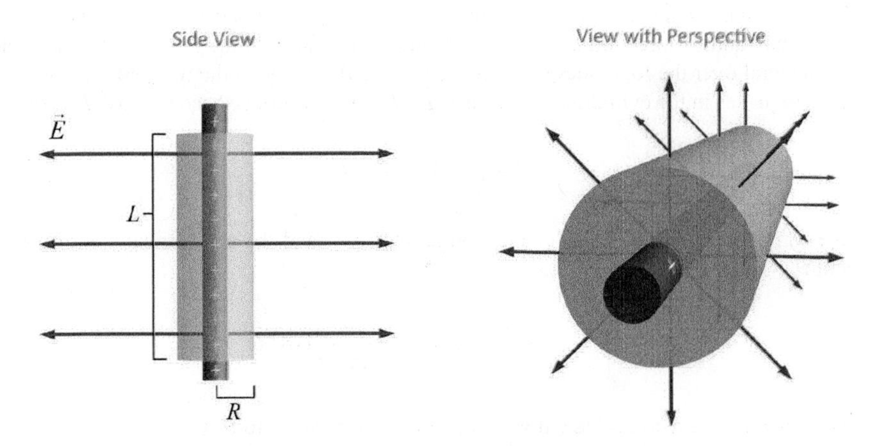

FIGURE 3.6 Infinite line of charge with linear charge density λ and Gaussian cylinder of radius R and length L.

This flux is easy to calculate since the electric field is everywhere perpendicular to the surface and the magnitude of the field is a constant over the surface. Consequently, the flux is just equal to EA. The surface area of the barrel is just equal to the circumference of the cylindrical shell multiplied by the length, which is equal to $2\pi RL$. Therefore, the flux is just equal to $EA = (\lambda / 2\pi\varepsilon_o R)(2\pi RL) = \lambda L / \varepsilon_o = Q / \varepsilon_o$, as promised.

We'll now verify that Gauss' law also holds if we choose the surface that encloses the line charge to be a box of length L and sides $2R$, as shown in Figure 3.7.

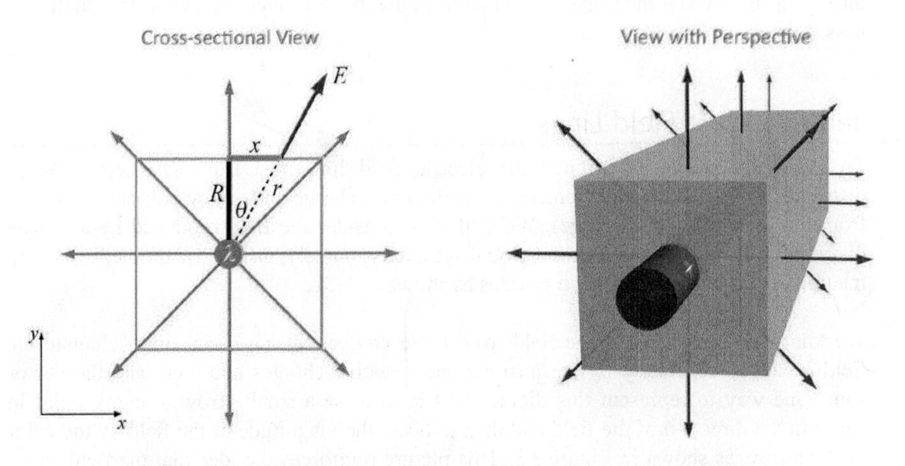

FIGURE 3.7 Infinite line of charge with linear charge density λ and Gaussian box with sides $2R$, $2R$, and length L.

Here, we must do an integral because the magnitude of $\vec{E}$ is not constant over the surface of the box and its direction is not perpendicular to the faces of the box. The plan is to do the integral over the four sides of the box since the flux through the two end surfaces will be zero just as in the cylindrical shell case. $\vec{E}\cdot d\vec{A}$ on this surface becomes just $E_y L\,dx$.

$$\Phi = \int \vec{E}\cdot d\vec{A} = \int E_y L\,dx$$

We can express E_y in terms of the charge density λ and the geometrical parameters R and θ.

$$E_y = \frac{\lambda}{2\pi\varepsilon_o r}\cos\theta = k\frac{\lambda}{2\pi\varepsilon_o R}\cos^2\theta$$

The integral is easy to evaluate if we change variables from x to θ.

$$x = R\tan\theta \qquad \Rightarrow \qquad dx = R\sec^2\theta\,d\theta$$

$$\Phi = \int\left(\frac{\lambda}{2\pi\varepsilon_o R}\cos^2\theta\right)LR\sec^2\theta\,d\theta = \frac{\lambda L}{2\pi\varepsilon_o}\int d\theta$$

Integrating $d\theta$ over the four sides of the box (0 to 2π), we arrive at the result that

$$\Phi = \lambda L / \varepsilon_o$$

This result for the electric flux using the box agrees exactly with the result we got using the cylindrical shell — as it must! The electric flux just corresponds to the number of lines through a surface and the same number of lines that pass through the cylindrical shell also pass through the box.

3.6 Electric Dipole Field Lines

To this point we've determined the electric field lines just from symmetric charge distributions (the spherical symmetry from the point charge and the cylindrical symmetry from the infinite line of charge). We will now consider the field produced by a charge distribution that does not possess a global symmetry, namely, an electric dipole, two equal magnitude but oppositely signed charges as shown.

We know how to calculate the field from these charges; at any point, just calculate the field produced separately by the positive and negative charges and then take the vector sum. One way to represent this dipole field is to draw a small arrow at every point in space in the direction of the field and then indicate the magnitude of the field by the color of the arrow, as shown in Figure 3.8. This picture reinforces the idea that the field exists everywhere and intuitively displays the main features of the field, its direction at any point, and where it is big and where it is small.

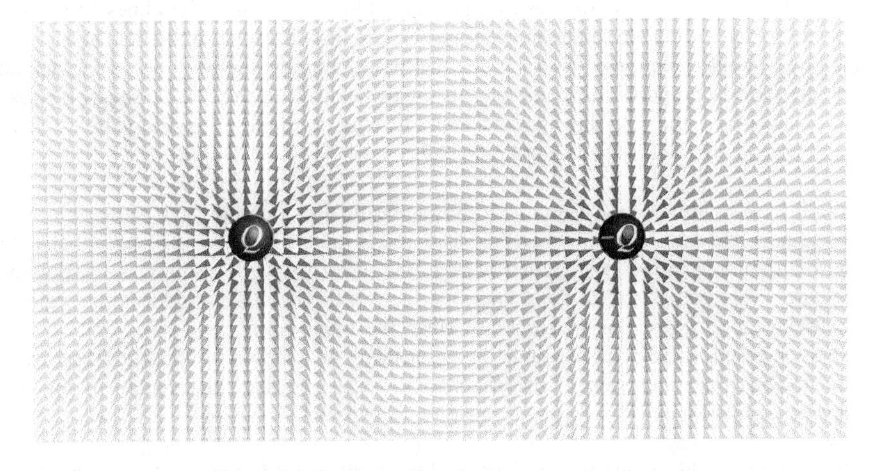

FIGURE 3.8 Arrow representation of field from electric dipole in which the direction of the field is indicated by the direction of the arrow and the magnitude of the field is indicated by the color of the arrow.

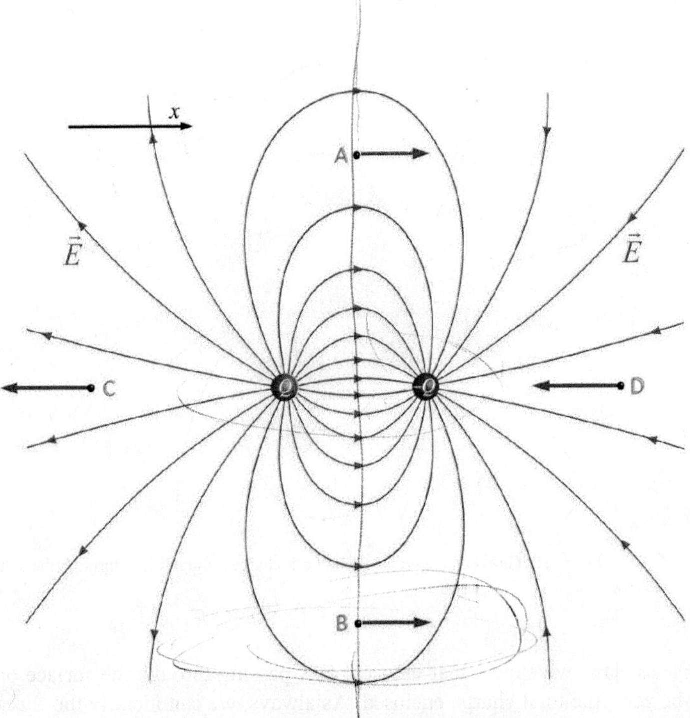

FIGURE 3.9 A two-dimensional representation of the electric field lines produced by an electric dipole. The direction of the field is indicated by the direction of the arrows and the magnitude of the field is indicated by the density of the lines.

How do we represent this field using electric field lines? We know field lines originate on positive charges and terminate on negative charges. The direction of the field at any point is given by the direction of the field line that passes through the point and the magnitude of the field is proportional to the density of the field lines that pass through a surface perpendicular to the field direction at that point. When we follow this procedure we obtain the curved field lines, as shown in Figure 3.9.

We'll now examine some features of the dipole field as inferred from these field lines. First, note that the direction of the field at A and B (pointing in the $+x$ direction) and at C and D (pointing in the $-x$ direction) agrees with what we've determined in the previous lecture. From the density of these lines, we see that the field is strongest in the region in between the charges; this makes sense since the fields from the positive and negative charges generally point in the same direction there, as opposed to opposite directions in the regions on either side of the charges. We also see that the strength of the field decreases as we move away from the charges, as expected from Coulomb's law.

We'll now use this dipole field to further illustrate the main features of Gauss' law. We'll start by investigating Gauss' law for a few different surfaces. Recall that Gauss' law says that the flux through a closed surface is proportional to the charge enclosed. Let's start with a sphere that encloses both charges, as shown in Figure 3.10.

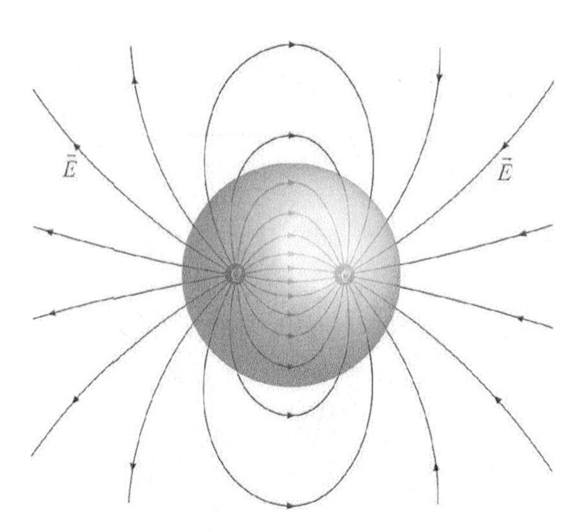

FIGURE 3.10 Gaussian sphere encloses both charges; the net flux through the sphere is zero.

From Gauss' law, we expect that the total flux passing through the surface of the sphere should be zero, the total charge enclosed. As always, we can identify the flux through the surface of this sphere as the total number of lines leaving the surface. On the left-hand side of Figure 3.10 we see a set of lines leaving the sphere as indicated by the outward-pointing arrows. Those same field lines curve around and then enter the surface of the sphere on the right side, as indicated by the inward-pointing arrows. The total flux through the surface is therefore zero. For the lines that leave the surface, the product $\vec{E} \cdot d\vec{A}$ is positive while for the lines that enter the surface, the product $\vec{E} \cdot d\vec{A}$ is negative. When we integrate over the entire surface of the sphere, we will get zero flux, in accordance with Gauss' law.

We'll now replace this large sphere with two smaller spheres, each one enclosing just one of the charges, as shown in Figure 3.11. Here, all of the field lines passing through the surface of the sphere on the left are leaving the sphere (the blue arrows), so the flux must be positive ($\vec{E} \cdot d\vec{A}$ is positive everywhere). Consequently, the charge enclosed by the sphere on the left must be positive. Similarly, all of the field lines passing through the surface of the sphere on the right are entering the sphere (the red arrows), indicating that the flux is negative ($\vec{E} \cdot d\vec{A}$ is negative everywhere) and, therefore, the charge inside must be negative. Since we have identical numbers of lines that leave the sphere on the left as that that enter the sphere on the right, we see that the magnitudes of the charges inside each sphere must be equal, in agreement with our initial assumption.

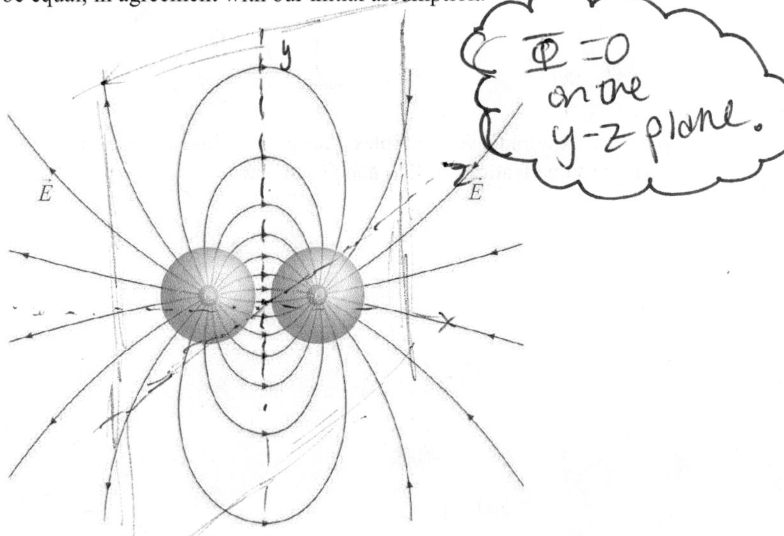

FIGURE 3.11 A Gaussian sphere on the left encloses the positive charge, while the one on the right encloses the negative charge. The flux through the left sphere is positive, while the flux through the right sphere is negative. The magnitudes of the fluxes are equal.

3.7 Summary

We began by constructing a geometric representation of the electric field produced by a single point charge. We identified the direction of the field at any point as the direction of the radial line from the charge that passes through the point. To get the magnitude, we noticed that the electric field decreases as $1/r^2$, while the surface area of a sphere increases as r^2. Consequently, we identified the magnitude of the electric field at any point as the density of radial lines passing through a surface perpendicular to the field at that point. We later learned that we can construct such “field lines” for the electric field from any arbitrary collection of charges.

We then introduced a new quantity, the electric flux, which corresponds to the number of field lines that pass through a surface. In particular, the electric flux through a surface is defined as

$$\Phi \equiv \int_{surface} \vec{E} \cdot d\vec{A}$$

This flux quantity was then used to develop an expression for Gauss’ law, one of the four fundamental equations that describe electromagnetic theory. In particular, Gauss’ law states that that the electric flux through any closed surface is proportional to the charge enclosed by that surface.

$$\oint_{surface} \vec{E} \cdot d\vec{A} = \frac{Q_{enclosed}}{\varepsilon_o}$$

We then explored two examples (the infinite line of charge and the electric dipole) to illustrate some features of flux and Gauss’ law.

MAIN POINTS

Electric Field Lines

Electric field lines represent the magnitude and direction of the electric field.

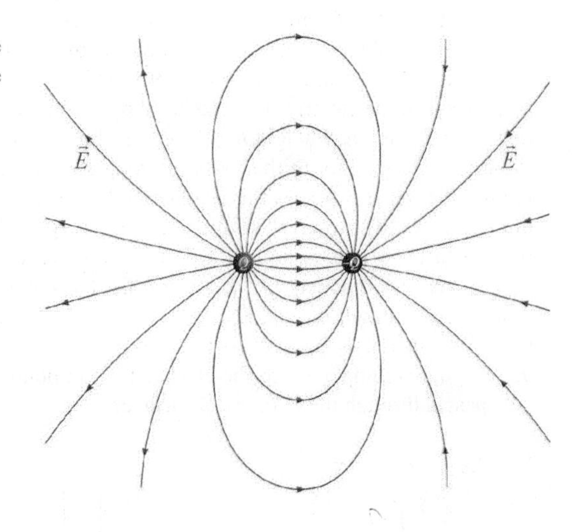

Direction:
Tangent to the field line
(from + charge to – charge)

Magnitude:
Density of field lines

Electric Flux

The electric flux through a surface is defined as the integral of $\vec{E}\cdot d\vec{A}$ over that surface. The geometric interpretation of the electric flux is that it simply counts the number of field lines that pass through the surface.

$$\Phi_E \equiv \int \vec{E}\cdot d\vec{A}$$

Gauss' Law

The electric flux through any closed surface is proportional to the charge enclosed by that surface.

$$\oint_{surface} \vec{E}\cdot d\vec{A} = \frac{q_{enclosed}}{\varepsilon_o}$$

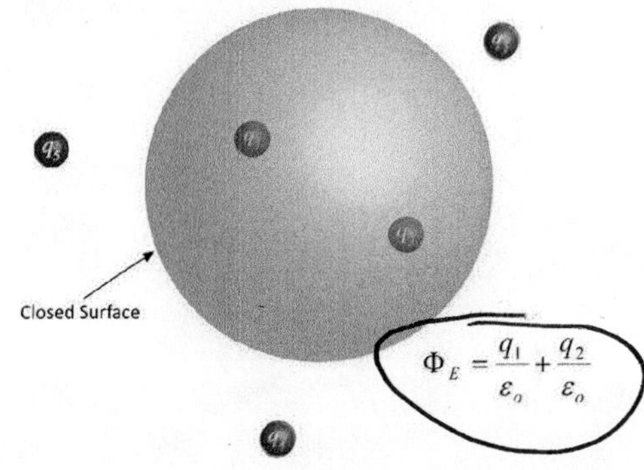

PROBLEMS

1. Electric Flux and Field from Lines of Charge: An infinite line of charge with charge density $\lambda_1 = -0.2\ \mu\text{C/cm}$ is aligned along the y-axis. (a) What is $E_x(\text{P})$, the value of the x-component of the electric field produced by the line of charge at point P, which is located at $(x, y) = (a, 0)$, where $a = 9.1$ cm? (b) What is $E_y(\text{P})$, the value of the y-component of the electric field produced by the line of charge at point P? (c) A cylinder of radius $a = 9.1$ cm and height $h = 9.4$ cm is aligned with its axis along the y-axis as shown. What is the total flux Φ that passes through the cylindrical surface? A positive number implies the net flux leaves the cylinder and a negative number implies the net flux enters the cylinder. (d) Another infinite line of charge with charge density $\lambda_2 = 0.6\ \mu\text{C/cm}$ parallel to the y-axis is now added at $x = 4.55$ cm as shown. What is the new value for $E_x(\text{P})$, the x-component of the electric field at point P? (e) What is the total flux Φ that now passes through the cylindrical surface? (f) The initial infinite line of charge is now moved so that it is parallel to the y-axis at $x = -4.55$ cm. What is the new value for $E_x(\text{P})$, the x-component of the electric field at point P? (g) What is the total flux Φ that now passes through the cylindrical surface?

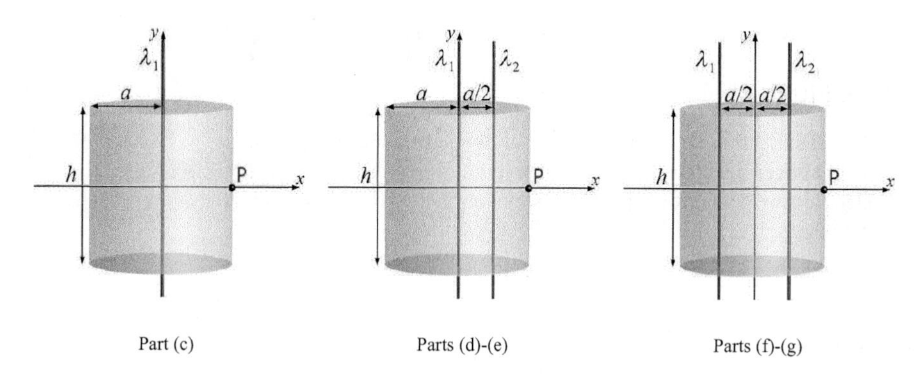

FIGURE 3.12 Problem 1

UNIT 4

GAUSS' LAW

4.1 Overview

We will now use Gauss' law to calculate the electric fields produced by a variety of charge distributions. The distinguishing feature of these examples is that in each case the charge distributions exhibit one of the three special spatial symmetries—spherical, cylindrical, or planar.

We will first investigate the electric field produced by a uniformly charged solid spherical insulator. We will then determine the electric field produced by an infinite solid cylindrical conductor that carries a uniform charge density. Finally, we will investigate the electric field produced by an infinite sheet that is uniformly charged.

We will conclude with a special example—two infinite sheets of charge that have equal but opposite signed uniform charge densities. Even though this example does not possess total planar symmetry, we can use the principle of superposition in conjunction with Gauss' law to make the calculation.

4.2 Gauss' Law and Symmetry

In the previous unit we introduced Gauss' law, which states that the total flux that passes through any closed surface is proportional to the electric charge enclosed by that surface. It is common to refer to this closed surface through which the flux is calculated as the **Gaussian surface.** We demonstrated the validity of Gauss' law by explicitly calculating the flux through some Gaussian surfaces and seeing it is proportional to the charge enclosed.

We will now use Gauss' law to calculate the electric fields produced by some special charge distributions. In general, these calculations would be extremely difficult because the electric field term occurs inside the integral in the Gauss' law equation.

For example, Figure 4.1 shows the electric dipole that we introduced in the previous unit. If we attempt to use a sphere centered on the dipole as the Gaussian surface, for example, we see that the magnitude of the field is not constant on the surface (e.g., it decreases as we move away from the axis of the dipole) and the direction of the field is also not constant (e.g., it is perpendicular to the surface along the axis and becomes more parallel to the surface as we move away). Consequently, there is no easy way to evaluate $\vec{E} \cdot d\vec{A}$ at all points on the surface in such a way that we can move E outside the integral, which would be necessary if we are to solve for E.

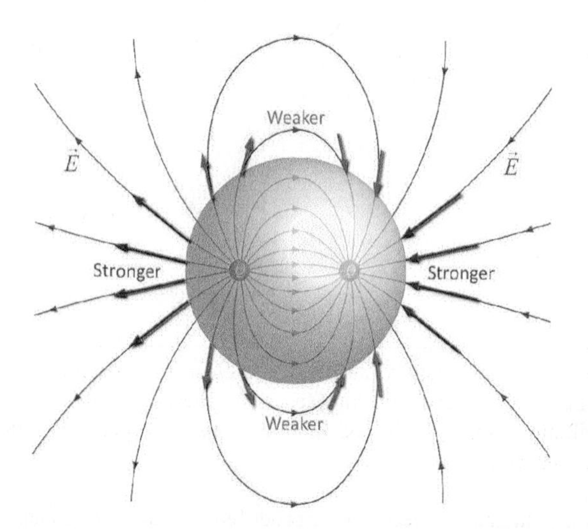

FIGURE 4.1 Electric field lines produced by an electric dipole. A Gaussian sphere centered on the dipole cannot be used to determine the electric field from Gauss' law since neither the direction nor the magnitude of the field at the surface of the sphere is constant.

On the other hand, if we replace the dipole by a single point charge, as shown in Figure 4.2, we have a completely different situation. Here, the electric field is constant at all points on the surface of the sphere and its direction is always perpendicular to the surface. Consequently, $\vec{E} \cdot d\vec{A}$ is just equal to E times dA at all points, and we can take E outside of the integral and solve for it.

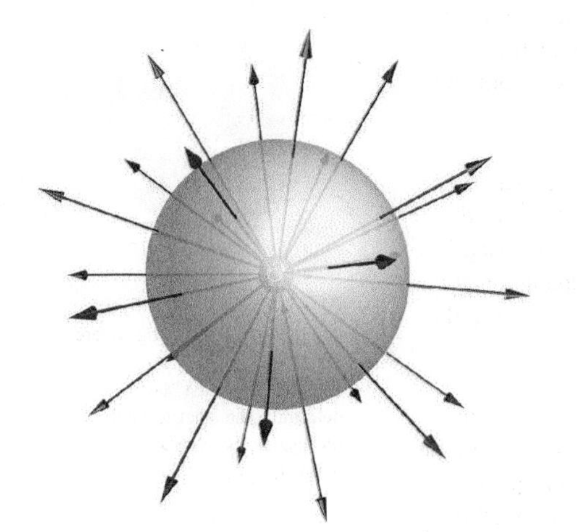

FIGURE 4.2 Electric field lines produced by a point charge. A Gaussian sphere centered on the charge can be used to determine the electric field from Gauss' law since the direction and the magnitude of the field at the surface of the sphere is constant.

$$\Phi \equiv \int_{surface} \vec{E} \cdot d\vec{A} = E \int_{surface} dA$$

In general, for cases that have a high degree of symmetry, it is possible to choose a Gaussian surface such that the electric field term can be moved outside of the integral. For example, if we can choose a surface such that the electric field has a constant magnitude at all points on the surface and is always perpendicular to the surface, then we can move the constant electric field term outside the integral and solve for it.

There are three such symmetries for which this procedure is possible, corresponding to our three spatial dimensions. We have the full three-dimensional symmetry of the sphere, the two-dimensional symmetry of the infinite cylinder, and the one-dimensional symmetry of the infinite plane. In each of these cases, we can construct Gaussian surfaces that allow us to solve for the electric field by moving its term outside of the integral in the Gauss' law equation.

4.3 Example: Spherical Symmetry

For our first example, we consider a solid insulating sphere that has a radius a and a uniform **volume charge density** ρ centered at the origin (Figure 4.3). In other words, inside the sphere the charge per unit volume (Q/V) is a constant equal to ρ and outside the sphere the charge density is zero. We've chosen an insulator because in our simple model of an insulator, charges cannot move; we can specify the charge distribution and it will remain fixed. Charge distributions on conductors are more complicated and we will delay discussing them until later in this unit.

We want to determine the electric field $\vec{E}$ from Gauss' law.

$$\oint_{surface} \vec{E} \cdot d\vec{A} = \frac{Q_{enclosed}}{\varepsilon_o}$$

In this case, the charge distribution possesses spherical symmetry. That is, the charge distribution itself is only a function of r, the distance from the center of the insulator.

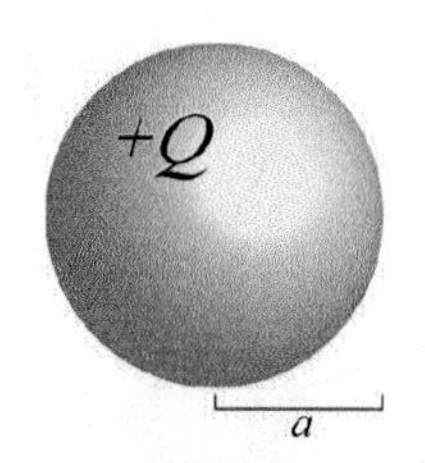

FIGURE 4.3 A uniformly charged solid insulating sphere with total charge Q and radius a. We will use Gauss' law to determine the electric field produced by this sphere at all points in space.

Since the charge distribution is spherically symmetric, the electric field produced by this charge distribution must also be spherically symmetric; its magnitude can only depend on the distance from the center of the insulator and its direction must be radial.

To calculate the value of the electric field at a distance r from the center of the sphere, we choose our Gaussian surface to be a concentric spherical shell with radius r. We know the direction of the field at all points on this Gaussian surface must be radial and therefore perpendicular to the surface. Consequently, the scalar product $\vec{E} \cdot d\vec{A}$ just becomes the simple product E times dA. Furthermore, since all points on the Gaussian surface are equidistant from the origin, we know the magnitude of the electric field at all points on this surface must be the same. Hence the electric field at the surface is a *constant* and can be taken outside the integral.

$$\oint_{surface} \vec{E} \cdot d\vec{A} = \oint_{surface} E\, dA = E \oint_{surface} dA$$

The remaining integral is easy to evaluate; it's just equal to the area of the Gaussian surface. Therefore, the left-hand side of Gauss' law simply reduces to the product of the value of the surface area of the sphere and the value of the electric field at a distance equal to the radius of that sphere.

$$E \oint_{surface} dA = E(4\pi r^2)$$

Gauss' law then becomes:

$$\oint_{surface} \vec{E} \cdot d\vec{A} = \frac{Q_{enclosed}}{\varepsilon_o} \quad \rightarrow \quad E(4\pi r^2) = \frac{Q_{enclosed}}{\varepsilon_o}$$

We'll first determine the electric field outside the insulator (i.e., for $r > a$). For all such values of r, the charge enclosed by the Gaussian surface is just the total charge of the insulator ($Q_{enclosed} = Q$). Therefore, we obtain the result that the electric field is proportional to the total charge of the insulator and inversely proportional to the distance from the center of the insulator.

$$E(4\pi r^2) = \frac{Q}{\varepsilon_o} \quad \Rightarrow \quad E = \frac{1}{4\pi\varepsilon_o}\frac{Q}{r^2}$$

In fact we see that the electric field E at a distance r from the center of the insulator is identical to that produced by a point charge Q located at the center of the insulator.

We'll now calculate the field inside the insulator (i.e., $r \leq a$). We can follow the same procedure as for the field outside the insulator. Namely, we can apply Gauss' law using a spherical Gaussian surface centered on the origin. Since the charge is uniformly distributed throughout the insulator, we know that the magnitude of the field must have the same value at all points on the spherical surface and the direction of the field must be perpendicular to this Gaussian surface everywhere.

The only change to our calculation is that the charge enclosed by the Gaussian surface is no longer the total charge of the insulator, i.e., the left-hand side of the Gauss' law equation remains the same

$$\oint_{surface} \vec{E} \cdot d\vec{A} = E(4\pi r^2)$$

but the right side changes; the charge enclosed by the sphere of radius r is less than the total charge Q. Consequently, we expect the field to be less than what it would be if it were just the continuation of the field for $r > a$.

To calculate the field inside the insulator, we need to determine the charge enclosed by a sphere of radius r, where $r < a$. Since the charge of the insulator is uniformly distributed, we can introduce the volume charge density of the insulator ρ, which is defined to be equal to the insulator's total charge divided by its volume.

$$\rho \equiv \frac{Q}{Volume} = \frac{Q}{\frac{4}{3}\pi a^3}$$

The charge enclosed by a sphere of radius r, then, is just equal to the product of this charge density and the volume of the sphere.

$$Q_{enclosed} = \rho\left(\frac{4}{3}\pi r^3\right) = Q\left(\frac{r}{a}\right)^3$$

Consequently, we obtain the result that the electric field inside the insulator increases linearly with r.

$$E = \frac{1}{4\pi\varepsilon_o}\frac{Q}{a^3}r$$

Figure 4.4 shows a plot of the magnitude of the electric field as a function of r. Here we see that the field outside the insulator ($r > a$) is identical to that produced by a point charge at the origin. Once we move inside the insulator ($r < a$), less charge is enclosed by our Gaussian surface, which means that the field cannot continue to increase as $1/r^2$. The exact deviation from the $1/r^2$ curve depends on the exact form of the charge distribution.

In this case, we assumed a uniform volume charge density that led to the linear behavior shown; a non-uniform but still spherically symmetric charge distribution would have led to a different radial dependence of the field inside the insulator.

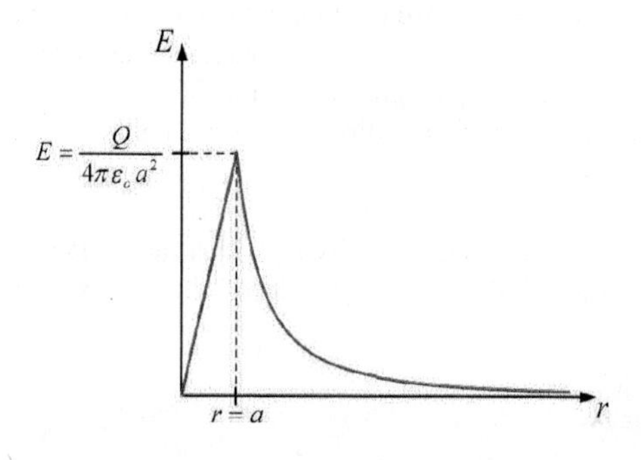

FIGURE 4.4 A plot of the electric field produced by a uniformly charged solid spherical insulator of radius a and total charge Q as a function of r, the radial distance from the center of the sphere.

4.4 Charges on Conductors

In our last example, the charged object, the sphere, is an insulator. Because charges cannot move in an insulator, the charge distribution has to be specified, namely that it is uniform throughout the volume of the spherical insulator.

If the sphere had instead been a solid *conductor*, any charge we put on it would be free to move. Before we can apply Gauss' law to this conductor (or any other charged conductor), we must first determine how the charge will be distributed. To answer this question, we first need to recall that charges are free to move in a conductor. If there is an electric field inside the conductor, the charges will experience a net force and move. Therefore, if the system is in static equilibrium, the electric field inside the conductor must be zero.

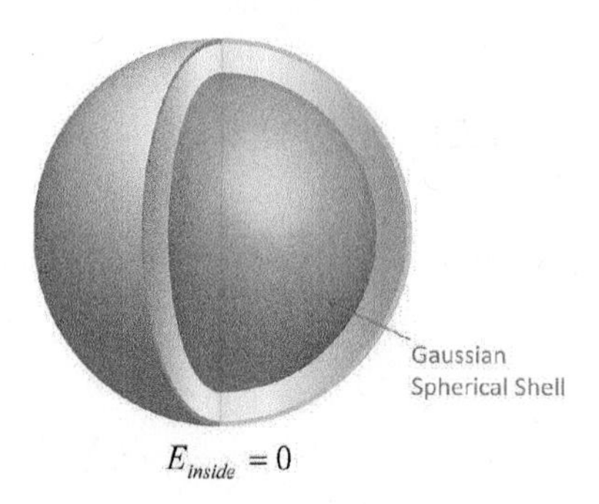

FIGURE 4.5 Applying Gauss' law to a charged, solid spherical conductor, we see $E = 0$ inside, which implies that $Q = 0$ inside.

But if the field is zero inside the conductor, we can apply Gauss' law to show that there can be no net charge inside the conductor. Consider a Gaussian surface whose center coincides with that of a spherical conductor and whose radius is just slightly smaller than the radius of the sphere (Figure 4.5). We know the electric field is zero everywhere on this surface so that the left-hand side of the Gauss' law equation (the flux Φ) is zero. Therefore, the right-hand side of Gauss' law (the enclosed charge or $Q_{enclosed}$) must also be zero. Therefore, all of the excess charge must reside on the surface of the spherical conductor. This result is a totally general one; it applies to all conductors independent of their shape.

4.5 Induced Charges on Conductors

We've just shown that all excess charge on a conductor must reside on its surface. Suppose we hollow out the solid spherical conductor from the last section to obtain a spherical shell with inner radius R_i and outer radius R_o, as shown in Figure 4.6. If we now place charge Q on the shell, where does it go? The arguments from the last section can be repeated here to show that this charge must be uniformly distributed on the outer surface of the shell.

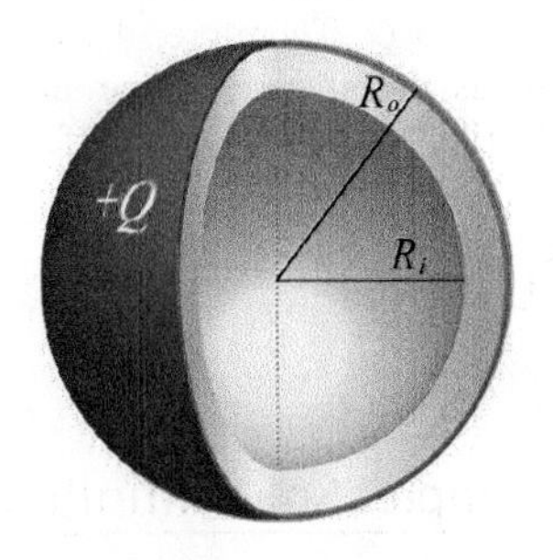

FIGURE 4.6 Conducting spherical shell (shown with a section removed) with inner radius R_i, outer radius R_o, and net charge Q. Applying Gauss' law, we determine that Q must be distributed uniformly on the outer surface (shaded blue for clarity).

Suppose we now place a point charge q_o at the center of this spherical shell. What will happen to the charge on the conducting shell? The charges on the conductor are certainly free to move, but what determines how they will move? We know the electric field in the conductor must be zero. Adding the point charge q_o would create a non-zero field in the conducting shell. Therefore, the charges in the conducting shell must move so as to cancel the field produced by the point charge inside the shell.

We can use Gauss' law to determine how the charge is distributed on the shell. Namely, we'll choose the Gaussian surface to be a sphere centered on the point charge with a radius between R_i and R_o, as shown in Figure 4.7. Since the E field is zero at all points on the surface of the Gaussian sphere, we know

$$\oint_{surface} \vec{E} \cdot d\vec{A} = 0$$

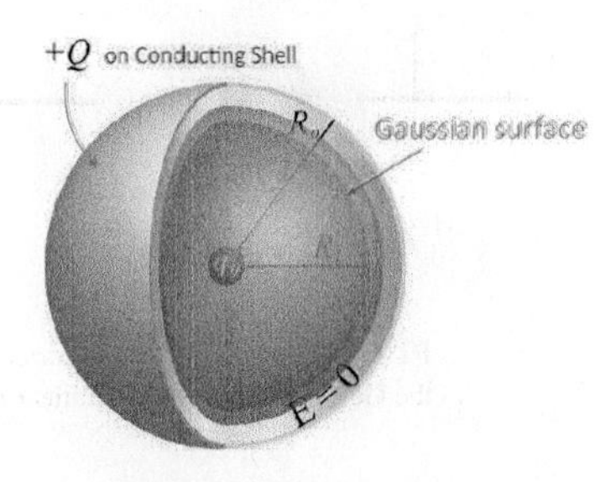

FIGURE 4.7 Gaussian surface with radius r ($R_i < r < R_o$) used to determine induced charge distribution on conducting spherical shell when a point charge (q_o) is placed at the center of the shell.

Consequently, the charge enclosed by the Gaussian sphere must also be zero. The charge enclosed is equal to q_o plus the induced charge on the inner surface of the shell. Therefore, we know this induced charge must be equal to $-q_o$. The spherical symmetry of the shell insures that this charge will be distributed uniformly on the surface. Therefore, we can determine that this **surface charge density**, σ_i, must be equal to the total induced charge divided by the surface area of the inner surface of the shell.

$$\sigma_i = -\frac{q_o}{4\pi R_i^2}$$

What about the outer surface of the shell? We can determine σ_{outer}, the charge density on the outer surface of the shell because we know that the total charge on the shell must be equal to Q. Therefore, the total charge on the outer shell is equal to the net charge Q minus the charge on the inner shell $(-q_o)$. Consequently, we determine the charge distribution on the outer surface:

$$\sigma_i = \frac{Q+q_o}{4\pi R_o^2}$$

4.6 Example: Solid Infinite Cylindrical Conductor

For our next example, we consider a solid infinite cylindrical conductor that has a radius a and uniform linear charge density λ. We can use Gauss' law to determine the electric field outside this cylindrical conductor. Since the charge must be distributed uniformly on the surface of the cylinder, we know the magnitude of the field produced by this charge can only depend on r, the two-dimensional distance from the axis of the cylinder, and its direction must always be perpendicular to this axis.

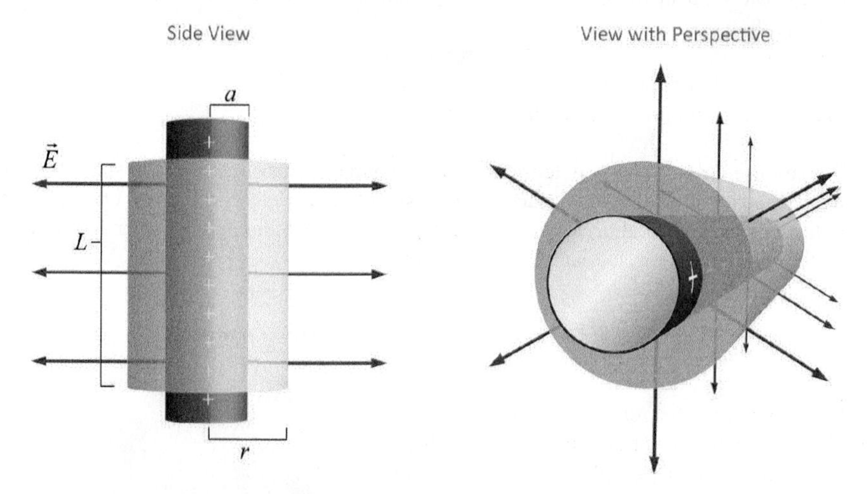

FIGURE 4.8 Solid cylindrical conductor of radius a and linear charge density λ and the Gaussian cylinder of radius r and length L.

To find the value of the field that is a distance r from the axis of the cylinder, we choose our Gaussian surface to be a cylinder of length L and radius r centered on the axis of the conductor (Figure 4.8). For the two end surfaces, E is parallel to the surface, and therefore perpendicular to dA. Consequently, $\vec{E}\cdot d\vec{A}$ over the ends is zero. Therefore, the only contribution to the flux Φ through the cylinder comes from the barrel surface. We know the direction of the field must be perpendicular to this surface and the magnitude of the field must be the same at all points on this surface. Consequently, the scalar product $\vec{E}\cdot d\vec{A}$ just becomes the arithmetic product EA and we can once again move the electric field term E outside the integral in the Gauss' law equation.

$$\oint_{surface} \vec{E} \cdot d\vec{A} = EA = E(2\pi rL)$$

The expression on the right-hand side of the Gauss' law equation is always $Q_{enclosed}$, the charge enclosed by this cylinder, divided by ε_o. In this case, the charge enclosed is just the linear charge density λ multiplied by the length L of the Gaussian cylinder.

$$Q_{enclosed} = \lambda L$$

Therefore, the electric field outside the charged cylindrical conductor is just given by

$$E(r) = \frac{\lambda}{2\pi\varepsilon_o r} \qquad (r > a)$$

We see then that the electric field *outside* a charged infinite cylindrical conductor is identical to the field produced by an infinite line of charge that has the same linear charge density. This result is not a coincidence. Since both of these charge distributions have cylindrical symmetry, the field at a distance r from the axis of symmetry is totally determined by the charge enclosed by a cylinder of radius r. The same charge enclosed, the same field.

This situation is just like that of our original example in which the field outside of the solid insulting sphere was found to be equal to that of a point charge located at the position of the center of the sphere. The same charge enclosed, the same field.

Since the electric field inside the conductor is zero, we can now plot the radial dependence of the field for all values of r, as shown in Figure 4.9. The field is zero inside the conductor, has its maximum value at the surface, and then decreases as $1/r$.

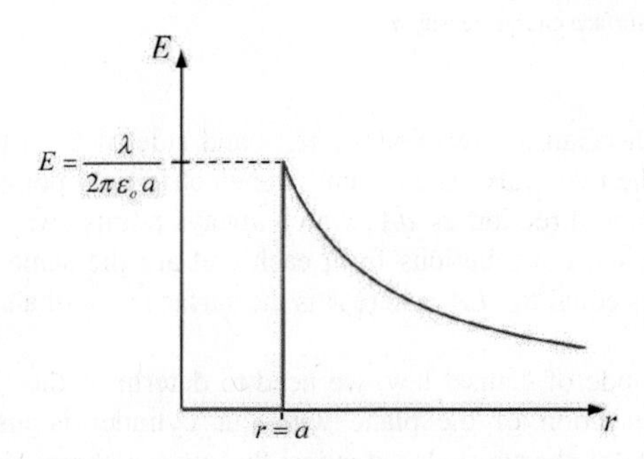

FIGURE 4.9 The radial dependence of the magnitude of the electric field produced by an infinite solid cylindrical conductor of radius a and linear charge density λ.

4.7 Infinite Sheet of Charge

For our next example, we consider an infinite sheet of charge that has a uniform surface charge density σ.

Here the charge density has planar symmetry, namely charge is uniformly distributed over a single plane. Consequently, we know that the direction of the electric field must be perpendicular to the surface. To determine the electric field at any point, we need to choose a Gaussian surface whose surfaces are either parallel or perpendicular to the plane so that $\vec{E} \cdot d\vec{A}$ will be either zero or just EA. With such a choice, we will be able to move the electric field term outside the integral in the Gauss' law equation and solve for it.

Let's choose a cylinder that cuts through the plane, as shown in Figure 4.10. Here the electric field is parallel to the barrel surface and perpendicular to the two end surfaces of the cylinder. Since the field may depend on the distance from the plane, we'll position our cylinder so that the two ends are equidistant from the plane.

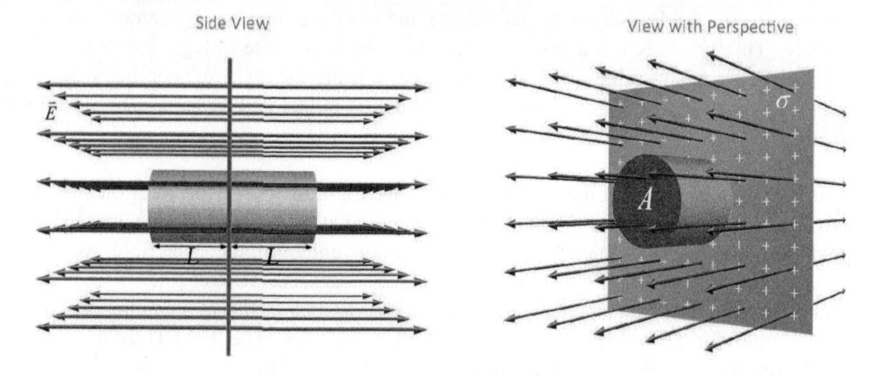

FIGURE 4.10 The Gaussian surface (a cylinder of length $2L$ and cross-sectional area A) used in Gauss' law to determine the electric field produced by an infinite sheet of charge with surface charge density σ.

We'll now apply the Gauss' law. For the left-hand side of the equation, $\vec{E} \cdot d\vec{A}$ will be non-zero only on the two ends. At each end, the electric field points away from the plane of charge, in the same direction as $d\vec{A}$, which always points away from the inside of the volume. Therefore, the contributions from each end are the same, leaving the total flux through the surfaces equal to $2EA$, where A is the surface area of a single end.

For the right-hand side of Gauss' law, we need to determine the charge enclosed by our cylinder. The intersection of the plane with our cylinder is just a circle of area A. Consequently, the total charge enclosed is just the surface charge density σ times the area A. Putting all this together, we obtain the final result:

$$E = \frac{\sigma}{2\varepsilon_o}$$

Perhaps this result seems surprising to you. The electric field from an infinite sheet of charge does not depend on the distance from the sheet. We can actually understand this result if we think about the field lines produced by an infinite sheet of charge. Because the sheet is infinite, the direction of the field must be perpendicular to the sheet; therefore, the field lines must also be perpendicular to the sheet. The magnitude of the field is given by the density of these lines. The density of these lines must just stay constant; there is no place for them to expand into. Consequently, the magnitude of the field produced by an infinite sheet of charge must be constant.

4.8 Superposition

We've now covered the three symmetries—spherical, cylindrical, and planar—for which we can use Gauss' law to calculate the electric fields. We'll conclude with an observation that the principle of superposition, which we have used a lot so far, comes in handy once again to let us use Gauss' law to calculate electric fields for cases that do not possess one of these global symmetries. In particular, we will now consider an example in which we have two parallel infinite sheets of charge (Figure 4.11). One sheet has a uniform surface charge density $+\sigma$, while the other sheet has a uniform surface charge density $-\sigma$.

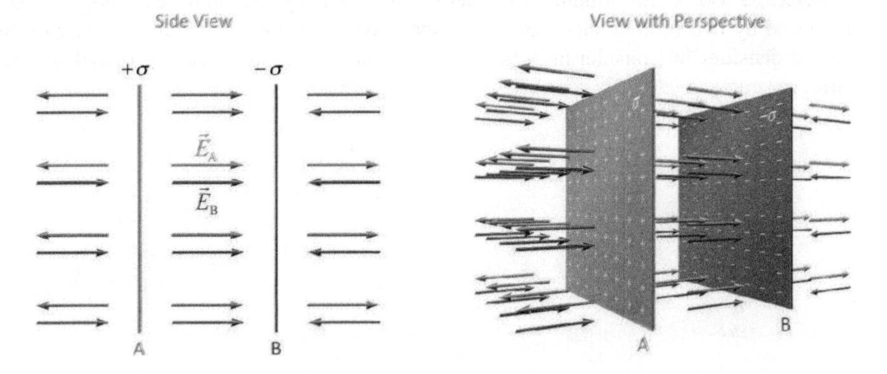

FIGURE 4.11 Use of superposition to determine the electric field produced by two infinite sheets of charge having equal magnitude but oppositely signed charge densities.

This charge distribution does not possess global planar symmetry because there are two planes. However, because electric fields obey the law of superposition, the field produced by this combination of planes must be identical to vector sum of the fields produced by each sheet separately.

There are three regions of space to consider, (1) in between the plates, (2) outside the positively charged sheet, and (3) outside the negatively charged sheet. We see the contributions from the two sheets exactly cancel in the two regions outside the sheets. Consequently, the electric field in these regions is zero. In the region between the two sheets, the fields add to give a total field that is twice the field from either sheet. Consequently, the magnitude of the field in this region is just equal to σ / ε_o, and the direction of the field is toward the negatively charged sheet.

4.9 Summary

We began by observing that because we live in a world that has three spatial dimensions, there are three kinds of symmetric charge distributions (spherical, cylindrical, and planar) that produce fields that can be calculated directly from the Gauss' law equation. We first considered a uniformly charged solid spherical insulator and found that the direction of the electric field is radial and the magnitude of the field increases linearly from zero while inside the insulator and then falls off as $1/r^2$ while outside the insulator.

We then discussed a uniformly charged infinite solid cylindrical conductor. The electric field is zero inside the conductor, which forces the charge to reside only on the outer surface. Once outside the conductor, the direction of the field is always perpendicular to the axis of the conductor, and the magnitude of the field decreases as $1/r$, just as in the case of the infinite line of charge. Finally, we found that the field produced by an infinite sheet of charge is directed perpendicularly to the sheet and has the same magnitude everywhere.

We also observed that this technique of using Gauss' law to calculate electric fields can be extended to some examples that do not possess any of these three global symmetries through the use of the principle of superposition. Namely, we could calculate the field produced by two infinite sheets of charge that have equal but oppositely signed uniform charge densities by considering it to be the superposition of the fields produced by each sheet separately.

MAIN POINTS

Use of Gauss' Law to Determine Electric Fields

Spherical Symmetry → Gaussian surface is a sphere centered on the ***point*** of symmetry.

Cylindrical Symmetry → Gaussian surface is a cylinder centered on the ***axis*** of symmetry.

Planar Symmetry → Gaussian surface is a cylinder axis of cylinder perpendicular to the ***plane*** of symmetry.

Spherical symmetry

Cylindrical symmetry

Planar symmetry

Gaussian Surfaces

Use of Gauss' Law to Determine Charge Densities on Conductors

Excess charge on conductor must reside on its surface.

Charges can be induced on surfaces of conductors in order to create $E = 0$ in the conducting material.

Connection between Symmetries and Dimensions

Spherical: Field lines expand fully into 3 Dimensions
Density of field lines $\propto 1/Area \sim 1/r^2 \Rightarrow \vec{E} \sim 1/r^2$

Cylindrical: Field lines expand into 2 Dimensions
Density of field lines $\propto 1/Area \sim 1/r \Rightarrow \vec{E} \sim 1/r$

Planar: Field lines cannot expand: only 1 Dimension
Density of field lines $\propto$ Constant $\Rightarrow \vec{E} \sim$ **constant**

PROBLEMS

1. Point Charge and Charged Sphere: A point charge $q_1 = -9$ μC is located at the center of a thick conducting spherical shell of inner radius $a = 3$ cm and outer radius $b = 4.4$ cm. The conducting shell has a net charge of $q_2 = 2.3$ μC. (a) What is $E_x(\text{P})$, the value of the x-component of the electric field at point P, located a distance 8.8 cm along the x-axis from q_1? (b) What is $E_y(\text{P})$, the value of the y-component of the electric field at point P, located a distance 8.8 cm along the x-axis from q_1? (c) What is $E_x(\text{R})$, the value of the x-component of the electric field at point R, located a distance 1.5 cm along the y-axis from q_1? (d) What is $E_y(\text{R})$, the value of the y-component of the electric field at point R, located a distance 1.5 cm along the y-axis from q_1? (e) What is σ_b, the surface charge density at the outer edge of the shell? (f) What is σ_a, the surface charge density at the inner edge of the shell? (g) For how many values of x: (4.4 cm $< x < \infty$) is it true that $E_x(x,0) = 0$: *none*, *one*, or *more than one*? (h) Define E_2 to be equal to the magnitude of the electric field at $r = 1.5$ cm when the charge on the outer shell (q_2) is equal to 2.3 μC. Define E_o to be equal to the magnitude of the electric field at $r = 1.5$ cm if the charge on the outer shell (q_2) were changed to 0. Compare E_2 and E_o: $E_2 < E_o$, $E_2 = E_o$, or $E_2 > E_o$.

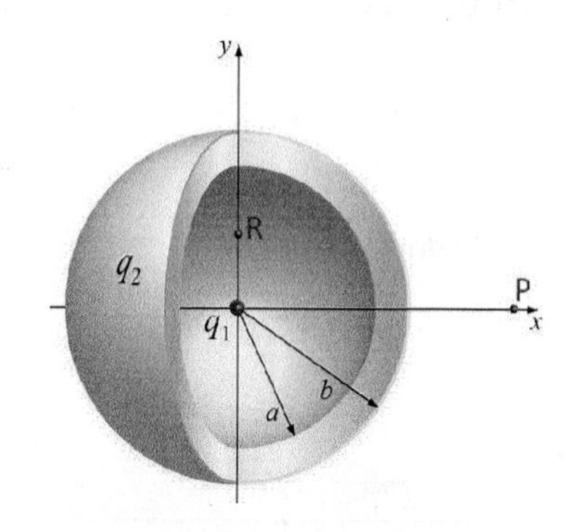

FIGURE 4.12 Problem 1. The spherical shell has been opened for clarity.

2. Line Charge and Charged Cylindrical Shell: An infinite line of charge with linear density $\lambda = 6.5$ μC/m is positioned along the axis of a thick, infinitely long, cylindrical insulating shell of inner radius $a = 2.6$ cm and outer radius $b = 4$ cm. The insulating shell is uniformly charged with a volume density of $\rho = -554$ μC/m^3. (a) What is λ_2, the linear charge density of the insulating shell? (b) What is $E_x(\text{P})$, the value of the x-component of the electric field at point P, located a distance 6.4 cm along the y-axis from the line of charge? (c) What is $E_y(\text{P})$, the value of the y-component of the electric field at point P, located a distance 6.4 cm along the y-axis from the line of charge? (d) What is $E_x(\text{R})$, the value of the x-component of the electric field at point R, located a distance 1.3 cm along a line that makes an angle of 30° with the x-axis? (e) What is $E_y(\text{R})$, the value of the y-component of the electric field at point R, located a distance 1.3 cm along a line that makes an angle of 30° with the x-axis? (f) For how many values of r, where 2.6 cm $< r <$ 4 cm, is the magnitude of the electric field equal to 0: *none*, *one*, or *more than one*? (g) If we were to double λ_1 ($\lambda_1 = 13$ μC/m), how would E,

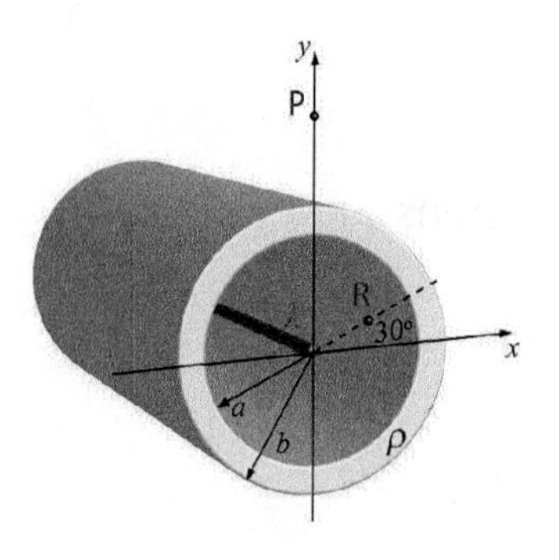

FIGURE 4.13 Problem 2

the magnitude of the electric field at point P, change?

(i) E would double.
(ii) E would increase by more than a factor of two.
(iii) E increases by less than a factor of two.
(iv) E decreases by less than a factor of two.
(v) E decreases by more than a factor of two.

(h) In order to produce an electric field of zero at some point $r > 4$ cm, how would λ_1 have to change?

(i) Change its sign and increase its magnitude.
(ii) Change its sign and decrease its magnitude.
(iii) Keep its sign the same and increase its magnitude.
(iv) Keep its sign the same and decrease its magnitude.

3. Infinite Charged Sheet and Infinite Conducting Slab: An infinite sheet of charge, oriented perpendicular to the x-axis, passes through $x = 0$. It has a surface charge density $\sigma_1 = -4.4\ \mu\text{C/m}^2$. A thick, infinite conducting slab, also oriented perpendicular to the x-axis, occupies the region between $a = 2.7$ cm and $b = 4.1$ cm. The conducting slab has a net charge per unit area of $\sigma_2 = 90\ \mu\text{C/m}^2$. (a) What is $E_x(\text{P})$, the value of the x-component of the electric field at point P, located a distance 6.5 cm from the infinite sheet of charge? (b) What is $E_y(\text{P})$, the value of the y-component of the electric field at point P, located a distance 6.5 cm from the infinite sheet of charge? (c) What is $E_x(\text{R})$, the value of the x-component of the electric field at point R, located a distance 1.35 cm from the infinite sheet of charge? (d) What is $E_y(\text{R})$, the value of the y-component of the electric field at point R, located a distance 1.35 cm from the infinite sheet of charge? (e) What is σ_b, the charge per unit area on the surface of the slab located at $x = 4.1$ cm? (f) What is E_x, the value of the x-component of the electric field at a point on the x-axis located at $x = 3.26$ cm? (g) What is σ_a, the charge per unit area on the surface of the slab located at $x = 2.7$ cm? (h) Where along the x-axis is the magnitude of the electric field equal to zero?

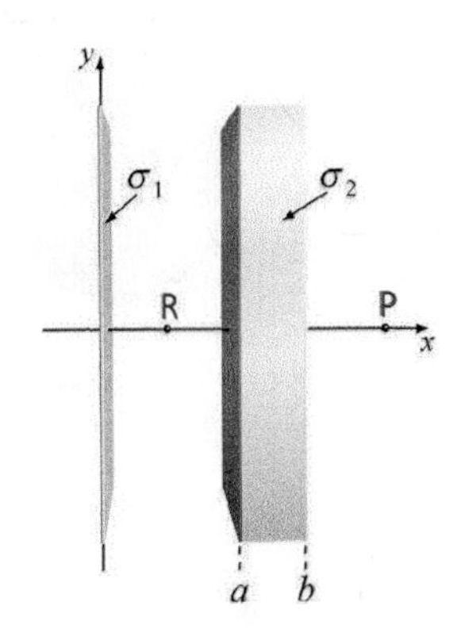

FIGURE 4.14 Problem 3

(i) $x < 0$
(ii) $0 < x < 2.7$ cm
(iii) $x > 4.1$ cm
(iv) None of these regions.

4. Spheres (INTERACTIVE EXAMPLE): An insulator in the shape of a spherical shell is defined by an inner radius $a = 4$ cm and an outer radius $b = 6$ cm and carries a total charge of $Q = +9\ \mu\text{C}$. (You may assume that the charge is distributed uniformly throughout the volume of the insulator.) What is E_y, the y-component of

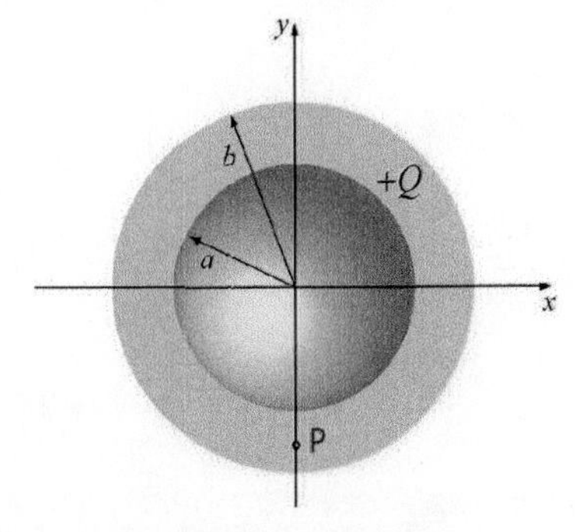

FIGURE 4.15 Problem 4

the electric field at point P, which is located at $(x, y) = (0, -5\text{ cm})$, as shown in the diagram?

5. Cylinder (INTERACTIVE EXAMPLE): Two coaxial cylindrical conductors are shown in perspective and cross-section above. The inner cylinder has radius $a = 2$ cm, length $L = 10$ m, and carries a total charge of $Q_{inner} = +8$ nC ($1\text{ nC} = 10^{-9}$ C). The outer cylinder has an inner radius $b = 6$ cm, outer radius $c = 7$ cm, length $L = 10$ m, and carries a total charge of $Q_{outer} = -16$ nC. What is E_x, the x-component of the electric field at point P, which is located at the midpoint of the length of the cylinders at a distance $r = 4$ cm from the origin and makes an angle of 30° with the x-axis?

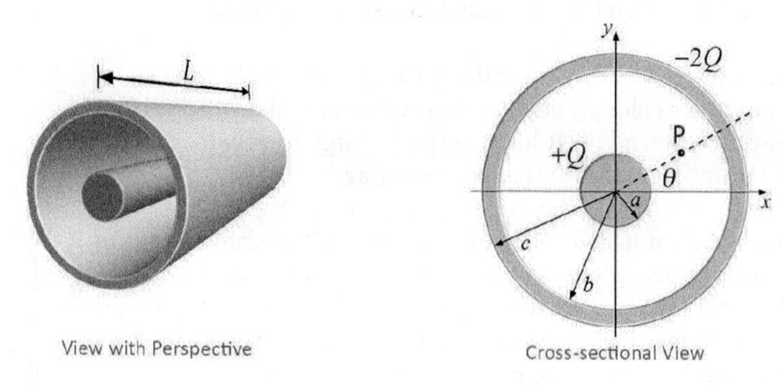

FIGURE 4.16 Problem 5

UNIT

5

ELECTRIC POTENTIAL ENERGY

5.1 Overview

We will now develop the concept of electric potential energy. We'll begin by explicitly showing the work done by the Coulomb force on an object as it moves between two points, and is independent of the path between those two points. This remarkable property of the Coulomb force identifies it as a conservative force and allows us to define an electric potential energy that is associated with this force.

We'll then present a couple of examples that illustrate the use of this potential energy. First, we will calculate the velocity a charged particle attains as it is released from rest in the electric field created by another point charge. Next, we'll calculate the amount of work required to assemble a system of point charges from infinity to a specific configuration and define that to be the electric potential energy of the system of charges.

5.2 The Coulomb Force Is a Conservative Force

As you may remember from mechanics, the concept of potential energy proved to be very useful for understanding certain physical situations. In particular, potential energies were defined for a few specific forces, namely the gravitational force and the force due to the compression or expansion of springs. These forces are called **conservative forces** because the work they do on an object as it moves between two points does not depend on the particular path between those two points (i.e., the work only depends on the initial and final position). We will now show that the Coulomb force is another such conservative force.

Let's begin by calculating the work done by the Coulomb force on a positive charge q_2 when it is moved from a distance r_B to a distance r_A from a fixed charge q_1, as shown in Figure 5.1. Since both charges are positive, the force on q_2 is repulsive and points to the right. As we move q_2 away from q_1, the magnitude of the force decreases. Thus, to calculate the work done by the Coulomb force on q_2 as it moves from r_B to r_A, we must do an integral.

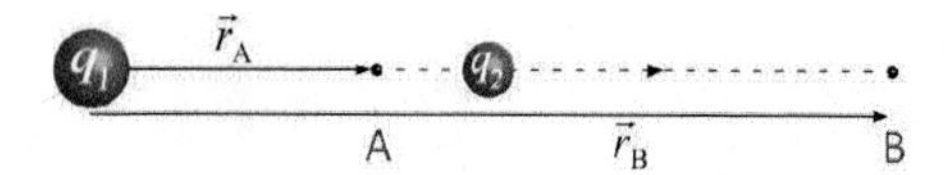

FIGURE 5.1 Positive charge q_2 is moved from point A to point B in the electric field produced by a fixed positive charge q_1.

In particular, the work done by the Coulomb force $\vec{F}_E$ on the moving particle as it travels from A to B is just given by the integral from r_A to r_B of $\vec{F}_E \cdot d\vec{r}$. Now since the vectors $\vec{F}_E$ and $d\vec{r}$ are always in the same direction, the dot product $\vec{F}_E \cdot d\vec{r}$ is just equal to the simple product $F_E dr$. We now use Coulomb's law to rewrite F_E in terms of the charges of the particles and the distance between them.

$$W_{AB} = \int_{r_A}^{r_B} \vec{F}_E \cdot d\vec{r} = \int_{r_A}^{r_B} F_E dr = \frac{1}{4\pi\varepsilon_o} \int_{r_A}^{r_B} \frac{q_1 q_2}{r^2} dr = \frac{q_1 q_2}{4\pi\varepsilon_o}\left(\frac{1}{r_A} - \frac{1}{r_B}\right)$$

The work done by the Coulomb force is proportional to the product of the charges ($q_1 q_2$) and to the difference of the inverse distances ($1/r_B - 1/r_A$).

It's always a good idea to see if your answer makes sense before moving on. In this case, we know the work done must be positive since the force and the displacement are in the same direction (both the force and the displacement point to the right). As we look at the expression for W_{AB}, we see it indeed will always be positive since r_A is always less than r_B.

In doing this calculation, we chose the path to be along the line connecting the two charges. We now want to demonstrate that we would have gotten the same answer for W_{AB}, no matter what path we had taken to get from A to B. Once we determine that the work done is independent of the path, we will have verified the claim that the Coulomb force is a conservative force.

In Figure 5.2, we show an alternate path (ACDB) that connects point A to point B. As q_2 is moved from A to C, the direction of the force is always perpendicular to the path. Consequently, the work done by the Coulomb force on q_2 as it is moved from A to C is zero. As q_2 is moved from C to D, the direction of the force is parallel to that of the path element $d\vec{r}$. Consequently, we will get the same value for the work done by the Coulomb force on q_2 as it is moved from C to D as we got when it was moved directly from A to B. Finally, as q_2 is moved from D to B, once again we see that the direction of the force is perpendicular to the path, resulting in the work done by the Coulomb force on q_2 being zero. Therefore, we see that the work done by the Coulomb force on q_2 as it is moved from A to B, either along the direct path AB or the alternate path ACDB, is the same!

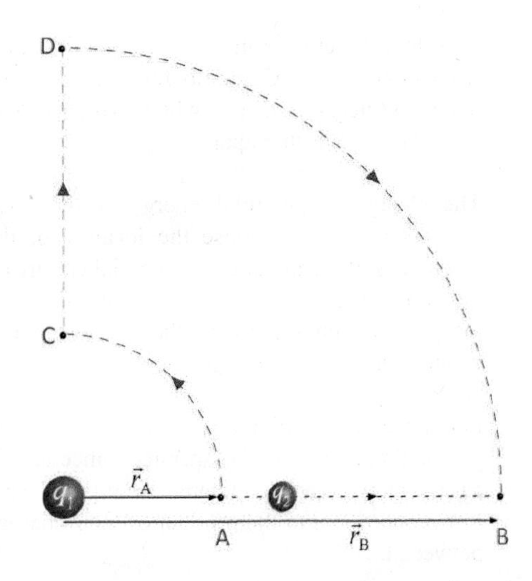

FIGURE 5.2 An alternate path connecting points A and B. The work done along this path is identical to that done along the direct path in Figure 5.1, thereby demonstrating that the Coulomb force is a conservative force.

This result is, in fact, perfectly general. As q_2 is moved along a path between any two points, we can always approximate any small part of that path as being the sum of two pieces, one perpendicular to the force (resulting in no work being done) and one parallel to the force (resulting in a difference of $(1/r)$ terms). When we sum all of these small pieces, the intermediate $(1/r)$ terms cancel and we are left with the difference of the $(1/r)$ terms for the endpoints, the same result as for the direct path.

5.3 Electric Potential Energy

Since we've just shown that the Coulomb force is a conservative force, we can define a potential energy associated with that force. Just as in mechanics, the change in potential energy of the object is defined as equal to minus the work done by the force on that object.

We can use the results from the last section to determine that the *change* in **electric potential energy** as a positively charged particle is moved from a distance r_A to a distance r_B from a fixed positively charged particle (see Figure 5.1) is just proportional to the product of the charges and the difference of the inverse distances.

$$\Delta U_{AB} = \frac{q_1 q_2}{4\pi\varepsilon_o}\left(\frac{1}{r_B} - \frac{1}{r_A}\right)$$

Note that the change in electric potential energy here is *negative* as it must be since the work done by the Coulomb force on the moving charge was *positive* (i.e., the electric potential energy is *larger* when two particles with like charges are *closer* together than when they are further apart).

The change in potential energy is the physically relevant quantity here as we are completely free to choose the location of the zero of potential energy. Recall that in mechanics, the gravitational potential energy of a mass *m* near the surface of the earth was given by *mgh*, where *h* was the height of the mass above a convenient, but arbitrary, point that we had chosen to be the zero of potential energy (e.g., the floor, the table, or whatever).

For the case of point charges, it is natural to choose r_o, the location of the zero of potential energy to be "at infinity" since as r_o goes to infinity, $1/r_o$ goes to zero. In this case, then, the electric potential energy of particle 2 that is a distance *r* from particle 1 is just proportional to the product of their charges and inversely proportional to the distance between them.

$$U_r \equiv \Delta U_{\infty r} = \frac{q_1 q_2}{4\pi\varepsilon_o r}$$

This simple equation captures the essential features of this situation.

If the charges of the two particles have the same sign, the forces are repulsive and the electric potential energy is then positive, while if the charges have opposite signs, the forces are attractive and the electric potential energy is then negative. In both cases, the closer *the charged particles are to each other*, the larger the magnitude of the electric potential energy.

5.4 Example: Speed as a Function of Distance

We'll now do an example that illustrates the use of this concept of electric potential energy. Suppose a positively charged particle q_2, initially located a distance *d* from a fixed positively charged particle q_1, is released from rest. Since q_1 exerts a repulsive force $\vec{F}$ on q_2, q_2 will accelerate to the right, as shown in Figure 5.3. The problem is to calculate *v*(*x*), the speed of q_2 as a function of *x*, its distance from q_1.

We have two choices: we know we can always use Newton's second law, but since here the force does not remain constant throughout the motion, we would need to integrate the acceleration to determine the change in speed. There is an easier way, though, since the force responsible for the motion, the Coulomb force, is a conservative force, mechanical energy is conserved throughout the motion.

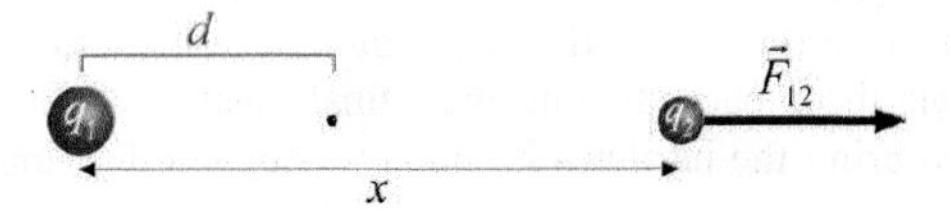

FIGURE 5.3 Positive charge q_2 is released from rest when it is a distance d from a fixed positive charge q_1. What is the speed of q_2 when it is a distance x from q_1?

Therefore, the total mechanical energy (the sum of kinetic energy plus potential energy) has the same value at all times during this process. Initially, the total mechanical energy is all potential energy. As time goes on, the potential energy decreases while the kinetic energy increases. In particular, the increase in kinetic energy is exactly equal to the decrease in potential energy. The increase in kinetic energy is just equal to $\frac{1}{2} m_2 v^2$, while the decrease in potential energy is given by the difference in the electric potential energies as calculated from the expression we obtained in the last section.

$$\frac{1}{2} m_2 v^2 = \frac{q_1 q_2}{4\pi\varepsilon_o}\left(\frac{1}{d} - \frac{1}{x}\right)$$

Simplifying, we see that the velocity v is proportional to the square root of $(x-d)/x$.

We have plotted in Figure 5.4 the speed of q_2 as a function of the separation between the two charges. Note the initial steep slope of the curve, corresponding to the large initial acceleration of the particle. As the separation between the particles increases, the acceleration decreases and the speed asymptotically approaches its maximum value, v_{max}, corresponding to the complete conversion of the initial electric potential energy into the final kinetic energy.

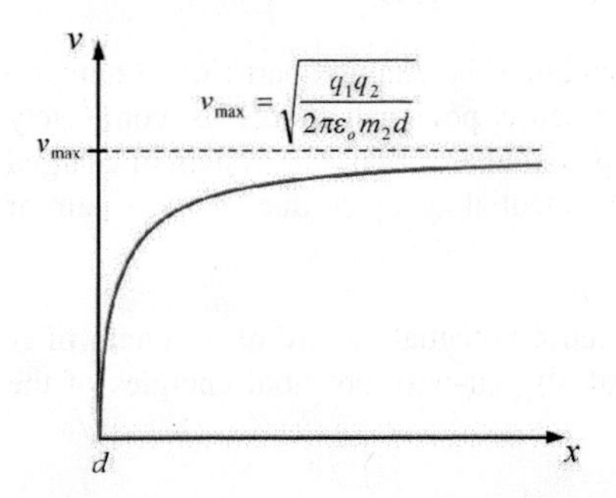

FIGURE 5.4 A plot of the speed of particle q_2 after it was released from rest (see Figure 5.3) as a function of x, its distance from q_1.

5.5 Example: Potential Energy of a System of Charges

Up to this point we have discussed the *electric potential energy of a single charged particle* in the electric field produced by another charged particle that is fixed in space. We will now discuss the *potential energy of the electric field* produced by a system of fixed charged particles.

We'll start with the system composed of three charged particles, as shown in Figure 5.5. We will *define* the electric potential energy of this system to be equal to the energy required to assemble these particles into their final positions from infinity. To make this calculation, we will bring the particles in, one by one, and determine the energy required at each step.

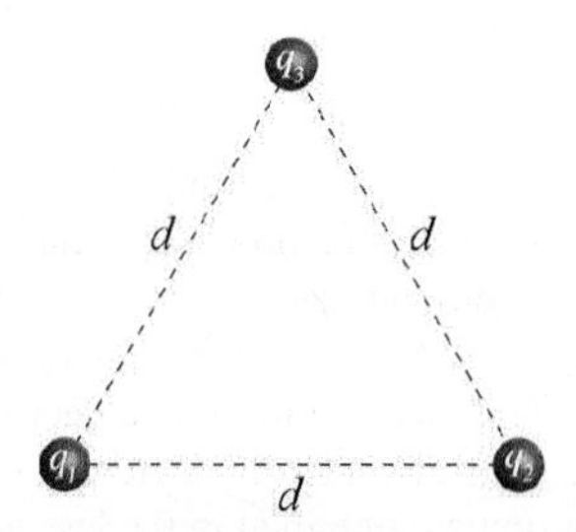

FIGURE 5.5 The system of three fixed charges fixed at the corners of an equilateral triangle of side *d*. We will calculate the potential energy of the electric field produced by these charges.

It takes no energy to position the first particle since all the other particles are at infinity. We now bring in q_2 from infinity to its final position. The electric potential energy of q_2 at this position is just equal to kq_1q_2/d. Therefore, the energy of the *system* of particles 1 and 2 is equal to kq_1q_2/d. We note that this energy is a positive number since these particles have like charges and exert repulsive forces on each other.

We now bring in q_3 from infinity to its final position. The electric potential energy of q_3 at this position is just equal to $kq_1q_3/d + kq_2q_3/d$. Therefore, the energy of the *system* of particles q_1, q_2, and q_3 is equal to $kq_1q_2/d + kq_1q_3/d + kq_2q_3/d$.

Although we performed this calculation by assembling the charged particles in numeric order, it is reassuring to see that the final electric potential energy is completely symmetric in the three charges. Indeed, the total potential energy of this system of charged particles is just the scalar sum of the electric potential energies due to each pair of particles.

We can extrapolate from this result that the electric potential energy of a system of *N* charged particles must just be the scalar sum of all pair-wise potential energies of the individual particles.

$$U_{system} = \sum_{pairs} U_{ij} = k\sum_{pairs}\frac{q_i q_j}{r_{ij}}$$

5.6 Summary

We began by demonstrating that the Coulomb force is a conservative force. By that, we mean that the work done by the Coulomb force on a charged particle as it is moved between two points is independent of the path taken between those two points.

We then proceeded to define a potential energy associated with the Coulomb force. In particular, we defined the change in electric potential energy of a charged particle moved from point A to point B to be the negative of the work done by the Coulomb force on that particle as it moves between the two points. Since the Coulomb force on the particle is just given by $\vec{F} = q\vec{E}$, we obtained the important relation that the change in potential energy ΔU_{AB} then becomes

$$\Delta U_{\mathrm{AB}} = -q\int_{\mathrm{A}}^{\mathrm{B}} \vec{E} \cdot d\vec{l}$$

where the path of integration can be any path connecting points A and B.

We then explored two examples. We first applied conservation of mechanical energy to a system composed of two charges, one fixed in place and one free to move, to determine the velocity of the moving charge as a function of its distance from the fixed charge. We then went on to define the potential energy of a system of point charges to be the negative of the work required to bring each charge, one by one, from infinity to their current positions. We arrived at the result that the potential energy of a system of point charges is just the scalar sum of the potential energies due to each pair of charges.

MAIN POINTS

The Coulomb Force Is a Conservative Force

The work done by the Coulomb force as a particle is moved between two points is independent of the path taken between those two points.

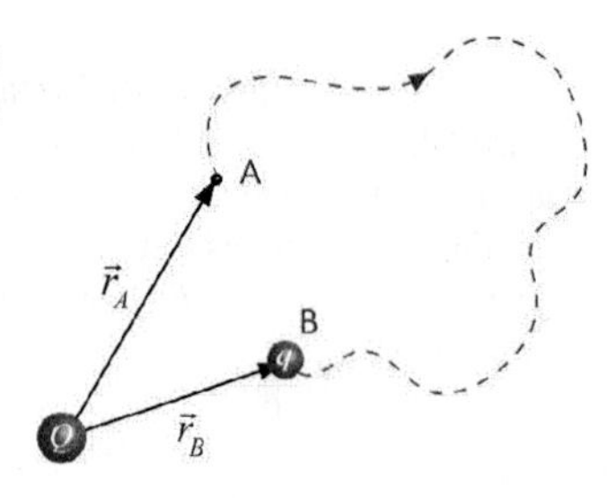

$$\vec{F}_E = \frac{1}{4\pi\varepsilon_o}\frac{Qq}{r^2}\hat{r}$$

An Electric Potential Energy is Associated with the Coulomb Force

The change in electric potential energy of a charged object as it is moved from point A to point B is defined as the negative of the work done by the Coulomb force on that object as it moves between the two points.

$$\Delta U_{A\to B} \equiv -W_{A\to B} = -\int_{A\to B} \vec{F}_E \cdot d\vec{l}$$

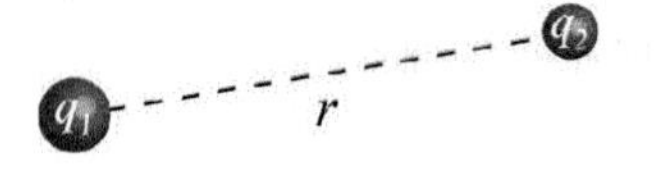

For the special case of point charges:

$$U_r \equiv \Delta U_{\infty\to r} = \frac{q_1 q_2}{4\pi\varepsilon_o r}$$

The Electric Potential Energy of a System of Fixed Charges

The electric potential energy of a system of fixed charged particles is just equal to the scalar sum of the electric potential energies due to each pair of particles.

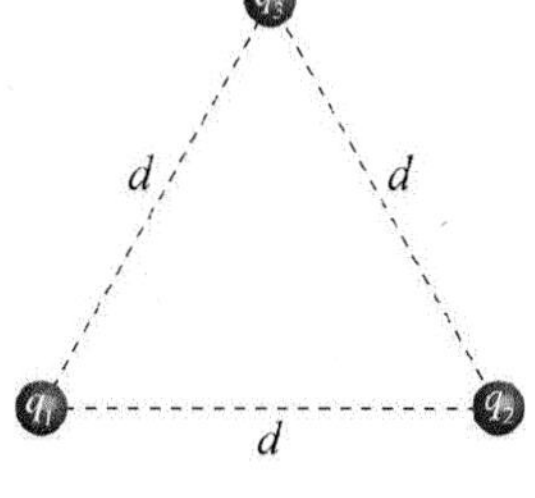

$$U_{System} = k\sum_{pairs}\frac{q_i q_j}{r_{ij}}$$

$$U_{System} = k\frac{q_1 q_2}{d} + k\frac{q_1 q_3}{d} + k\frac{q_2 q_3}{d}$$

PROBLEMS

1. Potential Energy of Point Charges: A point charge $q_2 = -3.5\ \mu\text{C}$ is fixed at the origin of a coordinate system as shown. Another point charge $q_1 = 4.7\ \mu\text{C}$ is initially located at point P, a distance $d_1 = 5.1\text{ cm}$ from the origin along the x-axis. (a) What is ΔU, the change in potential energy of charge q_1 when it is moved from point P to point R, located a distance $d_2 = 2$ cm from the origin along the x-axis as shown? (b) The charge q_2 is now replaced by two charges q_3 and q_4 that each have a magnitude of $-1.75\ \mu\text{C}$, half of that of q_2. The charges are located a distance $a = 1.2$ cm from the origin along the y-axis as shown. What is ΔU, the change in potential energy now if charge q_1 is moved from point P to point R? (c) What is the potential energy of the system composed of the three charges q_1, q_3, and q_4, when q_1 is at point R? Define the potential energy to be zero at infinity. (d) The charge q_4 is now replaced by charge q_5, which has the same magnitude, but opposite sign from q_4 (i.e., $q_5 = 1.75\ \mu\text{C}$). What is the new value for the potential energy of the system? (e) Charges q_3 and q_5 are now replaced by two charges, q_2 and q_6, having equal magnitude and sign ($-3.5\ \mu\text{C}$). Charge q_2 is located at the origin and charge q_6 is located a distance $d = d_1 + d_2 = 7.1\text{ cm}$ from the origin as shown. What is ΔU, the change in potential energy now if charge q_1 is moved from point P to point R?

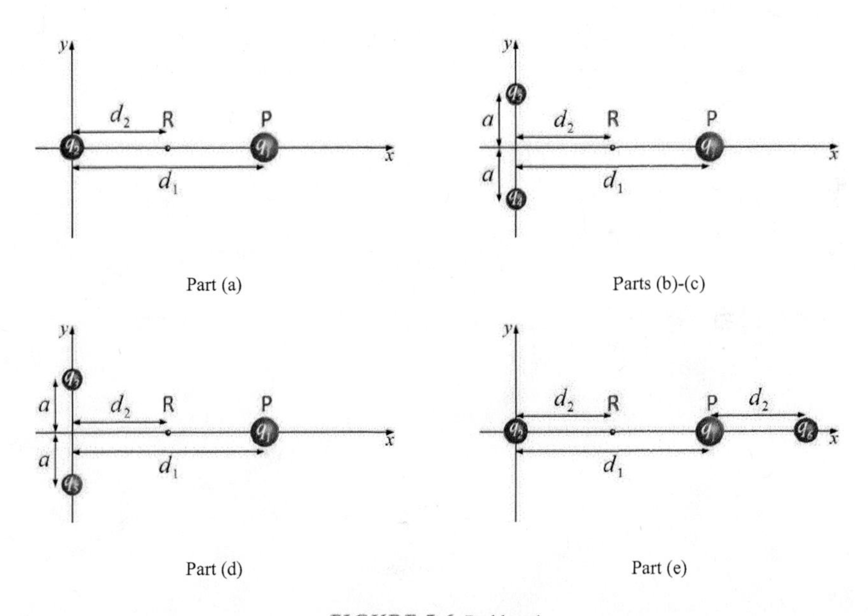

FIGURE 5.6 Problem 1

UNIT 6

ELECTRIC POTENTIAL

6.1 Overview

We will now develop the important concept of the electric potential. We'll begin by noticing that since the electric potential energy of a particle is proportional to its charge, we can define a new quantity, the **electric potential**, at any point in space as the *potential energy per unit charge of a test charge positioned at that point.* This procedure is exactly analogous to what we used in defining the electric field at any point in space as the *force per unit charge* at that point.

We'll then explore some of the consequences of this definition. In particular, we will draw an analogy to the gravitational force near the surface of the Earth to represent the electric potential as a "hill," where the steepness of the hill indicates the strength of the electric field. We'll also show how the potential (a scalar) or the electric field (a vector) can equally be used to describe all characteristics of the electrostatic environment.

Finally, we will present some examples that illustrate the calculation of the electric potential for several geometries. We will begin with a system of point charges, followed by a uniformly charged solid spherical insulator, and finally, a uniformly charged cylindrical shell.

6.2 Electric Potential Defined

We will now develop the concept of electric potential from the electric potential energy in exactly the same way we developed the concept of the electric field from the electric force. Namely, when we saw the Coulomb force on a charged particle at a point in space was proportional to its charge, we defined the electric field, a property of the space, as the *force per unit charge.*

We will now do exactly the same thing with the electric potential energy. Since the change in the particle's electric potential energy is proportional to its charge, we will define the **electric potential difference**, a property of the space, as the *change in electric potential energy per unit charge.*

We start from the definition of the electric potential energy in terms of the work done by the Coulomb force. The work done by the Coulomb force $\vec{F}$ as a charged particle moved from point A to point B is equal to the integral of $\vec{F} \cdot d\vec{l}$ from A to B.

$$W_{\mathrm{A}\to\mathrm{B}} = \int_{\mathrm{A}}^{\mathrm{B}} \vec{F} \cdot d\vec{l}$$

We can express the Coulomb force at any point P along the path as the product of the charge and the electric field at that point. From this expression for the work done by the field, we can obtain the expression for ΔU, the change in electric potential energy of the charge as it is moved from A to B.

$$\Delta U_{\mathrm{A}\to\mathrm{B}} = -q\int_{\mathrm{A}}^{\mathrm{B}} \vec{E} \cdot d\vec{l}$$

Note that the change in potential energy is proportional to the charge q. Therefore, we can divide ΔU by q to obtain our final expression for ΔV, the electric potential difference between A and B directly in terms of the integral of $\vec{E} \cdot d\vec{l}$ along any path connecting A and B.

$$\Delta V_{\mathrm{A}\to\mathrm{B}} \equiv \frac{\Delta U_{\mathrm{A}\to\mathrm{B}}}{q} = -\int_{\mathrm{A}}^{\mathrm{B}} \vec{E} \cdot d\vec{l}$$

For example, to determine the electric potential difference between A and B due to the electric field produced by a point charge at the origin, we need to integrate kq/r^2. The result is that the electric potential difference is proportional to the difference of the inverse distances of A and B from the point charge. This potential difference ΔV is the physically relevant quantity, but note that it is defined in terms of two points in space (A and B). It is common to introduce a quantity V, which is defined for each single point in space as the difference in electric potential between that point and some reference point

$\vec{r}_o$ called the zero of potential. That is, $V(r)$ is defined to be equal to the potential difference between $\vec{r}_o$ and $\vec{r}$:

$$V(r) = -\int_{\vec{r}_o}^{\vec{r}} \vec{E} \cdot d\vec{l}$$

where $\vec{r}_o$ is the point at which V is defined to be zero. For our example of the electric potential produced by a point charge at the origin, it is common to choose the zero of the potential to be at $r_o = \infty$, so that $V(r)$ becomes simply equal to kq/r.

The dimensions of potential (or potential difference) are Joules per Coulomb. The common unit for electric potential is the volt (V), defined as 1 Joule / Coulomb. I'm sure you're familiar with batteries, devices whose purpose it is to provide a constant potential difference between their terminals, as measured in volts.

6.3 Obtaining the Electric Field from the Electric Potential

We've just defined the electric potential in terms of the path integral of the electric field. This definition establishes an essential connection between the electric field and the electric potential. We can think of these quantities as being different encodings of all information necessary to describe a given electrostatic environment. How do we convert from one description to the other? The big idea is simply that if we can integrate the electric field to determine the electric potential, we can also do the inverse, which is to differentiate the electric potential to determine the electric field.

We'll start with the example of the point charge fixed at the origin that we discussed in the last section. Namely, choosing the zero of potential at $r_o = \infty$ yields the simple expression for the potential: $V(r) = kq/r$. Figure 6.1 shows a plot of this potential as a function of r, the distance from the point charge. We expect that the electric field produced by the point charge can be obtained from this expression by simply differentiating. In particular, we expect $E = -dV/dr$. When we differentiate the form for $V(r)$, we do indeed get back the familiar result for the E field produced by a point charge, namely $E = kq/r^2$.

To illustrate this result, the slopes of the tangent lines to the potential function (Figure 6.1(a)) are drawn at two values of r ($r = 1$ and $r = 2$) in the $V(r)$ vs. r plot. These slopes correspond to the values of the electric field function at the same values of r. Indeed, we see that the slope has decreased by a factor of 4, exactly the decrease expected for the electric field if r is increased by a factor of two.

When considering problems involving two or three dimensions, we need to differentiate the potential in two or three dimensions to yield all of the components of the electric field. This is called finding the *gradient* of the potential.

$$\vec{E} = -\vec{\nabla} V$$

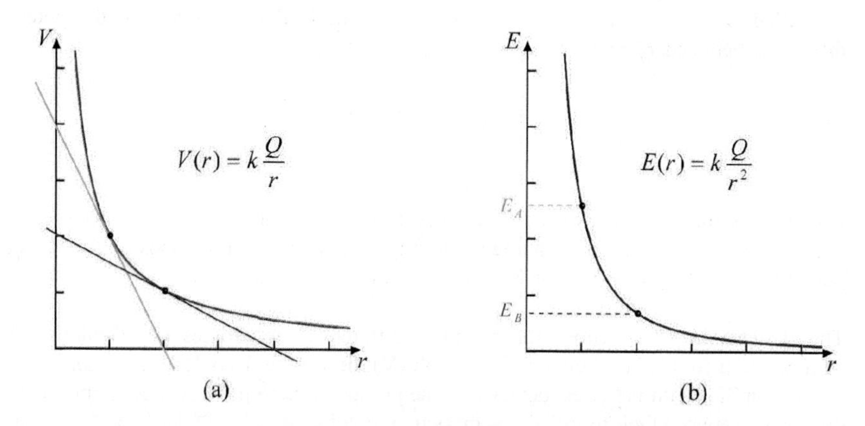

FIGURE 6.1 (a) The electric potential produced by a point charge as a function of the distance from the point charge. (b) The magnitude of the electric field produced by a point charge as a function of the distance from the point charge. The *slopes of the tangent lines* in plot (a) correspond to the *values* in plot (b): $E = -dV/dr$

A great way to visualize this in the case of two dimensions, x and y, is to plot the magnitude of the electric potential in the third dimension, z. The potential shown in Figure 6.2 is just that of our example of a point charge located at the origin in the x-y plane. As in the one-dimensional case, the steepness of the potential at any point tells us the magnitude of the electric field, but now that the slope at any point of the potential surface also has a direction, thereby giving us the direction of the field as well.

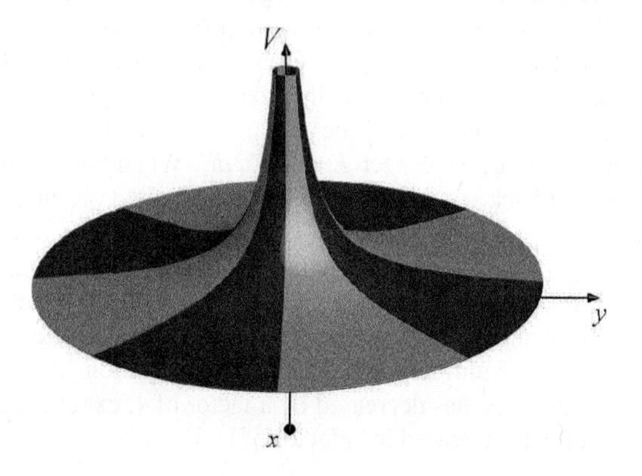

FIGURE 6.2 A plot of the electric potential V produced by a positive point charge at the origin as a function of two orthogonal spatial dimensions, x and y.

The mathematical form of the gradient depends upon the coordinate system used. In Cartesian coordinates, the form of the gradient is what you might expect, namely the generalization of the one-dimensional expression.

$$\vec{\nabla} V = \frac{\partial V}{\partial x}\hat{i} + \frac{\partial V}{\partial y}\hat{j} + \frac{\partial V}{\partial z}\hat{k} \qquad \text{(Cartesian)}$$

The gradient takes on a somewhat more complicated form in spherical and cylindrical coordinates.

$$\vec{\nabla} V = \frac{\partial V}{\partial r}\hat{r} + \frac{1}{r}\frac{\partial V}{\partial \theta}\hat{\theta} + \frac{1}{r\sin\theta}\frac{\partial V}{\partial \phi}\hat{\phi} \qquad \text{(Spherical)}$$

$$\vec{\nabla} V = \frac{\partial V}{\partial r}\hat{r} + \frac{1}{r}\frac{\partial V}{\partial \theta}\hat{\theta} + \frac{\partial V}{\partial z}\hat{k} \qquad \text{(Cylindrical)}$$

6.4 Equipotentials

We have just seen that we can visualize the electric potential created by a positive charge as a "hill," in that a change in electric potential is represented as a change in height. This representation is a powerful way of thinking about the electric potential. The "steepness" of the hill represents the strength of the electric field. This picture reinforces the fact that "absolute height" is not important, what matters is the change in height (i.e., the electric field is determined from this change in height, from the "steepness" of the curve that certainly does *not* depend on what particular height you call "zero"). We can choose "zero" to be anywhere; for a collection of point charges, we usually choose zero to be "at infinity," whereas geographic heights are usually given relative to sea level.

We can develop this graphical picture further by including a new quantity, an **equipotential**, the locus of all points having the same potential. Clearly, in our picture an equipotential results from intersecting a plane of constant height with our potential surface, as indicated by the equally spaced stripes from the side view of the potential plot shown in Figure 6.3.

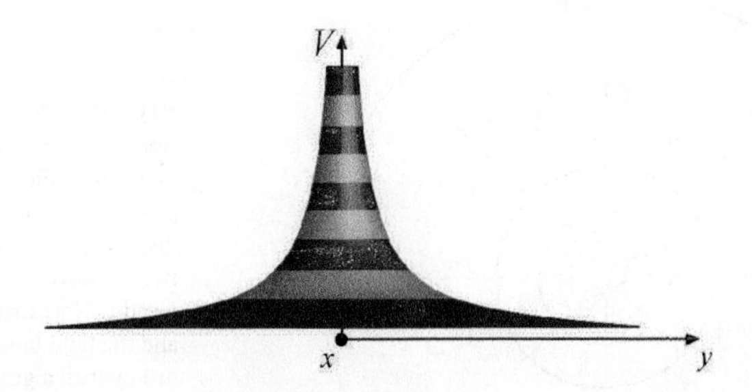

FIGURE 6.3 A plot of the electric potential V produced by a positive point charge at the origin as a function of two orthogonal spatial dimensions, x and y. The equally-spaced stripes correspond to equipotentials, the locus of all points in the x-y plane that have the same potential.

If we rotate the plot so that we are looking directly down at the potential, we see that the stripes no longer appear to be equally spaced, as shown in Figure 6.4. This view gives us the two-dimensional projection of these equipotentials in the x-y plane. Note that the spacing of these equipotentials (the red circles) increases as we increase r and move away from the origin (the location of the positive point charge that is creating this potential).

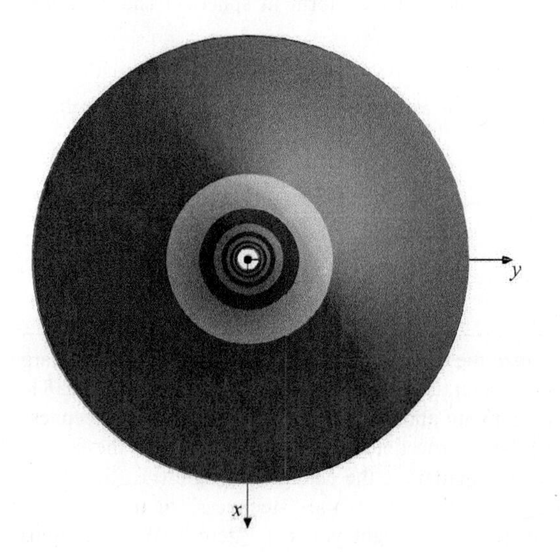

FIGURE 6.4 The projection of the electric potential V produced by a positive point charge at the origin onto a plane described by two orthogonal spatial dimensions, x and y. The stripes correspond to *equipotentials*, the locus of all points in the x-y plane that have the same potential.

We now superimpose the radial electric field lines produced by this positive charge in Figure 6.5. Note that these field lines are always perpendicular to the equipotentials. This is a completely general result; since to move between points in the direction of the electric field requires work to be done, the potential must change. Consequently, *an equipotential must everywhere be perpendicular to the electric field.*

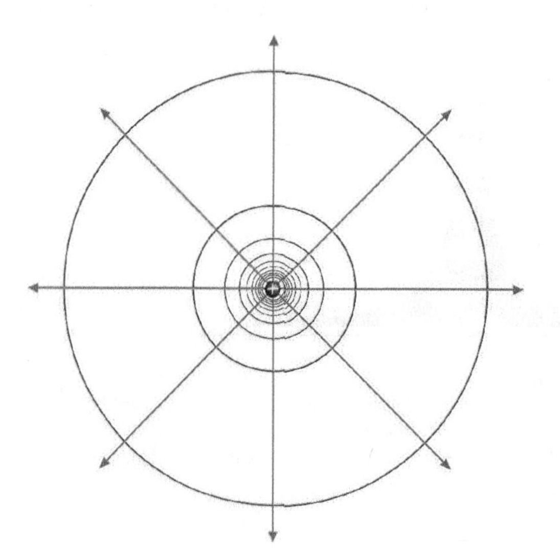

FIGURE 6.5 The equipotentials (red circles) and the electric field lines (blue lines) produced by a positive point charge at the origin displayed in a plane described by two orthogonal spatial dimensions, x and y. Note that the equipotentials and the field lines are everywhere orthogonal, a general result for any charge distribution.

Note that we have chosen to represent the equipotentials using constant spacing in potential just as a topographic map displays contours of constant elevation using constant spacing in height. Therefore, at places where the equipotentials are close together, the potential is changing rapidly (i.e., a strong electric field), while at places where they are far apart, the potential is relatively constant (i.e., a weak electric field). Since potential is a scalar, the potential created by a combination of charged particles is just the scalar sum of the potential created by each individual particle. Figure 6.6 shows the potential created by an electric dipole. Note that while the positive charge creates a hill, the negative charge creates a valley.

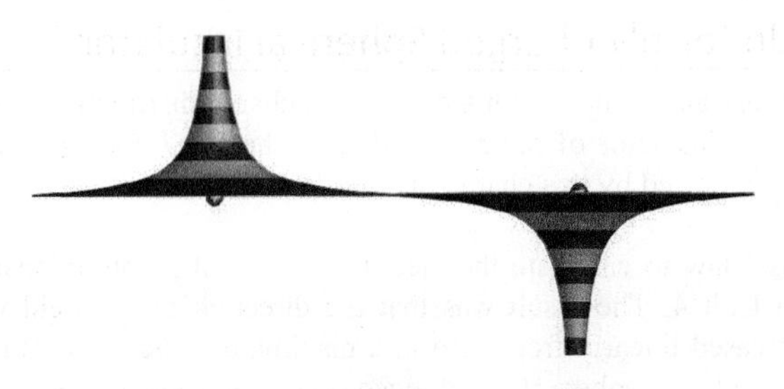

FIGURE 6.6 The electric potential produced by an electric dipole. The positive charge (blue) creates a hill, while the negative charge (red) creates a valley.

6.5 Example: Electric Potential from Collection of Point Charges

Our first example is the collection of point charges shown in Figure 6.7. Two positive charges and one negative charge, all having the same magnitude, are positioned at three corners of a square. The problem is to calculate the electric potential at point P, the fourth corner of the square.

Since we have a collection of point charges here, we can and we will define the zero of potential to be "at infinity." We will use the principle of superposition to calculate the total potential as the sum of the potentials created by each charge separately.

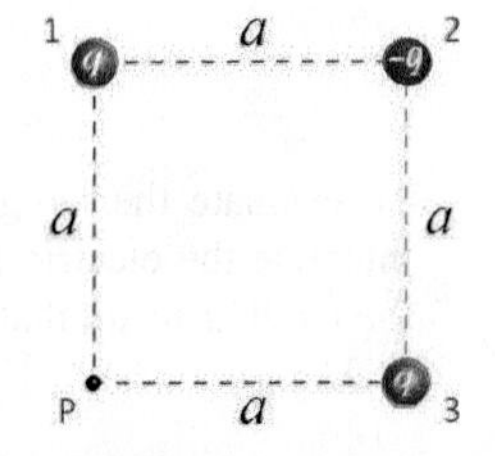

FIGURE 6.7

Since the potential is a scalar, the sum is just an arithmetic sum. There are no vectors here to worry about.

The electric potential due to a point charge is equal to kQ/r. Each positive charge is a distance a from point P; consequently, the contribution from each of these charges separately is just equal to $+kQ/a$. The negative charge is located a distance $a\sqrt{2}$ from point P so that its contribution is equal to $-kQ/a\sqrt{2}$. The total electric potential at point P then is just the sum of these three terms.

$$V_P = \sum_i V_i = k\frac{q}{a} - k\frac{q}{a\sqrt{2}} + k\frac{q}{a}$$

It's always a good idea to check to see if the result makes sense. Our final result predicts that the electric potential at point P is positive (since $2 > 1/\sqrt{2}$) and decreases as $1/a$. These predictions make sense since the potential must be positive since all charges have the same magnitude, but the *closest* charges are both positive and the contribution from each charge separately decreases as $1/a$ as the size of the square increases.

6.6 Example: Uniformly Charged Spherical Insulator

We'll now discuss an example with a continuous charge distribution, namely a uniformly charged spherical insulator of radius a and total charge Q. Our goal is to calculate the electric potential created by this charge distribution.

We used Gauss' law to calculate the electric field at all points in space for this charge distribution in Unit 4. The result was that the direction of the field was radial and its magnitude increased linearly from zero to a maximum value at the surface and then fell off as $1/r^2$ outside the sphere. Indeed, once outside the sphere, the field was equivalent to that of a point charge of magnitude Q located at the center of the sphere. The spatial dependence of the magnitude of the electric field for this charge distribution was shown in Figure 4.4.

To calculate the electric potential, we return to its definition, namely that

$$V(r) = -\int_{\vec{r}_o}^{\vec{r}} \vec{E} \cdot d\vec{l}$$

We choose our zero of potential to be "at infinity" so that $V(r)$ becomes

$$V(r) = -\int_{\infty}^{r=r} \vec{E} \cdot d\vec{l}$$

To evaluate the integral, we'll consider the two regions separately. First, for $r > a$, we integrate the electric field (which is proportional to Q/r^2) from infinity to r and obtain the familiar result that $V(r)$ is proportional to Q/r.

$$V(r) = \frac{Q}{4\pi\varepsilon_o r}$$

To determine the electric potential for $r < a$, we will need to do the integral in two pieces, since the forms for the electric field in the regions inside and outside the sphere are different. In particular, we will integrate first from infinity to a and then from a to r. In both regions, the electric field is parallel to r, so that $\vec{E} \cdot d\vec{l}$ becomes just $E\,dr$.

$$V(r) = -\int_{\infty}^{r=r} E\,dr = -\int_{\infty}^{r=a} E\,dr - \int_{r=a}^{r=r} E\,dr$$

We first integrate from infinity to $r = a$ to get the familiar $(1/(4\pi\varepsilon_o))\,(Q/a)$ result. We now add to this value the integral from a to r. This last integral $(\Delta V(a \rightarrow r))$ is proportional to $(a^2 - r^2)$. Therefore, we can put these pieces together to obtain $V(r) = V(a) + \Delta V(a \rightarrow r)$, which gives the final result:

$$V(r) = \frac{Q}{4\pi\varepsilon_o}\left[\frac{1}{a} + \frac{1}{2a^3}\left(a^2 - r^2\right)\right]$$

We plot $V(r)$ in Figure 6.8. We see that $V(r)$ is a continuous function that peaks at the center of the sphere. Once again, we see the potential represented as a "hill."

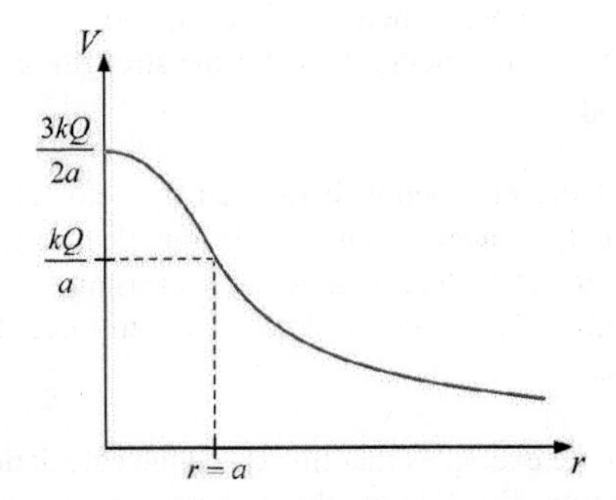

FIGURE 6.8 The electric potential V produced by a uniformly charged solid spherical insulator of radius a as a function of r, the distance from its center.

6.7 The Big Picture

Before concluding this unit, we will stop here to take stock of the big picture that we are developing.

We start from the idea that all electrostatic effects arise from the Coulomb force between charged objects. Since the Coulomb force on any charged particle is proportional to the charge of that particle, we can define a vector quantity, the electric field, as the force per unit charge at all points in space. Since this Coulomb force is a conservative force, we can also define a scalar quantity, the electric potential, as the electric potential energy per unit charge at all points in space. Although the electric field is a vector and the electric potential is a scalar, we can use either one of them to determine *all* electrostatic effects.

6.8 Summary

We began by defining the electric potential, at any point in space, as the potential energy per unit charge of a test charge positioned at that point, much in the same way that we defined the electric field at any point in space as the force per unit charge at that point. Since the potential energy was defined in terms of the work done by the Coulomb force, we can obtain the potential directly from the electric field, namely

$$V(r) = -\int_{\vec{r}_o}^{\vec{r}} \vec{E} \cdot d\vec{l}$$

where $\vec{r}_o$ is the point at which V is defined to be zero. We then went on to draw an analogy to the gravitational force near the surface of the Earth by representing the electric potential as a "hill," with the steepness of the hill indicating the strength of the electric field. In this picture, equipotentials, the loci of all points having the same potential, correspond to contours of constant height. These equipotentials are always perpendicular to electric field lines and their spacing indicates the strength of the electric field, how fast the potential is changing.

We then demonstrated the connection between the vector electric field and the scalar electric potential. Namely, that we could obtain the electric field from the potential by simply differentiating it. The electric field is a measure of how fast the potential is changing in any direction. We generalized this result to three dimensions by introducing the three dimensional derivative, the gradient.

Finally, we explored some examples that illustrated the calculation of the electric potential for both a discrete charge distribution (three charges at the corners of a square) and a continuous charge distribution (a uniformly charged solid spherical insulator). In this last example, we found that the potential resembled a "hill," where the potential decreases as the distance from the center of the distribution increases.

MAIN POINTS

Electric Potential

The electric potential difference between any two points is defined as the energy required to move a unit positive charge between the two points.

$$\Delta V_{A\to B} \equiv \frac{\Delta U_{A\to B}}{q} = -\int_{A}^{B} \vec{E}\cdot d\vec{l}$$

The electric potential at a given point, $\vec{r}$, is defined as the potential difference between that point and an arbitrary point, $\vec{r}_o$, chosen as the zero of electric potential.

$$V(\vec{r}) \equiv \Delta V_{\vec{r}_o\to\vec{r}} = -\int_{\vec{r}_o}^{\vec{r}} \vec{E}\cdot d\vec{l}$$

Equipotentials

An equipotential is defined as the locus of all points having the same potential. Equipotentials are always perpendicular to the electric field and their spacing indicates the strength of the electric field.

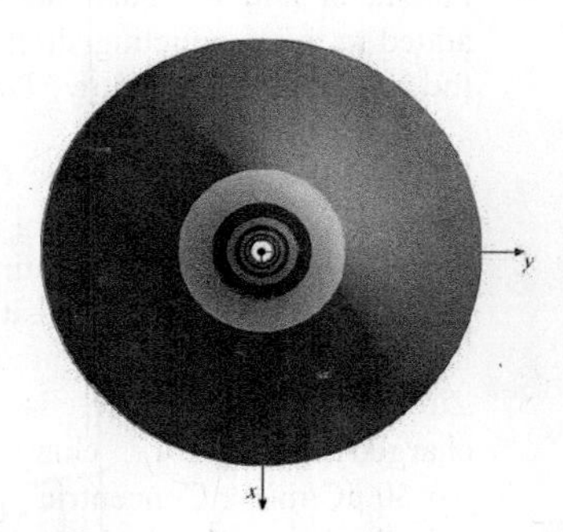

The Electric Field Can Be Obtained from the Electric Potential

The electric field (a vector) can be obtained from the electric potential (a scalar) at any point in space by applying the gradient function. The electric field is a measure of how fast the potential is changing.

$$\vec{E} = -\vec{\nabla} V$$

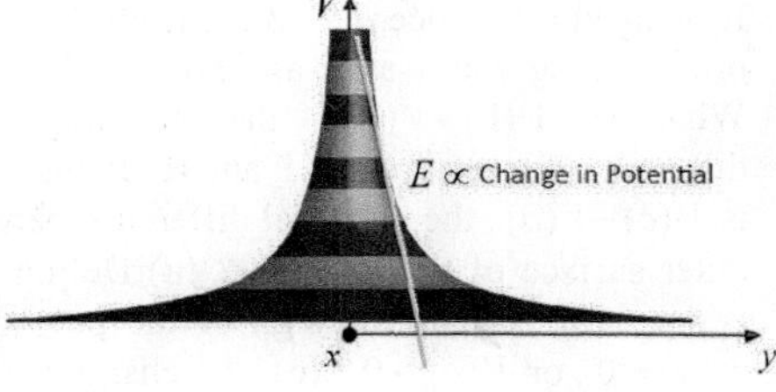

PROBLEMS

1. Potential of Concentric Spherical Insulator and Conductor: A solid insulating sphere of radius $a = 3.3$ cm is fixed at the origin of a coordinate system as shown. The sphere is uniformly charged with a charge density $\rho = -164\ \mu\text{C/m}^3$. Concentric with the sphere is an uncharged spherical conducting shell of inner radius $b = 10.9$ cm and outer radius $c = 12.9$ cm. (a) What is $E_x(\text{P})$, the x-component of the electric field at point P, located a distance of 27 cm from the origin along the x-axis as shown? (b) What is $V(b)$, the electric potential at the inner surface of the conducting shell? Define the potential to be zero at infinity. (c) What is $V(a)$, the electric potential at the outer surface of the insulating sphere? Define the potential to be zero at infinity. (d) What is $V(c) - V(a)$, the potential difference between the outer surface of the conductor and the outer surface of the insulator? (e) A charge $Q = 0.0212\ \mu\text{C}$ is now added to the conducting shell. What is $V(a)$, the electric potential at the outer surface of the insulating sphere, now? Define the potential to be zero at infinity.

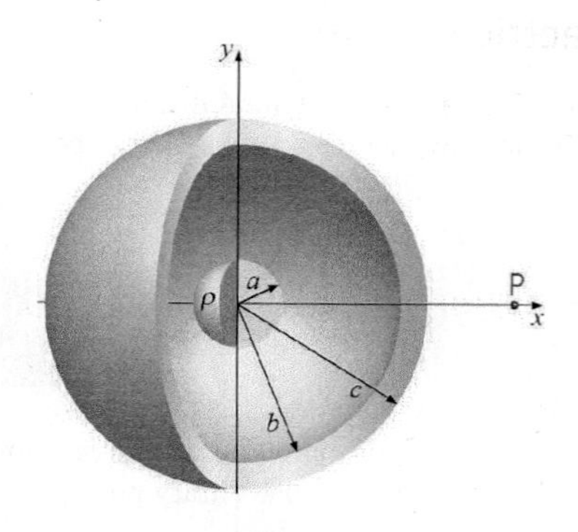

FIGURE 6.9 Problem 1. The insulator and conductor are shown with a section removed for clarity.

2. Potential of Concentric Cylindrical Insulator and Conducting Shell: An infinitely long solid insulating cylinder of radius $a = 4.2$ cm is positioned with its symmetry axis along the z-axis as shown. The cylinder is uniformly charged with a charge density $\rho = 30\ \mu\text{C/m}^3$. Concentric with the cylinder is a cylindrical conducting shell of inner radius $b = 14.7$ cm and outer radius $c = 19.7$ cm. The conducting shell has a linear charge density $\lambda = -0.38\ \mu\text{C/m}$. (a) What is $E_y(\text{R})$, the y-component of the electric field at point R, located a distance $d = 53$ cm from the origin along the y-axis as shown? (b) What is $V(\text{P}) - V(\text{R})$, the potential difference between points P and R? Point P is located at $(x, y) = (53\text{ cm}, 53\text{ cm})$. (c) What is $V(c) - V(a)$, the potential difference between the outer surface of the conductor and the outer surface of the insulator? (d) Defining the zero of potential to be along the z-axis ($x = y = 0$), what is the sign of the potential at the surface of the insulator: $V(a) < 0$, $V(a) = 0$, or $V(a) > 0$? (e) The charge density of the insulating cylinder is now changed to a new value, ρ', and it is found that the electric field at point P is now zero. What is the value of ρ'?

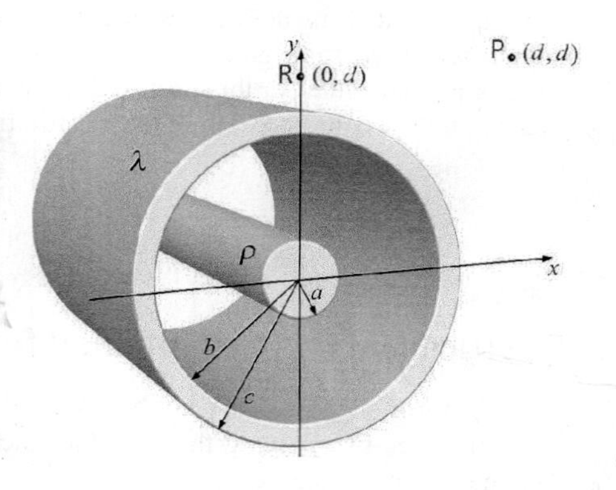

FIGURE 6.10 Problem 2

3. Potential of Infinite Sheets of Charge and Conducting Slab: An infinite sheet of charge is located in the *y-z* plane at $x = 0$ and has uniform charge density $\sigma_1 = 0.58\ \mu\text{C/m}^2$. Another infinite sheet of charge with uniform charge density $\sigma_2 = -0.51\ \mu\text{C/m}^2$ is located at $x = c = 40$ cm. An uncharged infinite conducting slab is placed halfway in between these sheets (i.e., between $x = 18$ cm and $x = 22$ cm). (a) What is $E_x(\text{P})$, the *x*-component of the electric field at point P, located at $(x, y) = (9\text{ cm}, 0)$? (b) What is σ_a, the charge density on the surface of the conducting slab at $x = 18$ cm? (c) What is $V(\text{R}) - V(\text{P})$, the potential difference between point P and point R, located at $(x, y) = (9\text{ cm}, -22\text{ cm})$? (d) What is $V(\text{S}) - V(\text{P})$, the potential difference between point P and point S, located at $(x, y) = (31\text{ cm}, -22\text{ cm})$? (e) What is $E_x(\text{T})$, the *x*-component of the electric field at point T, located at $(x, y) = (49\text{ cm}, -22\text{ cm})$? (f) Which of the following plots gives the correct *x*-dependence for the potential function between $x = 0$ and $x = 40$ cm?

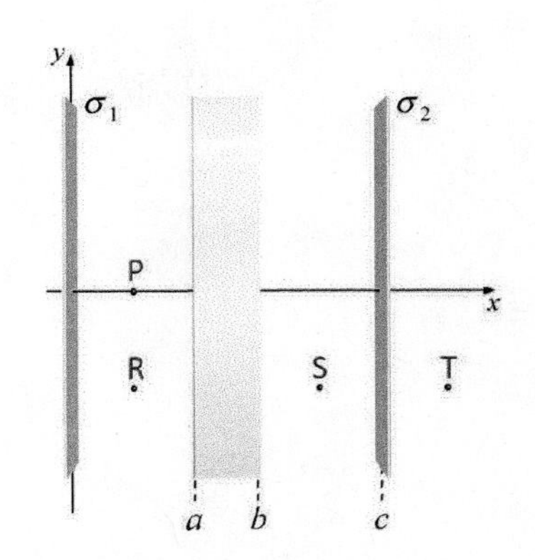

FIGURE 6.11 Problem 3

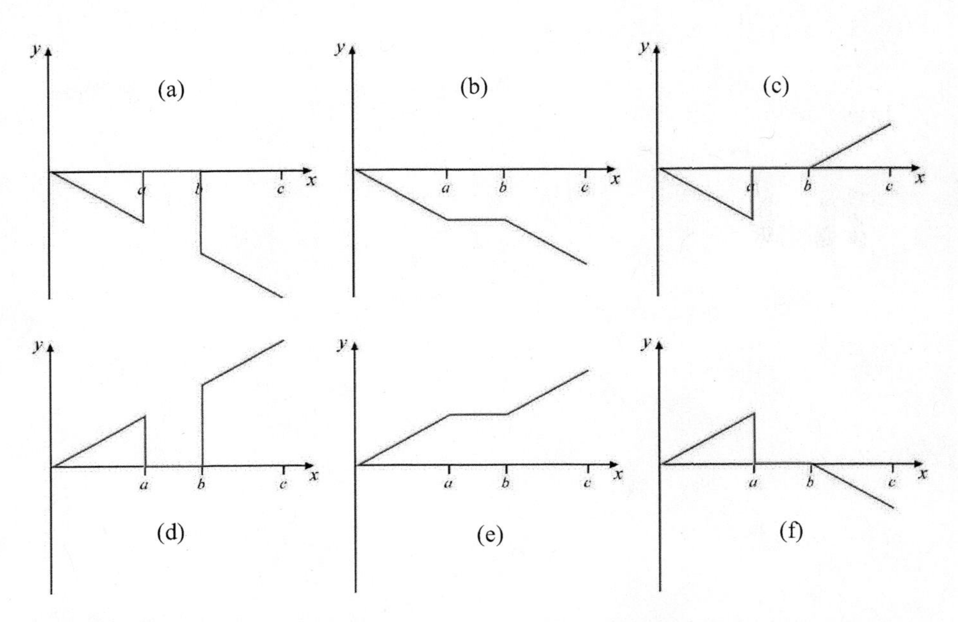

4. Spheres *V* (INTERACTIVE EXAMPLE): A solid metal sphere of radius $a = 2.5$ cm has a net charge $Q_{in} = -3$ nC ($1\text{ nC} = 10^{-9}$ C). The sphere is surrounded by a concentric conducting spherical shell of inner radius $b = 6$ cm and outer radius $c = 9$ cm. The shell has a net charge $Q_{out} = +2$ nC. What is V_o, the electric potential at the center of the metal sphere, given the potential at infinity is zero?

UNIT 7

CONDUCTORS AND CAPACITANCE

7.1 Overview

We will now extend our study of electric potential to include conductors. We will also introduce the concept of capacitance.

In Unit 6, we introduced the concept of electric potential, but all of the examples had fixed charge distributions. Once we consider conductors, the charges are free to move and their distribution cannot be arbitrarily specified. In fact, the concept of electric potential is extremely useful in describing the movement of these charges. We will discover that the surfaces of a conductor are equipotentials and that charges on conductors with an internal cavity will move to create a zero field in this cavity, thereby "shielding" that area from outside electrical effects.

We will then introduce the concept of capacitance in the context of moving electric charge between two spatially separated conductors. We will determine the capacitance for two specific examples, parallel plates, and concentric spheres. We will also discuss the energy that is stored in a capacitor when it is charged.

7.2 Conductors are Equipotentials

Our first task is to demonstrate the important result that conductors are equipotentials, i.e., that every point on a conductor, on the surface or inside, is at the same potential.

The electric potential difference between any two points is found by integrating the electric field between them. Since the field is zero everywhere in the conductor, this integral will also be zero. Consequently the potential difference between any two points in the conductor is zero, which means that the entire conductor itself is an equipotential.

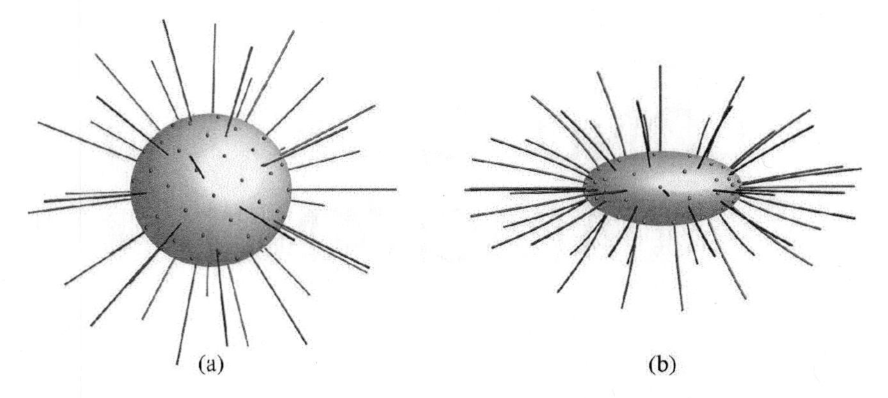

FIGURE 7.1 (a) The electric field lines produced by a spherical conductor. (b) The electric field lines produced by a non-spherical conductor. The field lines in (b) are no longer spherically symmetric, but they are everywhere perpendicular to the surface, since the surface is an equipotential. Note that the charge distribution is also no longer uniform.

Figure 7.1(a) shows a spherical conductor that carries an excess charge $+Q$. This charge distributes itself uniformly on the surface and produces the spherically symmetric field shown.

Note that the field lines leaving the conductor are all perpendicular to the surface, in agreement with what we learned in the last unit that equipotentials are always perpendicular to the field lines. When we change the shape of the cross-section of the conductor so that it is no longer circular as in Figure 7.1(b), we will find that the charges redistribute themselves so that the charge density is no longer uniform and the field is no longer spherically symmetric. Since the conductor is still an equipotential, however, we see that, at its surface, the field lines are still perpendicular to the surface.

7.3 Equipotential Example

We'll now look at an example that reinforces the idea of conductors as equipotentials. Consider the case shown in Figure 7.2 where we have two conducting spheres, each with net charge $+Q$ and separated by a large distance. R_B, the radius of sphere B, is four times R_A, the radius of sphere A.

FIGURE 7.2 Two conducting spheres, each with net charge $+Q$ separated by a large distance. The radius of the sphere B is four times that of the sphere A.

If the spheres are separated by a distance that is large compared to the radius of either sphere, we can neglect the contribution of the field produced by B near the surface of A. Consequently, we can calculate the potential at the surface of A by integrating the field produced by A only. The field produced by A for all points beyond its surface is equivalent to the field produced by a point charge $+Q$ located at the center of A. Consequently, when we integrate this field from infinity to the surface of A we obtain the familiar result for the electric potential due to a point charge.

$$V_A = \frac{1}{4\pi\varepsilon_o}\frac{Q}{R_A}$$

Repeating this operation for sphere B, we obtain the expected result that the potential is proportional to $1/R_B$ so that the potential at the surface of sphere B is ¼ of that at the surface of A.

$$V_B = \frac{1}{4\pi\varepsilon_o}\frac{Q}{R_B} = \frac{1}{4}V_A$$

What will happen if we connect the spheres by a thin wire? Charge will probably flow, but how much will flow and in what direction? The answer can be determined by realizing that connecting the two spheres by the wire has served to make them, in essence, a single conductor. Therefore, charges will flow until both spheres come to the same potential.

Which way will the charges flow? Since A is originally at a higher potential than B, the charge will flow "downhill" from A to B until the spheres come to the same potential. To determine the final charges on A and B, we set their potentials to be equal and determine that the final charge on sphere B will be four times the final charge on sphere A. Since the total charge must remain the same, the final charge on B will increase to eight-fifths of its original charge while the final charge on A will decrease to two-fifiths of its original charge.

Note that even though sphere B has four times the charge of sphere A, the charge density on sphere A is still larger (four times larger in fact) than the charge density at the surface of sphere B, since the surface area is proportional to the square of the radius of the sphere. This result reflects a general feature of conductors that the charge density is largest at the points of largest curvature (the smallest radius), producing the largest electric fields at those points.

7.4 Charge Distribution on Conductors

We'll now look at another important example, an uncharged spherical conducting shell with a positive charge placed in the center of the cavity, as shown in Figure 7.3.

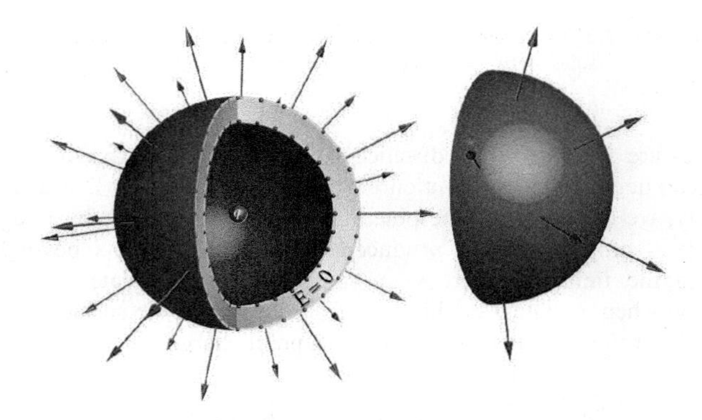

FIGURE 7.3 A positive point charge at the center of an uncharged spherical conducting shell. Negative charges are induced on the inner and surface of the shell leaving the outer shell with a net positive charge (shaded red and blue, respectively) such that $E = 0$ in the conducting material.

The presence of the positive charge will cause the charges in the conducting shell to move so as to preserve the zero electric field inside the shell. Consequently, both surfaces of the shell will acquire an induced charge. In fact, we considered this exact problem in Unit 4 and determined, using Gauss' law, that the total induced surface charge on the inner surface was equal to $-Q$, while the total induced charge on the outer surface was equal to $+Q$. These induced charges were distributed uniformly on the surfaces. Since we have complete spherical symmetry, the electric field outside the conducting shell is spherically symmetric and is equivalent to that produced by the induced charge on the outside surface alone.

We now want to see what changes when we move the point charge inside the cavity to the right, as shown in Figure 7.4. We know the charges on the conductor must move in order to preserve the zero electric field inside the shell. The magnitude of the induced charge on the inner surface cannot change since the field being zero everywhere inside the shell

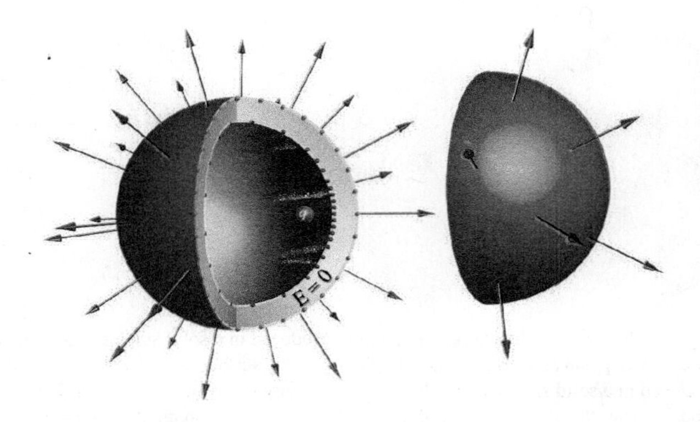

FIGURE 7.4 The point charge from Figure 7.3 is moved off-center. The magnitude of the induced charge on the inner surface remains the same, but its distribution changes in order to produce $E = 0$ in the conducting material. The induced charge distribution on the outer surface remains the same and produces a spherically symmetric field outside the shell.

requires that the induced charge on the inner surface and the point charge must sum to zero. The distribution of the induced negative charge will change, however, with more charge moving closer to the point charge

The perhaps surprising result is that the distribution of the positive charge on the outside surface does not change! The negative charges on the inner surface of the shell produce a field everywhere in the shell (and outside the shell) that exactly cancels the field produced by the point charge. Therefore, the positive charges on the outer surface of the shell are blissfully ignorant of all this action. The field outside the shell remains the same, independent of the exact location of the point charge inside the cavity.

7.5 Shielding in a Conductor with a Cavity

We've just explored what happens to the charges in a spherical conducting shell when a point charge is moved around inside its cavity. One important result was that the field outside the shell did not change as the point charge was moved around inside the cavity. We'll now examine what happens to the charges in a conductor when we move a point charge around in the region outside the conductor. We will find a similar result.

We'll start with an example of an uncharged solid spherical conductor. We now place a point charge close to the surface of the conductor. What will happen? We know that the field inside the conducting material must be zero. The charges on the surface of the conductor must separate so that the field they create inside the conductor exactly cancels the field produced by the point charge, as shown in Figure 7.5(a).

We'll now create a cavity in this conductor by cutting out a piece, as shown in Figure 7.5(b). How will the field change? The answer is it will not change at all! The field inside the cavity must be zero. We removed a chunk of conductor with zero net charge.

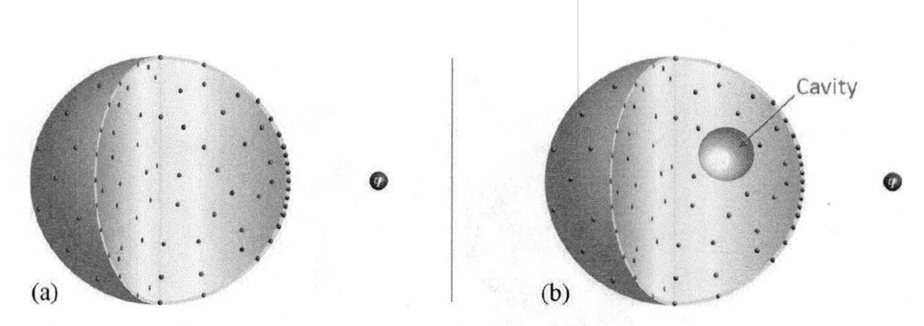

FIGURE 7.5 (a) The charge distribution produced in a solid spherical conductor when an external positive charge q is brought near the surface. (b) The charge distribution produced in a solid spherical conductor with a cavity when an external positive charge q is brought near the surface. Note that the charge distribution in the conductor does not change.

The charges on the surface of the conductor can't move because they are already positioned to cancel the field from the external point charge at every point inside the surface. Therefore, the field in the region of the cavity does not change; it remains zero.

Another way to think about this result is in terms of electric field lines. Electric field lines must start and stop on charges and they point "downhill." Since the cavity is empty, any field lines in the cavity must start and stop on charges on the inner surface of the conductor. But we know the conductor is an equipotential. Therefore, it would be impossible to have any field lines in the cavity since they would have to point "downhill," but the entire conductor is "flat," i.e., an equipotential.

We have just arrived at another very important result: that the charges on a conductor move to exactly cancel the electric field created by an external charge everywhere inside the conductor, including cavities enclosed by the conductor. This feature has great practical significance. It is the principle that makes it possible to "shield" sensitive equipment from external electric fields by placing the equipment inside a cavity of a conductor.

7.6 Electric Potential and Capacitance

We'll now move on to the study of a device often used in electrical circuits, the capacitor. A **capacitor** is an object with two spatially separated conductors. When charges are moved from one conductor to the other, these conductors will end up with an equal but opposite net charge that then creates an electric field in the space between the conductors. Consequently, a non-zero potential difference develops between the conductors. Since the electric field is proportional to the separated charge, the change in potential is also proportional to the amount of separated charge. We define the **capacitance** C of the pair of conductors as the ratio of the separated charge to the potential difference it produces.

$$C \equiv \frac{Q}{V}$$

The bigger the value of C, the more charge must be separated to produce a given potential difference. The dimensions of this quantity are coulombs per volt. The common unit for capacitance is the farad (F), defined as 1 coulomb / volt.

Our first example will be to calculate the capacitance of two conducting sheets, each having surface area A, separated by a distance d, as shown in Figure 7.6. If we move a total charge $+Q$ to the top plate, we will leave the bottom plate with charge $-Q$. This separation of charge creates an electric field in the region between the plates. The direction of this field points from the top plate to the bottom plate. Now, if d, the separation between the plates, is small compared to L, the length of the smallest side of the plates, the electric field between the plates will be approximately uniform. We can apply the principle of superposition to obtain the value for the $\vec{E}$ field between the plates. Namely, we know the magnitude of the field produced by each plate separately is equal to $\sigma / 2\varepsilon_o$, where σ is the magnitude of the surface charge density on each plate (Q / A). Since the directions of the fields are the same, we determine that the magnitude of the electric field between the plates is given by $E = \sigma / \varepsilon_o$.

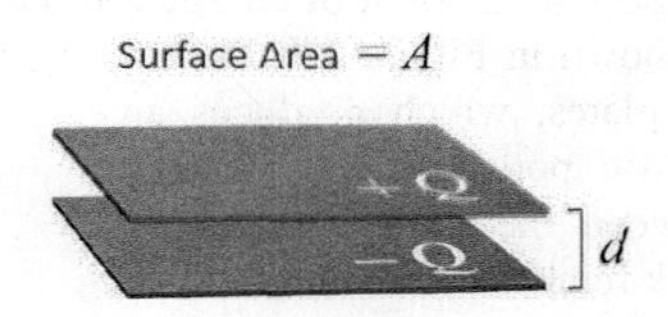

FIGURE 7.6 A parallel-plate capacitor having plates of area A separated by a distance d. Equal amounts of positive and negative charge are placed on the plates.

To find the capacitance, we need to calculate the potential difference that this separation of charge Q creates. To determine this potential difference, we need to integrate the electric field from the bottom plate to the top plate. We choose a path perpendicular to the sheets so that $\vec{E} \cdot d\vec{l}$ becomes $E\,dy$. Since E is a constant along this path, we can take it outside the integral and obtain

$$\Delta V = Ed = \frac{\sigma}{\varepsilon_o} d = \frac{Q}{\varepsilon_o A} d$$

As promised, this potential difference is proportional to Q, the amount of charge separated. The capacitance is simply the inverse of this coefficient.

$$C \equiv \frac{Q}{\Delta V} = \frac{\varepsilon_o A}{d}$$

Note that the capacitance C only depends on the geometry of the capacitor; in this case it is proportional to its area, and inversely proportional to the separation of its plates.

We can follow this same procedure to determine the capacitance of any pair of conductors and we will always find that the capacitance can depend only on the geometry of the system of conductors.

7.7 Capacitors Store Energy

When used in electrical circuits, the main functions of capacitors are to store and to release energy. We'll now calculate the amount of energy stored in a capacitor when there is a potential difference V between its plates by simply calculating the amount of work required to separate the charges. We will perform this calculation for the simplest geometry, the parallel-plate capacitor, but the result we obtain will be completely general.

Suppose an amount of charge q has already been moved from the bottom plate to the top, as shown in Figure 7.7. This charge will create an electric field E in the region between the plates, which produces an electric potential difference V between the plates. Now, the work required to move the next increment of charge dq from the bottom plate to the top is certainly positive and is just equal to the force times the distance:

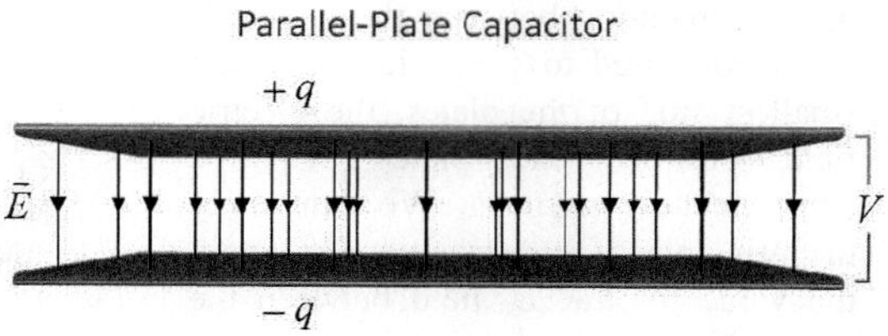

$$dW = (dq\ E)d$$

FIGURE 7.7 A parallel-plate capacitor in which a charge $+q$ has been moved from the bottom plate to the top plate, thereby creating a constant electric field in the region between the plates.

We can rewrite this expression in terms of the potential since the electric field is just equal to the potential difference divided by the separation.

$$dW = V\ dq$$

We can now identify this increment of work required to move dq from the bottom plate to the top as the increase in the potential energy of the system.

$$dU = dW = V\ dq$$

This expression holds true for any geometry since it represents the definition of the change in potential energy.

Once we have moved dq from the bottom plate to the top, the total separated charge increases, which leads to an increase in the potential difference between the plates. Consequently, moving the next increment of charge from the bottom plate to the top will require even more work. Therefore, to find the total energy stored in a capacitor when it is charged from 0 to Q, we need to integrate the work done as q increases from 0 to Q. We can do this integral by rewriting the potential V in terms of the charge q and the capacitance C:

$$U = \int_0^Q V\ dq = \int_0^Q \frac{q}{C}\ dq = \frac{1}{2}\frac{Q^2}{C}$$

We can use the definition of the capacitance to express the result in several equivalent ways. Namely,

$$U = \frac{1}{2}\frac{Q^2}{C} = \frac{1}{2}QV = \frac{1}{2}CV^2$$

7.8 Energy Is Stored in the Electric Field

We've just identified the energy stored in a capacitor as the change in potential energy of the system as the charges are moved from one plate to the other. We now want to ask exactly where that energy is stored.

Although the stored energy certainly depends upon the location of the separated charges, it is very instructive to rewrite this energy in terms of the electric field created by these charges. To simplify this task, we'll first choose a capacitor with a constant electric field. We know that the parallel-plate capacitor, with small separation relative to the size of the plates (shown in Figure 7.7) satisfies this condition.

We'll start with the expression of the stored energy: $U = CV^2/2$. Now we know the potential difference is just the product of the electric field strength and the separation between plates ($V = Ed$). We also know how to express the capacitance in terms of the area and separation of the plates ($C = \varepsilon_o A/d$).

Substituting these relations into our original expression for the stored energy, we obtain

$$U = \frac{1}{2}\varepsilon_o E^2 Ad$$

Since the volume of the region between the plates is just the product of the area of the plates and their separation, we can divide the total energy stored by this volume and obtain the **energy density** u in the region between the plates:

$$u = \frac{U}{Ad} = \frac{1}{2}\varepsilon_o E^2$$

The amazing thing is that although we used the constant field of the parallel-plate capacitor to obtain this result, it is totally general; all electric fields have an energy density of $\varepsilon_o E^2/2$.

7.9 Summary

The calculation of electrostatic phenomena is complicated for conductors since the charges are free to move in conductors. In fact, they will always move so as to ensure that the electric field in all conductors is zero and that the electric field at any surface of the conductors is perpendicular to that surface. Consequently, all points on a conductor must be at the same electric potential.

We also investigated the use of cavities in conductors to provide electric shielding. In particular, we discovered that a point charge placed in a cavity of a spherical conductor can induce a non-uniform charge distribution on the surface of the cavity, but the induced charge distribution on the outer surface of the conductor will be uniform, independent of the exact location of the point charge inside the cavity. Consequently, to the world outside the conductor, the field appears to be that created by a charged spherical conductor, independent of the location of the charge within the cavity.

Furthermore, if the point charge is outside the conductor, charges on the surface of the conductor will redistribute themselves so as to produce a zero electric field at all points inside the conductor's outer surface, including inside the cavity. Consequently, the cavity defines an environment that is shielded from the effects of charges external to the conductor.

Finally, we defined the capacitance C of a system composed of two spatially separated conductors (called a capacitor) in terms of the potential difference V between the conductors that is produced when each conductor carries an equal amount Q of oppositely signed excess charge ($C \equiv Q/V$). We also determined that a charged capacitor stores energy ($U = QV/2$) in the electric field created between its plates. The energy density in this region is proportional to the square of he electric field ($u = \varepsilon_o E^2/2$).

MAIN POINTS

Charge Distributions on Conductors

Charges in a conductor will always move so as to create a zero electric field at all points in the conductor. All conductors are equipotentials and the electric field at any surface of a conductor is always perpendicular to that surface.

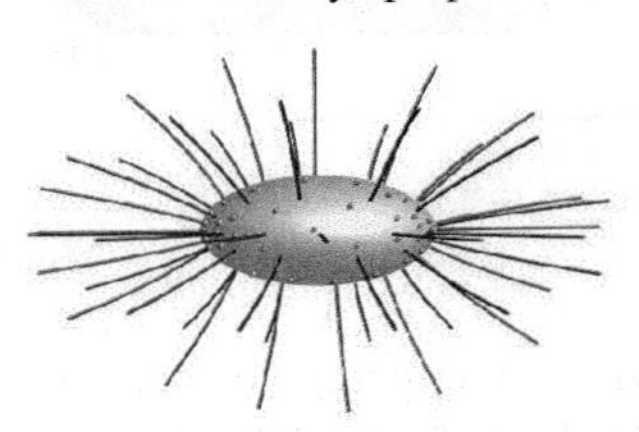

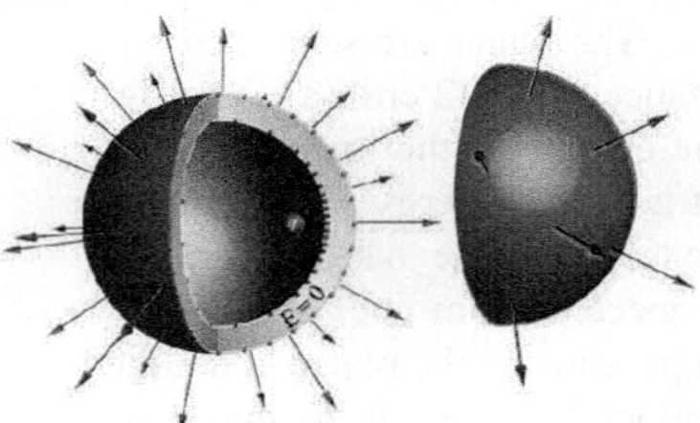

Capacitance

The capacitance C of a system, composed of two spatially separated conductors (called a capacitor), is defined as the ratio of the separated charge to the potential difference between the conductors.

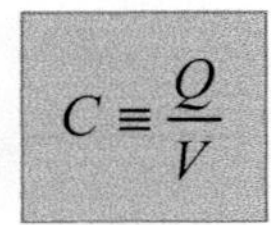

$$C \equiv \frac{Q}{V}$$

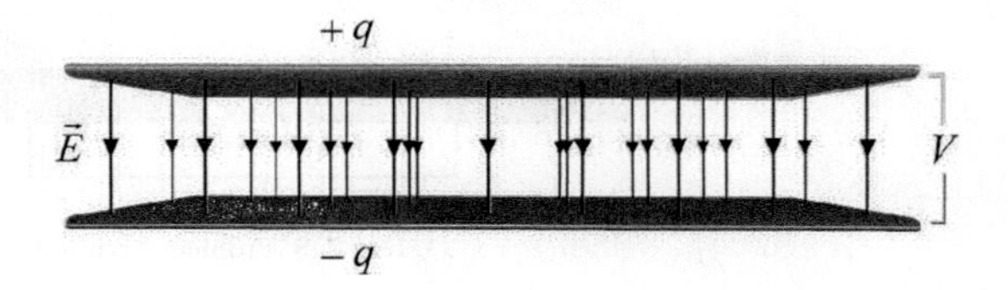

Energy in Capacitors

Energy is stored in electric fields with an energy density proportional to the square of the field.

$$u_E = \frac{1}{2}\varepsilon_o E^2$$

In capacitors, the energy is proportional to the product of the electric charge and the voltage.

$$U = \frac{1}{2}QV$$

PROBLEMS

1. Parallel-Plate Capacitor and Battery: Two parallel plates, each having area $A = 3{,}722\ \text{cm}^2$, are connected to the terminals of a battery of voltage $V_b = 6\ \text{V}$ as shown. The plates are separated by a distance $d = 0.42$ cm. (a) What is Q, the charge on the top plate? (b) What is U, the energy stored in the capacitor? (c) The battery is now disconnected from the plates and the separation of the plates is doubled ($d = 0.84$ cm). What is the energy stored in this new capacitor? (d) What is E, the magnitude of the electric field in the region between the plates? (e) Compare V, the magnitude of the new potential difference across the plates, to V_b, the voltage of the battery: $V < V_b$, $V = V_b$, or $V > V_b$? (f) Two uncharged parallel plates identical to the original plates are now connected to the initial pair of plates as shown.

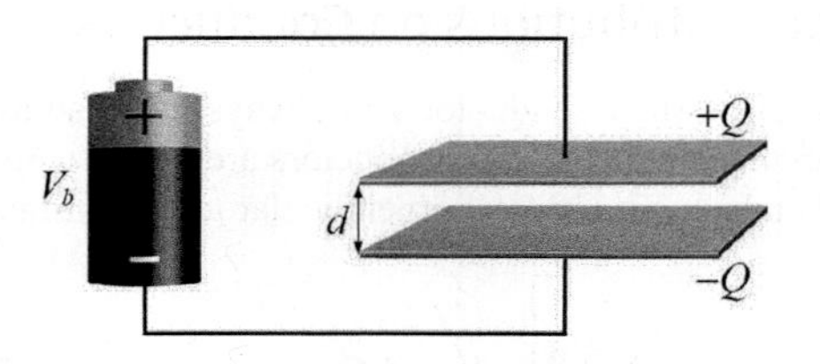

FIGURE 7.8 Problem 1

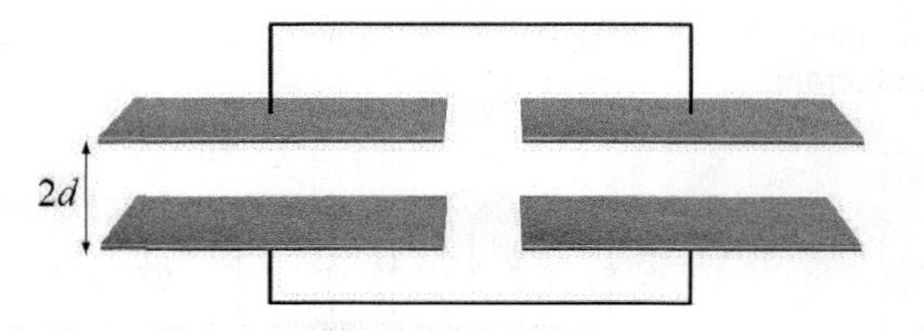

FIGURE 7.9 Problem 1 part (f)

How will the electric field, E, and potential difference across the plates, V, change, if at all?

(i) Both E and V will remain the same.
(ii) E will decrease and V will increase.
(iii) E will increase and V will decrease.
(iv) Both E and V will decrease.
(v) Both E and V will increase.

PART II: DC CIRCUITS

UNIT

CAPACITORS

8.1 Overview

We will now continue our study of capacitance by focusing on physical capacitors and their combinations in circuits. We'll start by exploring the relationships between the relevant physical quantities using the parallel-plate capacitor as an example. We'll then extend our treatment of real materials by introducing a simple model for a dielectric material, characterized by a dielectric constant. We will then discuss the effects of inserting a dielectric material between the plates of a capacitor. This exercise will require us to extend our understanding of electric fields to environments containing real materials. We will then determine the equivalent capacitance of various combinations of capacitors in circuits, introducing the central ideas of series and parallel connections.

8.2 Example: Parallel-Plate Capacitor

Consider a parallel-plate capacitor that has a net charge of $+Q$ on the top plate and $-Q$ on the bottom plate. What would happen if the plates were pulled a bit farther apart (i.e., d increases slightly)?

To analyze this situation, we must first determine what does not change when we pull the plates apart. Since the plates are isolated and not connected to anything else, the separated charge Q must stay the same; there's no way for charge to leave or enter the plates. Consequently, the electric field between the plates cannot change either since it is proportional to the charge density.

The potential difference between the plates will increase. Why? The potential difference is determined by integrating the electric field between the two plates. If the field remains the same but the distance increases, V must also increase.

What about the capacitance? Since the charge remains the same but the potential difference increases, we can see from the definition of capacitance ($C \equiv Q/V$) that it must decrease. This result is completely expected since the capacitance is totally determined by the geometry and we know that the capacitance of a parallel-plate capacitor is inversely proportional to the separation of the plates ($C = \varepsilon_o A/d$).

What happens to the energy stored in the capacitor as the plates are separated? The attractive Coulomb force does negative work as the plates are separated. Therefore, the potential energy of the capacitor must increase since the change in potential energy is defined as minus the work done by the electric field. This result is also confirmed from thinking about the energy in the field: The energy density is proportional to the square of the electric field, which remains the same, but the volume of the region between the plates has increased, leading to an increase in the energy stored.

8.3 Example: Adding a Conductor to a Parallel-Plate Capacitor

To this point we have restricted our discussion of capacitors to cases where there was nothing between the two separated conductors. Most real capacitors are not constructed in this way; they usually have the space between the conductors completely filled with a dielectric material. In order to understand why this practice is advantageous, we need to discuss electric fields in matter.

As a first step in that direction, we begin with an impractical but illustrative example. Consider a parallel-plate capacitor with positive charge $+Q$ on the top surface and negative charge $-Q$ on the bottom surface, satisfying the usual requirement $d << L$ with empty space between the two plates. Now we place a conducting block with the same area but thickness d_c / d between the plates, as shown in Figure 8.1. What will happen to the charges in this conductor?

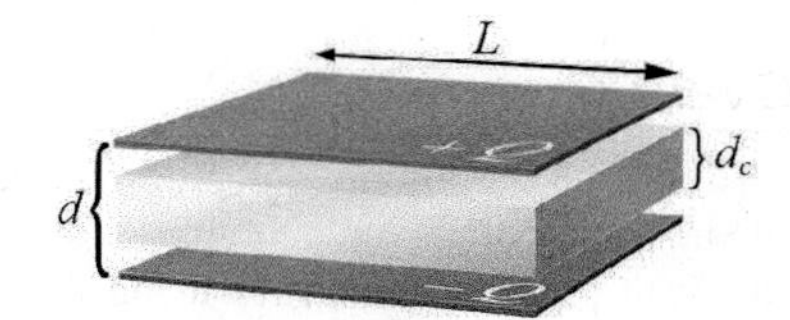

FIGURE 8.1 A parallel-plate capacitor with a solid conducting block of thickness d_c inserted in the region between the plates.

Clearly they will move so as to create a zero electric field inside the conductor. To cancel the electric field produced by the plates of the capacitor inside the block, an equal amount of negative charge ($-Q$) must be induced on the top of the block, while an equal amount of positive charge ($+Q$) must be induced on the bottom of the block.

This induced charge, while creating $E = 0$ inside the block, has no effect in the region outside the block since the field produced by the induced positive charges exactly cancels the field produced by the induced negative charges. Consequently, the fields in between the plates of the capacitor above and below the block remain the same. Since the fields remain the same above and below the block but the field in the block is zero, the potential difference, calculated by integrating the field between the two plates, must decrease. Since the charge on the plates remains constant, this decrease in potential difference leads to an increase in the capacitance of the capacitor. In particular, the capacitance ($C = \varepsilon_o A/(d - d_c)$) is increased by a factor of $d/(d - d_c)$.

Although adding a conducting block into a parallel-plate capacitor would have no practical value, it does provide a clean illustration of the important effect placing an object inside the space between the plates can have. Indeed you can think of this block as having reduced the average electric field throughout the volume to $E_{reduced} = E_o(d - d_c)/d$. We'll now discuss the case in which we fill the region between the plates with a dielectric, which will achieve a similar result.

8.4 Dielectrics

In our discussions of materials to date in this course, we have restricted ourselves to two types of idealized materials, conductors in which charges are totally free to move, and insulators in which charges cannot move at all. In the real world, the mobility of charges in most materials fall somewhere in between these two models. A **dielectric** is an insulating material in which the distribution of charges within individual molecules can be altered by an applied electric field. As we will now show, the addition of a dielectric material between the plates of a capacitor can significantly increase its capacitance.

We can understand the effect of adding a dielectric to a capacitor by considering the molecular dipole moment that is induced when the dielectric is placed in an external electric field. Namely, the presence of this field causes the centers of the positive and negative charge distributions within the molecule to separate somewhat, creating an electric dipole. The amount of this separation is determined by the molecular structure of the dielectric, but the negative center moves toward the positive plate of the capacitor while the positive center moves toward the negative plate of the capacitor. One way to visualize this separation is shown in Figure 8.2 where we show the individual molecules rotated to align with the applied electric field.

The effect of this orientation, therefore, is to create a layer of negative charge close to the positive plate, a layer of positive charge close to the negative plate, and a region between these layers that has no net charge. The field created in the dielectric by these two layers needs to be added to the field created in the dielectric by the charges on the plates. Since the field produced by the layers of charge in the dielectric is in the opposite direction to the field produced by the plates, the ultimate effect of this orientation then is to reduce the

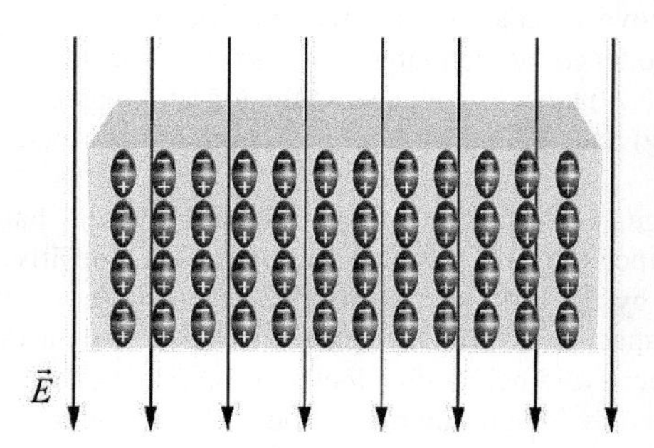

FIGURE 8.2 A visualization of the separation of charge in a dielectric in a parallel-plate capacitor. Individual molecules are shown rotated to align with the applied electric field.

electric field in the dielectric, just as we found in the previous case when we inserted a conducting slab in the region between the plates. This reduced field then leads to a lower potential difference for the same amount of charge on the plates, which in turn leads to an increased capacitance.

The factor by which the capacitance increases is called the **dielectric constant** κ of the material. For example, $\kappa = 2.6$ for polystyrene. The dielectric constant reflects the molecular structure of the material and can be quite different for different materials.

8.5 Example: Inserting a Dielectric

We will now consider what happens if we insert a dielectric with dielectric constant κ into a parallel-plate capacitor while it is connected to a **battery**. For this example, we'll consider the battery to be an ideal battery, which is defined to be a voltage source (i.e., we'll assume the electric potential difference across the terminals of the battery remains the same, no matter what devices are connected to it). This situation is illustrated in Figure 8.3.

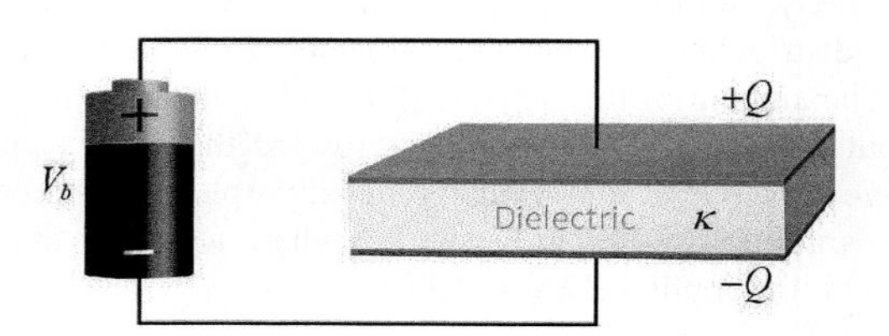

FIGURE 8.3 A dielectric with dielectric constant κ is inserted into a parallel-plate capacitor that is connected to an ideal battery of voltage V_b.

The capacitance is totally determined by the geometry and the dielectric constant. In this case, the capacitance will increase, namely $C = \kappa C_o$. After all, that was the purpose of introducing the dielectric in the first place.

The remaining properties (the electric potential difference, the charge, and the energy) are not determined totally by the geometry and the dielectric constant. To determine them, we must figure out what important feature of the situation *stays the same*. In this case, since the capacitor remains connected directly to the battery, the *electric potential difference* between the plates stays the same. The electric potential difference across the capacitor is just the electric potential supplied by the battery.

Once we know the potential difference remains constant, we can determine what happens to all of the other properties of the capacitor. Since V remains constant while C increases, we know from the definition of capacitance that Q, the amount of charge on each plate of the capacitor, must increase. In this case, the battery provides the force to move more charge from the bottom to the top plate.

The energy stored in the capacitor also increases, since the potential difference V remains the same, but the amount of separated charge Q increases. Again, this energy must be provided by the battery.

8.6 Capacitors in Parallel

We will now consider what happens when two capacitors are connected together. There are two distinct ways we can connect a pair of capacitors. We can either connect the two positive plates together and the two negative plates together (called the **parallel combination**) or we can connect the positive plate of one capacitor to the negative plate of the other capacitor (called the **series combination**). In either case we end up with a combination that we can consider as a single capacitor whose capacitance we now want to calculate.

To determine this equivalent capacitance, we'll start by connecting the leads of the parallel combination of a pair of capacitors to a battery, as shown in Figure 8.4. Since the electric potential difference across each individual capacitor will just be equal to the battery voltage, we can use the definition of capacitance to determine the charge on each capacitor. In particular, Q_1, the charge on C_1, is equal to C_1V_b. Similarly, $Q_2 = C_2V_b$.

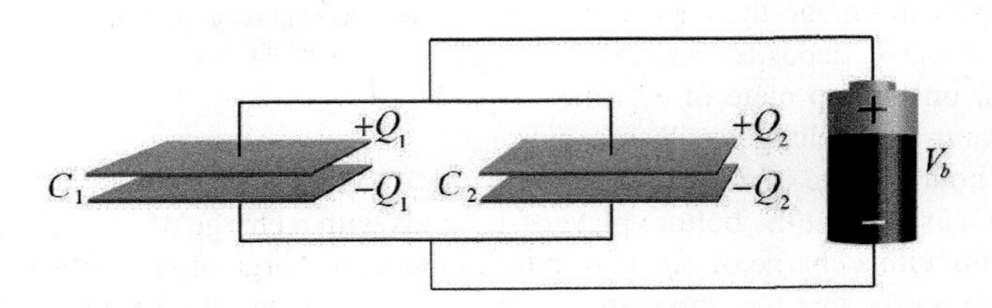

FIGURE 8.4 Capacitors C_1 and C_2 are connected in parallel and then to an ideal battery of potential difference V_b.

This combination of the capacitors then *behaves identically* to a single equivalent capacitor whose potential difference is V and whose charge is the sum of the charges on the two capacitors.

Consequently, the capacitance of this "equivalent capacitor" is just equal to the total charge on both capacitors divided by the voltage across them. Namely, $C_{equivalent} = (Q_1 + Q_2)/V_b$. We can express this result in terms of the original capacitances by applying the definition of capacitance to determine that the equivalent capacitance is just the sum of the individual capacitances: $C_{equivalent} = C_1 + C_2$.

We can see how this general result is actually produced in the specific case of two parallel-plate capacitors. When connected in parallel, the combination looks like a single capacitor that has an area $A_1 + A_2$ and a separation d. Since the capacitance of a parallel-plate capacitor is directly proportional to the area of the capacitor, we see that the equivalent capacitance must just be the sum of the individual capacitances, $C_1 + C_2$.

In conclusion, for two capacitors connected in parallel, the equivalent voltage is the same as the voltage across either capacitor, while the equivalent capacitance is given by the sum of the individual capacitances. Consequently, the charge on the equivalent capacitor is also given by the sum of the charges on the individual capacitors.

8.7 Example: Capacitors in Series

We will now connect the two capacitors in series and will find the equivalent capacitance of this combination. How do we start? Figure 8.5 shows the series combination of C_1 and C_2 connected to the terminals of an ideal battery. When we made this connection for the parallel combination, we noted that the potential difference across C_1 and C_2 had to be equal, and that fact led us directly to the equivalent capacitance.

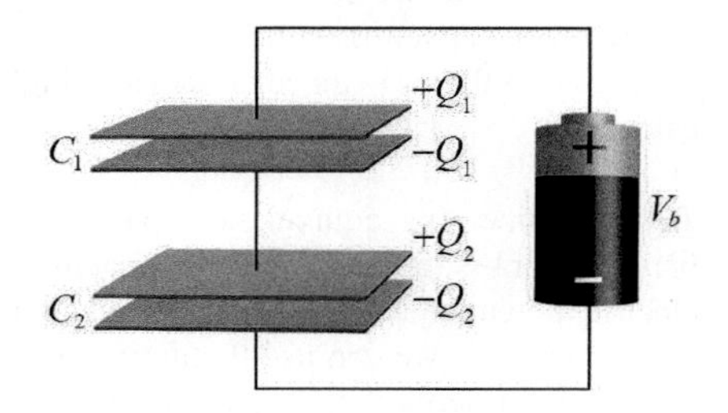

FIGURE 8.5 Capacitors C_1 and C_2 are connected in series and then to an ideal battery of potential difference V_b.

In this case, we cannot say that the potential differences V_1 and V_2 are equal. In fact, all we know is that $V_1 + V_2 = V_b$, the potential difference across the terminals of the battery. Let's look at what happens when the battery starts to charge the capacitors. As the battery deposits an initial charge $+q$ on the top plate of C_1, the same quantity of charge will move from the bottom plate of C_1 to the top plate of C_2, such that the bottom plate of C_1 ends with a charge of $-q$ and the top plate of C_2 ends up with a charge of $+q$. This in turn induces a charge of $-q$ on the bottom plate of C_2, which means that the same amount of charge flows into the bottom of the battery as flowed out of the top; this makes sense since the battery remains neutral. From this process, we see that the separated charge on each capacitor will be the same.

The potential difference across the series combination is equal to the sum of the voltages across each capacitor, i.e., $V = Q/C_1 + Q/C_2$. The total charge on the combination is Q, the same charge that is on each of the individual capacitors. Applying once again the definition of capacitance, we obtain $1/C_{equivalent} = V/Q = 1/C_1 + 1/C_2$. In other words, the inverse of the equivalent capacitance of a pair of capacitors connected in series is equal to the sum of the individual inverse capacitances.

We can see how this general result is actually produced in the specific case of two parallel-plate capacitors. When connected in series, the combination looks like a single capacitor of area A with separation $d_1 + d_2$. Since the inverse capacitance of a parallel-plate capacitor is directly proportional to the separation of the plates ($1/C = d/\varepsilon_o A$), we see that the inverse equivalent capacitance must just be the sum of the individual inverse capacitances ($1/C_1 + 1/C_2$).

In conclusion, for two capacitors connected in series, the charge on each capacitor is the same and is the charge on the equivalent capacitor, while the equivalent voltage is given by the sum of the individual voltages. Consequently, the equivalent inverse capacitance is also given by the sum of the individual inverse capacitances.

8.8 Example: Combination of Capacitors

Three capacitors are connected to a battery as shown in Figure 8.6. We want to determine the final charge on each of the capacitors. How do we start? Any network of capacitors can be described in terms of series and parallel combinations of capacitors. Once we make these identifications, we can reduce the given network to a single equivalent capacitance whose equivalent charge we can determine. Once we know the equivalent charge we can determine the charges on the parallel and series combinations that comprise the single equivalent capacitance.

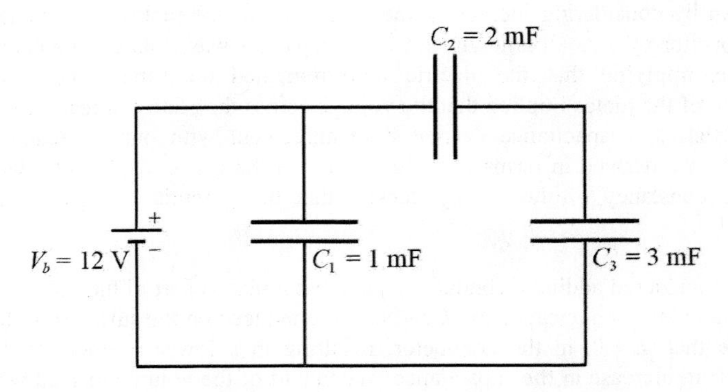

FIGURE 8.6 Capacitors $C_1 = 1$ mF, $C_2 = 2$ mF, and $C_3 = 3$ mF are connected as shown to an ideal battery of potential difference $V_b = 12$ V.

To start we'll note that C_2 and C_3 are connected in series. We can therefore replace these two capacitors by an equivalent capacitor C_{23}, where $(1/C_{23}) = (1/C_2) + (1/C_3)$. Substituting in the values for C_2 and C_3, we obtain C_{23} = 1.2 mF.

We now see that this equivalent series capacitance is connected in parallel with C_1. Therefore, we can replace C_1 and C_{23} by an equivalent capacitor $C_{123} = C_1 + C_{23}$. Substituting in the values for C_1 and C_{23}, we obtain C_{123} = 2.2 mF.

We have thus reduced this circuit to a single capacitor, C_{123} = 2.2 mF, connected across a battery of potential difference V_b = 12 V. Therefore, the charge on C_{123} must just be $Q_{123} = C_{123}V_b = 26.4$ mC.

Knowing this charge on the equivalent capacitance of the network, we can now reverse our procedure to determine the charges on the individual capacitors. Since C_1 and C_{23} are in parallel, we know the voltages across them are equal and in this case, equal to the battery voltage. Consequently, the charges on these capacitors will just be given by the product of the battery voltage and the appropriate capacitance; i.e., $Q_1 + C_1V_b = 12$ mC, and $Q_{23} = C_{23}V_b = 14.4$ mC. Now we are done since C_2 and C_3 are in series, $Q_2 = Q_3 = Q_{23} = 14.4$ mC.

It's always a good idea to check your answers. Here, we have two checks we can perform using the voltages. First, we expect V_1, the voltage across C_1, to be the battery voltage V_b = 12 V. From our results, we can calculate $V_1 = Q_1 / C_1$ = 12 mC / 1 mF = 12 V. Finally, we expect the voltages across C_2 and C_3 to sum to the battery voltage. From our results, we can calculate $V_2 = Q_2 / C_2 = 7.2$ V and $V_3 = Q_3 / C_3 = 4.8$ V, giving the expected $V_2 + V_3 = 12$ V.

8.9 Summary

We began by considering increasing the separation of the plates of an isolated parallel-plate capacitor by a small amount. Since the capacitor was isolated, the charge remained the same, implying that the electric field remained the same. The increase in the separation of the plates implied that the voltage across the plates increased. Consequently, we saw that the capacitance decreased in agreement with our expectations from the expression we derived in terms of the geometry of the capacitor. Finally, we determined from the constancy of the energy density that the potential energy of the capacitor increased.

We then considered adding a conducting slab that replaced part of the volume between the plates of a parallel-plate capacitor. Charges were induced on the surfaces of this conductor to ensure that $\vec{E} = 0$ in the conductor, resulting in a lower average field throughout, leading to an increase in the capacitance. When part of the volume instead was filled with a dielectric material, we found the capacitance increased. We understood this increase in a similar way, in which the molecules of the dielectric were oriented by the applied voltage resulting in a separation of the positive and negative centers of charge, creating a field in the opposite direction of that created by the applied charge on the plates, thus reducing the average field throughout.

We then obtained rules for finding the equivalent capacitance of both series and parallel combinations of capacitors. With these rules in hand, we are able to analyze any combination of capacitors.

MAIN POINTS

Dielectric Constant

The ratio of the capacitance when filled with a dielectric to that when filled with air is equal to κ, the dielectric constant of the material.

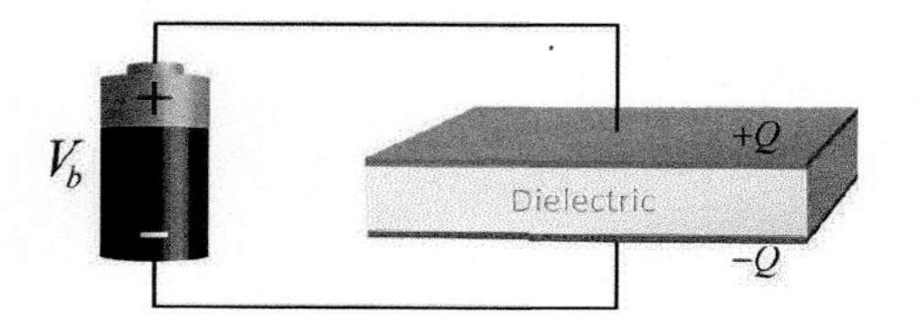

$$\kappa = \frac{C_{new}}{C_{original}}$$

Capacitors in Parallel

The equivalent capacitance of two capacitors connected in parallel is equal to the sum of the individual capacitances.

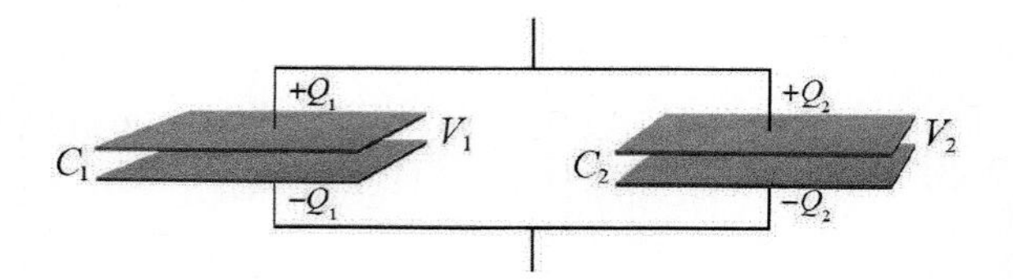

$$C_{equivalent} = C_1 + C_2$$
$$V_{equivalent} = V_1 = V_2$$
$$Q_{equivalent} = Q_1 + Q_2$$

Capacitors in Series

The equivalent inverse capacitance of two capacitors connected in series is equal to the sum of the individual inverse capacitances.

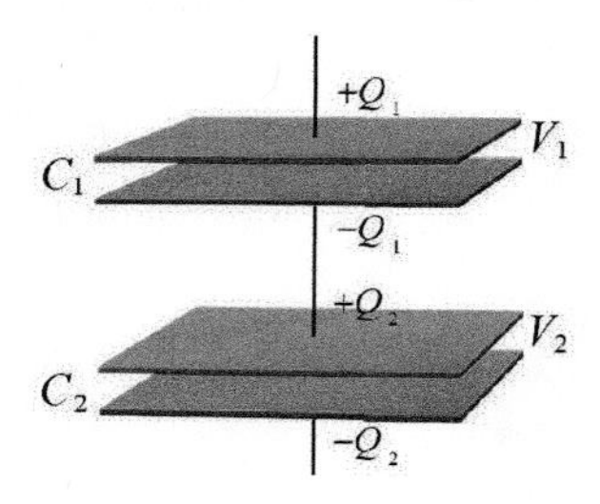

$$\frac{1}{C_{equivalent}} = \frac{1}{C_1} + \frac{1}{C_2}$$
$$V_{equivalent} = V_1 + V_2$$
$$Q_{equivalent} = Q_1 = Q_2$$

PROBLEMS

1. Parallel-Plate Capacitor and Dielectric: Two parallel plates, each having area $A = 2{,}198\ \text{cm}^2$, are connected to the terminals of a battery of voltage $V_b = 6\ \text{V}$ as shown. The plates are separated by a distance $d = 0.39$ cm. You may assume (contrary to the drawing) that the separation between the plates is small compared to the size of the plates. (a) What is C, the capacitance of this parallel-plate capacitor? (b) What is Q, the charge stored on the top plate of the capacitor? (c) A dielectric having dielectric constant $\kappa = 4$ is now inserted in between the plates of the capacitor as shown. The dielectric has area $A = 2{,}198\ \text{cm}^2$ and thickness equal to half of the separation (0.195 cm). What is the charge on the top plate of this capacitor? (d) What is U, the energy stored in this capacitor? (e) The battery is now disconnected from the capacitor and then the dielectric is withdrawn. What is V, the voltage across the capacitor?

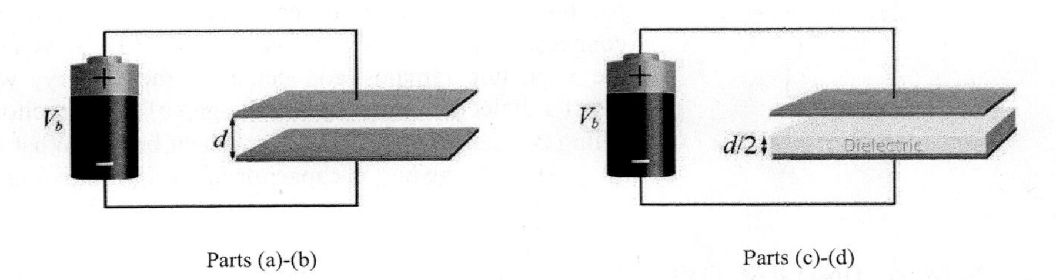

FIGURE 8.7 Problem 1

2. Concentric Cylindrical Conducting Shells: An infinitely long solid conducting cylindrical shell of radius $a = 4.9$ cm and negligible thickness is positioned with its symmetry axis along the z-axis as shown. The shell is charged, having a linear charge density $\lambda_{inner} = -0.43\ \mu\text{C/m}$. Concentric with the shell is another cylindrical conducting shell of inner radius $b = 15.6$ cm and outer radius $c = 19.6$ cm. This conducting shell has a linear charge density $\lambda_{outer} = 0.43\ \mu\text{C/m}$. (a) What is $E_x(\text{P})$, the x-component of the electric field at point P, located a distance $d = 9.2$ cm from the origin along the x-axis as shown? (b) What is $V(c) - V(a)$, the potential difference between the two cylindrical shells? (c) What is C, the capacitance of a one-meter length of this system of conductors? (d) The magnitudes of the charge densities on the inner and outer shells are now changed (keeping $\lambda_{inner} = -\lambda_{outer}$) so that the resulting potential difference doubles ($V_{ca,new} = 2V_{ca,initial}$). How does C_{new}, the capacitance of a one-meter length of the system of conductors when the charge density is changed, compare to C, the initial capacitance of a one-meter length of the system of conductors: $C_{new} < C$, $C_{new} = C$, or $C_{new} > C$? (e) What is $\lambda_{outer,new}$?

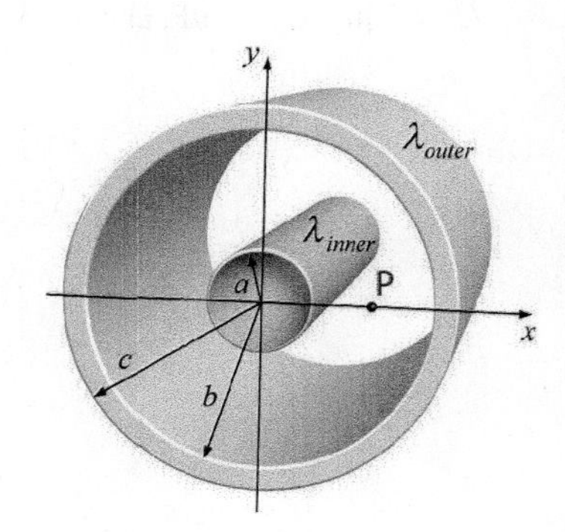

FIGURE 8.8 Problem 2

3. Circuit with Capacitors and a Battery: A circuit is constructed with five capacitors and a battery as shown. The values for the capacitors are: $C_1 = C_5 = 4.8\ \mu\text{F}$, $C_2 = 2.0\ \mu\text{F}$, $C_3 = 6.6\ \mu\text{F}$, and $C_4 = 4.8\ \mu\text{F}$. The battery voltage is $V_b = 12\ \text{V}$. (a) What is C_{ab}, the equivalent capacitance between points a and b? (b) What is C_{ac}, the equivalent capacitance between points a and c? (c) What is Q_5, the charge on capacitor C_5? (d) What is Q_2, the charge on C_2? (e) What is Q_1, the charge on capacitor C_1? (f) What is V_4, the voltage across capacitor C_4?

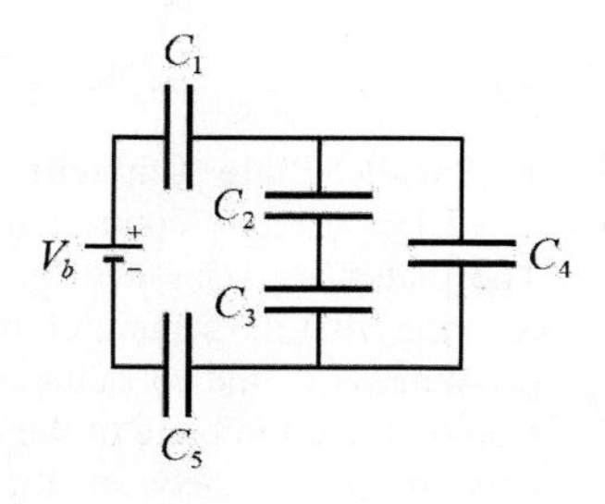

FIGURE 8.9 Problem 3

4. Physical Capacitors (INTERACTIVE EXAMPLE): An air-gap parallel-plate capacitor of capacitance $C_o = 20\ \text{nF}$ is connected to a battery with voltage $V_b = 12\ \text{V}$. While the capacitor remains connected to the battery, we insert a dielectric ($\kappa = 2.6$) into the gap of the capacitor, filling one half of the volume as shown below. What is Q_{final}, the charge on the capacitor in the final situation?

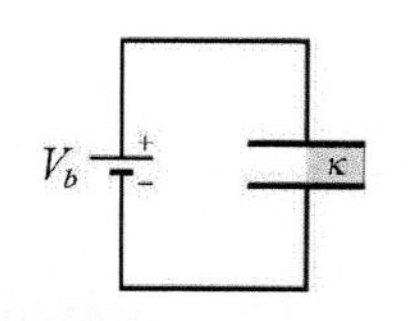

FIGURE 8.10 Problem 4

5. Network (INTERACTIVE EXAMPLE): In this circuit, what is the potential difference across capacitor C_4? Use the following values in your calculation: $V_b = 12\ \text{V}$, $C_1 = 3\ \mu\text{F}$, $C_2 = 2\ \mu\text{F}$, $C_3 = 2\ \mu\text{F}$, $C_4 = 1\ \mu\text{F}$, and $C_5 = 4\ \mu\text{F}$.

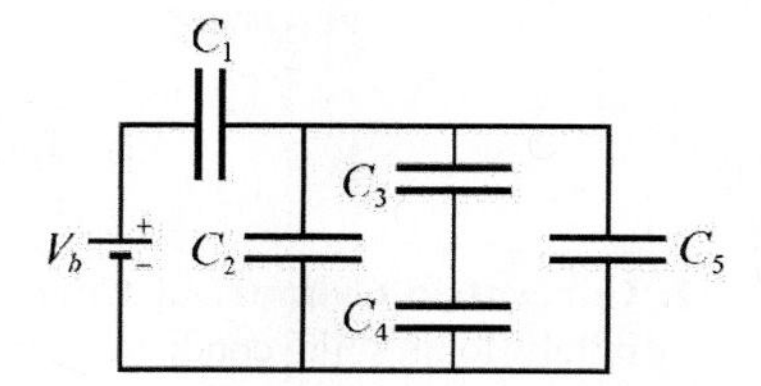

FIGURE 8.11 Problem 5

UNIT 9

ELECTRIC CURRENT

9.1 Overview

Our study, to this point, has been restricted to **electrostatics**: the determination of electric fields produced by stationary charge distributions. We now want to remove this restriction to study electric charges in motion. In particular, we will introduce the idea of electric current and use it to explore properties of electric circuits.

We'll start by defining electric current and developing a qualitative microscopic view of the motion of charge carriers in conductors. We will then use Ohm's law, an empirical relation between current density and the applied electric field in a conductor, to obtain a more quantitative description of the current. We will then introduce the concept of resistance and discuss the use of resistors in electric circuits. We will close with a brief description of the energy flow in circuits composed of batteries and resistors.

9.2 A Qualitative Description of Electric Current

Before we introduce the concept of the electric current, we'll first describe the microscopic motions of charges in a conductor.

Consider a piece of copper wire. This wire is electrically neutral, but because copper is a good conductor, its outermost electrons can easily become dissociated from the atom and are able to move freely through the wire. The positive ions left behind then form a rigid lattice through which these conduction electrons move. These electrons move with speeds determined by the temperature. For example, the average speed of electrons at room temperature is about 10^5 m/s. The average velocity of these electrons is zero, however, since their directions are completely random.

Now suppose we attach the ends of the wire to the terminals of a battery, as shown in Figure 9.1. The battery produces a constant electric potential difference, sometimes called an ***emf***, between the ends of the wire. Therefore, a net electric field must exist in the wire itself. This field, which is directed parallel to the wire from the positive terminal to the negative terminal, creates a force on the conduction electrons. This force leads to a non-zero average velocity of the conduction electrons, which results in an electric current.

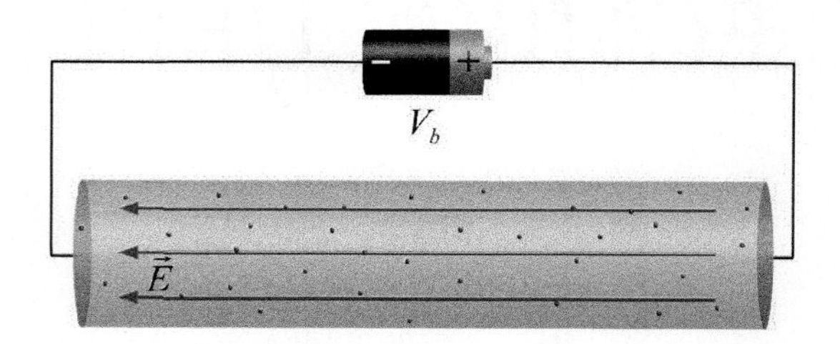

FIGURE 9.1 A battery produces an electric field in a copper wire. The field creates a non-zero average velocity of the conduction electrons in the copper, thereby creating an electric current in the wire.

We define the **electric current** I in the wire as the amount of charge that passes through a cross-section of the wire per unit time (i.e., $I \equiv dq / dt$). The SI unit for current is the ampere (A) and is equal to 1 coulomb per second.

9.3 A Quantitative Description of Electric Current

We'll now introduce a more quantitative description of the motion of charges in simple circuits. We'll start by defining the **current density** J as the total current I in the wire divided by its cross-sectional area A. We can represent this current density in terms of the motion of the charge carriers, the conduction electrons. Namely, the current density is given by

$$J = eN_e v_{drift}$$

where e is the charge on a single electron, N_e is the number of conduction electrons per unit volume, and v_{drift} is the average velocity (or drift velocity) of the conduction electrons.

We know the charge on an electron is 1.6×10^{-19} C. We can determine the density of charge carriers using Avogadro's number and some information about the particular conductor. For example, if we assume a sample of copper has one conduction electron per atom, we can use its density (8,920 kg/m^3) and its atomic weight (63 g / mol) to determine $N_e = 8\times10^{28}$ electrons / m^3.

The one remaining unknown quantity is the drift velocity. This quantity is not simply a property of the material; the drift velocity is the quantity that determines how the current density changes when different electric fields are applied to the conductor. To understand this dependence, we appeal to the empirical result that for the vast majority of substances, over a wide range of field strengths, the current density J is proportional to E, the electric field that gives rise to it. The constant of proportionality, σ, is called the **conductivity** of the material. As the name implies, σ is large for good conductors (6×10^7 Ω-m for copper) and small for good insulators (~10^{-12} Ω-m for glass). This empirical result ($J = \sigma E$) is called **Ohm's law**; materials for which it applies are called *ohmic materials*.

We can now get an estimate for v_{drift} by applying Ohm's law for this wire to obtain that v_{drift} is proportional to the product of the conductivity σ and the applied electric field E.

$$v_{drift} = \frac{J}{eN_e} = \frac{\sigma}{eN_e}E$$

Let's put in some typical numbers to determine the magnitude of v_{drift}. For a length $L = 50$ m of 24 gauge copper wire ($d = 0.5$ mm) connected to a one-volt battery, we obtain a drift velocity of 0.15 mm/second. This velocity is incredibly small; remember that typical thermal speeds of the conduction electrons are on the order of 10^5 m/s. Therefore, the picture we have arrived at is that the conduction electrons are zipping around rapidly at high speeds, but the applied electric field from the battery creates a tiny average velocity for the electrons directed along the wire toward the positive terminal of the battery. We get sizable currents from these electrons with a tiny average velocity because of the large numbers of conduction electrons present in the wire.

9.4 Resistance

Now that we have developed the microscopic picture of how an electric current is produced in a conductor when a potential difference is applied to its ends, we will introduce a more macroscopic description of currents and voltages in real circuits.

We'll start with **resistors**, devices used in electric circuits to control voltages and current flow. There are many kinds of resistors, but the common property they all share is that they obey Ohm's law: that the current density is proportional to the applied electric field.

Consider a cylindrical resistor of length L, cross-sectional area A, and conductivity σ, as shown in Figure 9.2. If we connect this resistor across the terminals of a battery of *emf* V_b, an electric field $E = V_b / L$ will be created throughout the resistor. From Ohm's law, we see that this electric field will produce a current density $J = \sigma V_b / L$. Defining the **resistance** R of this resistor to be the ratio of the applied voltage V_b to the resulting current I (i.e., $R \equiv V_b / I$), we find

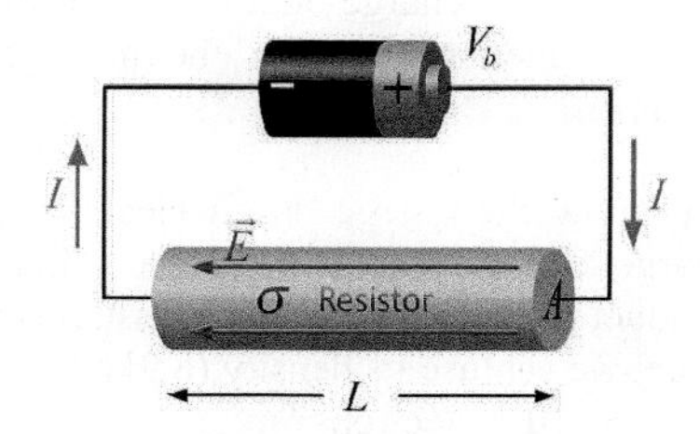

FIGURE 9.2 A resistor of length L, cross-sectional area A, and conductivity σ is connected to the terminals of a battery of *emf* V_b. An electric field $E = V_b / L$ is created in the resistor, which produces a current density $J = \sigma E$ in the resistor.

$$R = \frac{1}{\sigma}\frac{L}{A}$$

The constant of proportionality is $1/\sigma$; this quantity is usually denoted by ρ and is called the **resistivity**. The SI unit of resistance is called the ohm (Ω) and is equal to 1 volt per ampere.

Note that we initially used the term Ohm's law to describe the relationship between the current density and the applied electric field ($J = \sigma E$). When we talk about circuits composed of devices, it is usual to use the phrase Ohm's law to describe the relationship between the current and voltage across a resistor:

$$V = IR$$

These different expressions are really equivalent since we used $R = V / I$ to define the resistance based on the assumption that the current density is proportional to the applied field.

9.5 Resistors in Series

We will now discuss constructing circuits that are composed of combinations of resistors and batteries. Perhaps the first point to make is that we will use wires to make the connections between the circuit elements and we will assume that these wires are *ideal* in the sense that they are sufficiently short so that the voltage drop across any connecting wire is negligible. This assumption is easily justified; for example, even a one-meter long piece of 24-gauge wire ($d = 0.5$ mm) has a resistance of less than a tenth of an ohm. Resistors typically used in circuits are much larger, ranging from hundreds of ohms to millions of ohms. *Consequently, we will assume the resistance of all connecting wires to be zero.* In other words, we will treat connecting wires as equipotentials, so that all points on a given wire are at the same potential.

Just as was the case for capacitors, we can combine two resistors either in series or in parallel. We'll start with the series combination of R_1 and R_2 connected to the terminals of a battery, as shown in Figure 9.3. Current will flow through this complete circuit; the same current that flows through R_1 will also flow through R_2, let's call it I. Using Ohm's law, we see that the potential difference (or voltage drop) across each resistor is just equal to the product of this current and the individual resistance ($V_1 = IR_1$ and $V_2 = IR_2$).

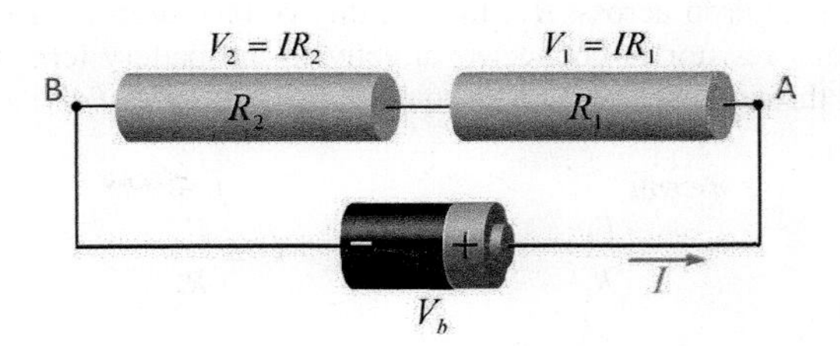

FIGURE 9.3 Resistors R_1 and R_2 are connected in series to a battery of *emf* V_b, causing current I to flow in the complete circuit.

Since the potential difference between any two points (e.g., A and B) is independent of the path, we see that V_{AB} is equal to the battery voltage, but it also must equal the sum of the voltages across each resistor ($V_{AB} = V_1 + V_2$). Therefore, if we replaced the series combination of R_1 and R_2 with a single resistor $R = R_1 + R_2$, we would obtain exactly the same current I. In other words, the equivalent resistance of two resistors connected in series is just the sum of the individual resistances.

$$R_{equivalent} = R_1 + R_2$$

Physically this makes sense since, for example, if the resistors had the same resistivity and the same area, connecting them in series effectively produces a single resistor with a length equal to the sum of the lengths of the individual resistors. Since the resistance is proportional to the length, the new resistor would have a resistance equal to the sum of the initial resistances.

9.6 Resistors in Parallel

We'll now discuss the parallel combination of R_1 and R_2 connected to the terminals of a battery, as shown in Figure 9.4. Once again, current flows through each of the resistors. In this case, however, the current that flows through the first resistor will not necessarily be equal to the current that flows through the second resistor. Clearly, the current leaving one terminal of the battery must split, with some going through the first resistor and some going through the second resistor.

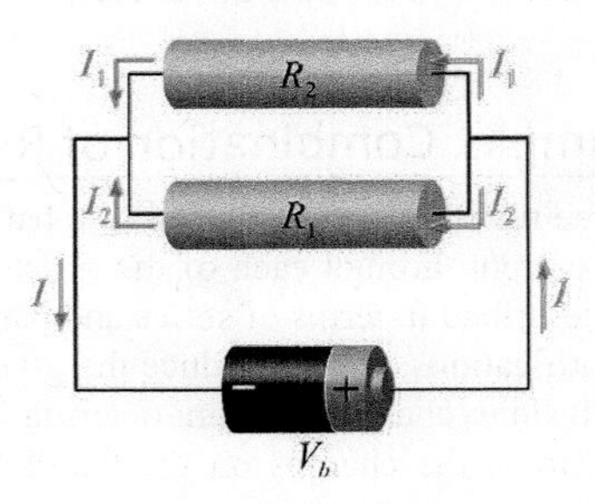

FIGURE 9.4 Resistors R_1 and R_2 are connected in parallel to a battery of *emf* V_b, causing current I to flow through the battery. This current splits, with I_1 going through R_1 and I_2 going through R_2.

What determines how much current goes through R_1 and how much goes through R_2? The key here is that the voltage drop across R_1 must be the same as the voltage drop across R_2. In fact, this voltage drop is just V_b, the *emf* of the battery since each resistor is connected directly to the battery terminals. Therefore, the current passing through each resistor is equal to the battery *emf* divided by the individual resistance.

$$I_1 = \frac{V_b}{R_1} \qquad I_2 = \frac{V_b}{R_2}$$

The total current passing through the battery is just the sum of individual currents passing through each resistor ($I = I_1 + I_2$).

To find the equivalent resistance of the parallel combination of the two resistors, we need to determine what single resistor when connected to the terminals of the battery would draw the same current as that drawn by the parallel combination of the two resistors. The single resistor will draw a current, which is equal to the battery voltage divided by the resistance ($I = V_b / R_{equivalent}$). We now substitute our value for the total current passing through the parallel combination into this equation and see that the battery voltage cancels and we are left with a simple expression for the inverse of the equivalent resistance.

$$\frac{V_b}{R_{equivalent}} = \frac{V_b}{R_1} + \frac{V_b}{R_2} \qquad \Rightarrow \qquad \frac{1}{R_{equivalent}} = \frac{1}{R_1} + \frac{1}{R_2}$$

In particular, we see that the inverse of the equivalent resistance is just equal to the sum of the individual inverse resistances.

Physically this makes sense since, for example, if the resistors had the same resistivity and the same length, connecting them in parallel effectively produces a single resistor with an area equal to the sum of the areas of the individual resistors. Since the reciprocal resistance is proportional to the area, the new resistor would have a reciprocal resistance equal to the sum of the initial reciprocal resistances.

9.7 Example: Combination of Resistors

Three resistors are connected to a battery, as shown in Figure 9.5. We want to determine the current through each of the resistors. How do we start? Any network of resistors can be described in terms of series and parallel combinations of resistors. Once we make these identifications, we can reduce the given network to a single equivalent capacitance whose equivalent charge we can determine. Once we know the equivalent charge we can determine the charges on the parallel and series combinations that comprise the single equivalent capacitance.

To start we'll note that R_2 and R_3 are connected in series. We can therefore replace these two resistors by an equivalent resistor R_{23}, where $R_{23} = R_2 + R_3$. Substituting in the values for R_2 and R_3, we obtain $R_{23} = 50\ \Omega$.

We now see that this equivalent series capacitance is connected in parallel with R_1. Therefore, we can replace R_1 and R_{23} by an equivalent resistor R_{123}, where

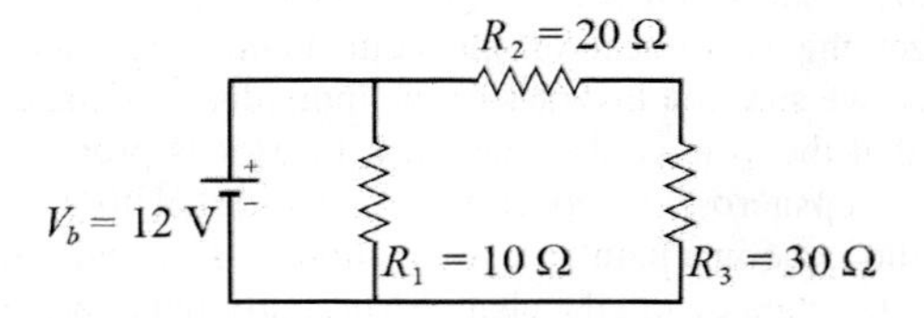

FIGURE 9.5 Resistors $R_1 = 10\ \Omega$, $R_2 = 20\ \Omega$, and $R_3 = 30\ \Omega$ are connected as shown to an ideal battery of potential difference $V_b = 12$ V.

$(1/R_{123}) = (1/R_1) + (1/R_{23})$. Substituting in the values for R_1 and R_{23}, we obtain $R_{123} = 8.33\ \Omega$.

We thus have reduced this circuit to a single resistor, $R_{123} = 8.33\ \Omega$., connected across a battery of potential difference $V_b = 12$ V. Therefore, the current through R_{123} must just be $I_{123} = V_b / R_{123} = 1.44$ A.

Knowing the current through the equivalent resistance of the network, we can now reverse our procedure to determine the currents through the individual resistors. Since R_1 and R_{23} are in parallel, we know the voltages across them are equal and in this case, equal to the battery voltage. Consequently, the current through these resistors will just be given by the battery voltage divided by the appropriate resistance; i.e., $I_1 = V_b / R_1 = 1.2$ A, and $I_{23} = V_b / R_{23} = 0.24$ A). Now we are done since R_2 and R_3 are in series, $I_2 = I_3 = I_{23} = 0.24$ A.

It's always a good idea to check your answers. Here, we have two checks we can perform using the voltages. First, we expect V_1, the voltage across R_1, to be the battery voltage $V_b = 12$ V. From our results, we can calculate $V_1 = I_1 R_1 = (1.2\text{ A})(10\ \Omega) = 12$ V. Finally, we expect the voltages across R_2 and R_3 to sum to the battery voltage. From our results, we can calculate $V_2 = I_2 R_2 = 4.8$ V and $V_3 = I_3 R_3 = 7.2$V, giving the expected $V_2 + V_3 = 12$ V. Note that the sum of the currents through R_1 and R_{23} is equal to the current through R_{123}, as it must be.

9.8 Comparison with Capacitors

Let's try to make sense of these expressions for the equivalent resistance of two resistors in series and in parallel in terms of the results we previously determined for the series and parallel combinations of two capacitors.

You may remember that when we combined capacitors in series, it was their reciprocal values ($1/C$) that were summed to get the reciprocal of the equivalent capacitance. When we combine resistors in series, we simply sum the individual resistances. This difference can be understood in that the voltage drop across a resistor is proportional to the resistance, while the voltage drop across a capacitor is proportional to 1 / capacitance. Since the voltage drops add in a series combination, the resistances add, but the reciprocal capacitances add.

You may remember that when we combined capacitors in parallel, we summed the capacitances to get the reciprocal of the equivalent capacitance. When we combine resistors in parallel, we sum the inverses of the individual resistances. This difference can be understood in that the voltage drop across a resistor is proportional to the resistance, while the voltage drop across a capacitor is proportional to 1 / capacitance. Since the voltage drops are the same in a parallel combination, it is the currents through the resistors (proportional to 1 / resistance) and the charges (proportional to capacitance) that add.

9.9 Power

Our final topic for this unit is a discussion of energy flow in circuits. Consider the simple circuit shown in Figure 9.6, consisting of a battery of *emf* V connected to a single resistor of resistance R. The big idea here is that the battery supplies the energy that the resistor dissipates. Let's see how this works.

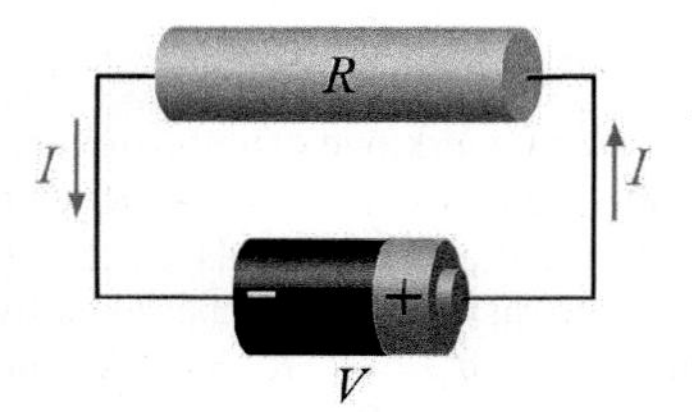

FIGURE 9.6 A resistor R is connected across the terminals of a battery of *emf* V. A current $I = V/R$ flows in the completed circuit. Energy supplied by the battery is dissipated in the resistor.

We'll start by determining the rate at which energy is transferred to the resistor, i.e., we want to calculate the **power** P, which is the time rate of change of the energy that is dissipated in the resistor.

$$P \equiv \frac{dU}{dt}$$

We can calculate this power as follows: In a time dt, an amount of charge $dq = I\,dt$ is moved through a potential difference V. Therefore, the change in potential energy is just equal to the product of the charge and the voltage.

$$dU = dq\,V = (I\,dt)V$$

The rate of energy change (the power) then becomes just the product of the current and the voltage.

$$P \equiv \frac{dU}{dt} = IV$$

This expression, $P = IV$, is valid for any device through which current I flows across a potential difference V. For a resistor, we know that the voltage drop is just equal to the product of the current and the resistance ($V = IR$). Consequently, we obtain that the power dissipated in the resistor is given by the product of the resistance and the square of the current.

$$P = I^2 R$$

The SI unit for power is the watt (W), which equals 1 volt-ampere.

We say that energy is dissipated in the resistor because the charge carriers leave the resistor with less potential energy than they had when they entered the resistor (i.e., $\Delta U = q\,\Delta V < 0$). We can understand this dissipation in terms of the electrons losing energy as they undergo collisions on their path through the resistor. Energy is transferred in this process from the electrons to the lattice of the resistor. This energy transfer results in an increase in the thermal energy of the resistor, i.e., the resistor heats up.

What happens at the battery? The current does flow through the battery. There is a net negative charge flow from the positive terminal to the negative terminal. Therefore the negative charges are moving from a higher potential to a lower potential and experience an increase in potential energy. Where does this energy come from? It must come from the battery itself. The battery is a source of (usually chemical) energy. It is this energy contained in the battery itself that allows the negative charge to "flow upstream" from the positive to the negative terminal. We have called the constant potential across the terminals of a battery, independent of the current being drawn, its *emf.* This abbreviation stands for "electromotive force," a historical reference to the "force" that allows the charge carriers to overcome the force from the electric field inside the battery. Of course, this electromotive force is not a force at all; it is the constant potential difference that appears between the terminals of the battery. For this reason, we only use the abbreviation *emf.* The rate at which the energy of the charge carriers increases as they pass through the battery is just the product of the current I with the *emf* V.

In this circuit then, the battery supplies energy at exactly the same rate as is dissipated in the resistor. If we were to add a capacitor to this circuit, the power discussion becomes somewhat more complicated; we will defer this discussion to Unit 11.

9.10 Summary

We began by defining the electric current in a conductor as the amount of charge that passes through a cross-section of the conductor per unit time (i.e., $I \equiv dq/dt$). We developed a microscopic view of current in which the charge carriers are dissociated electrons in the conductor. These conduction electrons move randomly at high speeds ($\sim 10^5$ m/s), but when a potential difference is introduced across the conductor (for example, by connecting it to the terminals of a battery), the resultant electric field gives these electrons a non-zero average velocity, which is the source of the electric current. This average velocity, the drift velocity, is quite small (generally millimeters per second), but the large number of conduction electrons in the material (typically $\sim 10^{29}\,/\,\text{m}^3$) can give rise to sizeable currents.

We introduced Ohm's law, the empirical result that, for a wide range of materials over a wide range of field strengths, the current density is proportional to the electric field that gives rise to it. The constant of proportionality is called the conductivity of the material. We then used this conductivity to determine the resistance of any particular conductor. The resistance was defined as the ratio of the voltage drop across the conductor to the current that passes through it. For a cylindrical conductor, we found that the resistance was proportional to its length and inversely proportional to its cross-sectional area. The constant of proportionality is 1 / (the conductivity), which we call the resistivity.

We then discussed the use of physical resistors in circuits. In particular, we determined that for resistors in series, we simply add the individual resistances to obtain the equivalent resistance, while for resistors in parallel, we need to add the inverse resistances to obtain the inverse of the equivalent resistance.

Finally, we explored the energy flow in a simple circuit composed of a single resistor connected to the terminals of the battery. When the charge carriers (the electrons) pass through the resistor, they move from a lower potential to a higher potential. Since the charge of the electron is negative, though, that means that these electrons lose potential energy. In particular, the energy they lose ends up as additional thermal energy in the resistor, i.e., the resistor heats up. The rate at which this energy is dissipated in the resistor is called the power, and is given by I^2R. On the other hand, when the electrons pass through the battery, they move from a higher potential to a lower potential, gaining energy. This energy is supplied by the battery itself. The rate at which the battery supplies this energy is just equal to the *emf* of the battery multiplied by the current through the battery.

MAIN POINTS

Definition of Electrical Current

The electric current in a conductor is the amount of charge that passes through a cross-section of the conductor per unit time.

$$I \equiv \frac{dq}{dt}$$

Ohm's Law and Resistance

Ohm's Law: For a wide range of materials over a wide range of field strengths, the current density is proportional to the electric field that gives rise to it.

$$J = \sigma E \qquad V = IR$$

Resistance: The resistance is equal to the ratio of the voltage drop across the resistor to the current that flows through it.

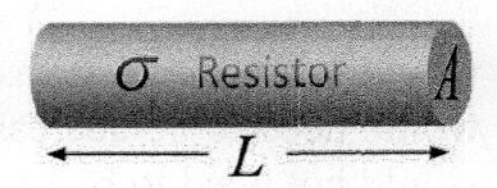

$$R = \frac{L}{\sigma A}$$

Resistors in Series and Parallel

Series: The equivalent resistance is equal to the sum of the individual resistances.

$$R_{equivalent} = R_1 + R_2$$

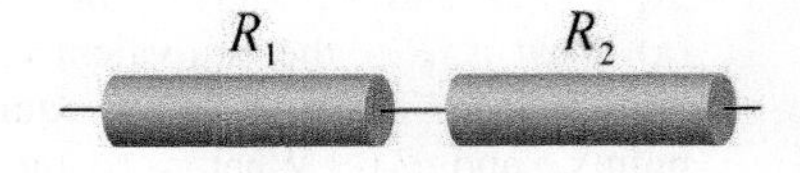

Parallel: The inverse of the equivalent resistance is equal to the sum of the inverses of the individual resistances.

$$\frac{1}{R_{equivalent}} = \frac{1}{R_1} + \frac{1}{R_2}$$

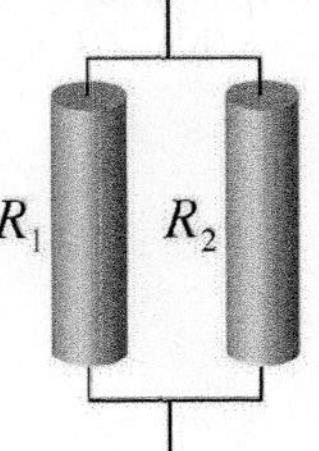

Power

Power is the time rate of change of the energy of a circuit component, and it is always equal to the product of the voltage drop across the component and the current that flows through the component.

$$P = IV$$

PROBLEMS

1. Circuit 1 with Resistors and a Battery: A circuit is constructed with five resistors and a battery as shown. The battery voltage is $V_b = 12\text{ V}$. The values for the resistors are $R_1 = 53\ \Omega$, $R_2 = 126\ \Omega$, $R_3 = 157\ \Omega$, and $R_4 = 96\ \Omega$. The value for R_X is unknown, but it is known that I_4, the current that flows through resistor R_4, is zero. (a) What is I_1, the magnitude of the current that flows through the resistor R_1? (b) What is V_2, the magnitude of the voltage across the resistor R_2? (c) What is I_2, the magnitude of the current that flows through the resistor R_2? (d) What is R_X, the value of the unknown resistor R_X? (e) What is V_1, the magnitude of the voltage across the resistor R_1? (f) If the value of the resistor R_2 were doubled, how would the value of the resistor R_3 have to change in order to keep the current through R_4 equal to zero?

(i) R_3 would need to be increased.
(ii) R_3 would need to be decreased.
(iii) R_3 would not need to be changed.
(iv) There is no change that could be made to R_3 to keep the current through R_4 equal to zero.

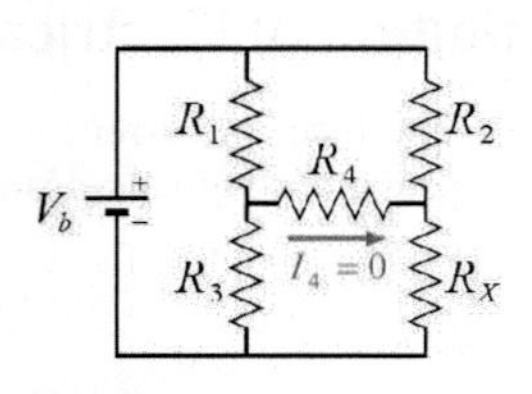

FIGURE 9.7 Problem 1

2. Circuit 2 with Resistors and a Battery: A circuit is constructed with five resistors and a battery as shown. The values for the resistors are: $R_1 = R_5 = 71\ \Omega$, $R_2 = 117\ \Omega$, $R_3 = 47\ \Omega$, and $R_4 = 88\ \Omega$. The battery voltage is $V_b = 12\text{ V}$. (a) What is R_{ab}, the equivalent resistance between points a and b? (b) What is R_{ac}, the equivalent resistance between points a and c? (c) What is I_5, the current that flows through resistor R_5? (d) What is I_2, the current that flows through resistor R_2? (e) What is I_1, the current that flows through the resistor R_1? (f) What is V_4, the voltage across resistor R_4?

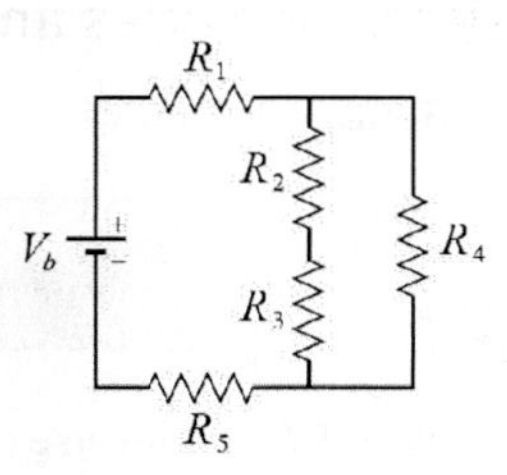

FIGURE 9.8 Problem 2

UNIT

10

KIRCHHOFF'S RULES

10.1 Overview

In the last unit we introduced the idea of electric current, spending most of our time building a microscopic picture of charges in motion in conductors. We will now take a more macroscopic perspective in analyzing voltages and currents in simple circuits.

We'll start by introducing the two rules we will need to perform all of the analysis, namely Kirchhoff's voltage and current rules. We will discuss conventions for current directions and voltage signs and then apply these rules to a few circuits. We will close by modeling a real battery as an ideal battery in series with an internal resistance.

10.2 Devices Review

Before we get to the analysis of simple circuits, we'll first provide a brief review of the devices we have available to us to create these circuits.

First, we have the **battery**. Ideal batteries maintain a constant potential difference across their terminals. Sometimes this potential difference is referred to as the *emf* of the battery. The symbol used for batteries in circuits is shown in Table 10.1; the longer line represents the positive terminal. Batteries provide the energy needed to move charge through the circuit, to establish electric currents.

Next, we have **capacitors**. Capacitors store charge. The symbol used for capacitors in circuits represents the two plates of a parallel-plate capacitor, as shown in Table 10.1. Capacitors are electrically neutral, but the charge is separated so that one plate has an excess of negative charge and the other plate has an equal amount of excess positive charge. The voltage drop across a capacitor is equal to the excess charge Q on one plate divided by C, its capacitance. Although charges do not jump across the plates of a capacitor, we can talk about the current through the capacitor when it is placed in a circuit as the rate of change of excess charge on its plates.

TABLE 10.1 Symbols and voltages for devices used in simple circuits.

Device	Symbol	Voltage
Battery	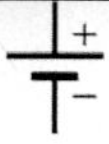	V_b, $\mathcal{E}$
Capacitor	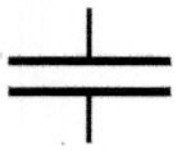	Q/C
Resistor		IR
Switch		closed = 0 open = ∞

Next, we have **resistors**. The voltage drop across a resistor is proportional to the current that flows through it. The symbol used for resistors in circuits is shown in Figure 10.1.

Finally, we have the **switch**. This device is used to add or remove other devices from the circuit. The symbol used for switches in circuits reflects the two positions (open or closed) of the switch, as shown in Table 10.1. When the switch is open, no current flows through it. When the switch is closed, current passes freely through the switch and the voltage across the switch is zero.

10.3 Kirchhoff's Rules

We will begin our study of circuit analysis by introducing the two rules we need to do all the work. The good news is that there is nothing really new going on in these rules; you already know the underlying principles.

The first rule is essentially a statement of the conservation of energy and is known as Kirchhoff's voltage rule. This rule states that if you start at any point on a circuit and follow that circuit around a complete loop, returning to where you started, the net change in the electric potential is zero. We can express this rule in the form of an equation:

$$\sum \Delta V_n = 0$$

where the sum over ΔV_n represents the algebraic sum of all changes in potential encountered while traversing a complete circuit loop. This rule certainly makes sense: The potential difference between a point and itself must be zero; that's the whole point of defining the potential as a property of the space. If you start climbing from any point on a hill and eventually return to that same point, your elevation has not changed!

The second rule is essentially a statement of charge conservation and is referred to as Kirchhoff's current rule. This rule states that at any "node" in a circuit (i.e., a place where two or more wires meet), the sum of the currents flowing into the node is equal to the sum of the currents flowing out of the node.

$$\sum I_{in} = \sum I_{out}$$

This rule certainly makes sense: Any charge that flows into a node must also flow out since there is no place for the charge to build up in the wire. These two rules are all that we need to analyze simple circuits. Our next task will be to apply these rules to analyze some example circuits.

10.4 Conventions

In order to apply these two rules to analyze circuits, we must first address the issue of sign conventions. In particular we will need to address conventions for the *direction* of *current* flow and the *sign* of *voltage* changes.

In all circuits discussed so far, we have been careful never to associate a direction with the symbol I for current to avoid possible confusion with the convention we will now introduce. Consider the circuit shown in Figure 10.1 in which a battery with *emf* V_b is connected to two resistors, R_1 and R_2, in series. This is a *one-loop circuit*; there is only one current whose magnitude is given by the battery voltage divided by the total resistance.

$$I = \frac{V}{R_1 + R_2}$$

We know that the charge carriers in the circuit are electrons and they are moving from lower potential to higher potential through the resistors, i.e., they move from c to b to a (*counterclockwise*). It might seem natural to represent the current I with an arrow pointing in the counterclockwise direction. We will instead follow the usual convention of representing the current in this circuit as pointing in the *clockwise* direction. In other words, we define the direction of the current as the direction of the flow of net *positive* charge. The main reason we choose this convention is of course historical, but it also seems to minimize the amount of extra negative signs needed to account for the negative charge of the electron. If we take the current to be the direction of positive charge flow, then the natural motion of charges is "downhill" rather than "uphill."

To apply Kirchhoff's voltage rule, we need to adopt a convention for the sign of the voltage changes ΔV as we go around a loop. Since the voltage changes eventually sum to zero, it doesn't matter if we choose voltage drops or voltage gains to be positive; we just need to be consistent. For this course, we will consistently define ΔV to be *positive* when we encounter a voltage *drop*. We'll now introduce some examples in which we will apply Kirchhoff's rules using these conventions.

10.5 Single Loop Example

We'll start with the example we discussed shown in Figure 10.1. We can actually solve this circuit by simply replacing the two resistors with one equivalent resistor and then applying Ohm's law to obtain the result that the current is just equal to the battery voltage divided by the equivalent resistance of the two resistors in series (i.e., $I = V_b/(R_1 + R_2)$). We will go ahead, though, and apply Kirchhoff's rules to illustrate the use of the conventions we discussed in the last section.

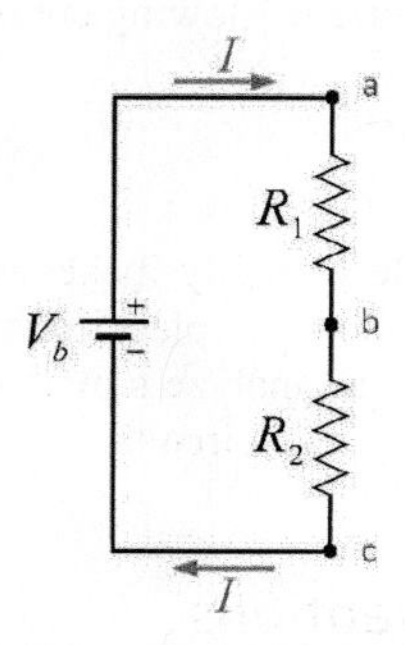

FIGURE 10.1 A circuit composed of two resistors (R_1 and R_2) connected to a battery of *emf* V_b. The current convention we adopt is shown: The direction of the current *I* is that of the *positive* charges.

The first step is to identify and label all currents. The circuit has no nodes; there is only one current and we'll call it *I*. What do we choose for its direction? In this case we can see that the net positive charge flow will be clockwise (e.g., from *a* to *b* to *c* through the resistors), so we will choose this direction for the current.

Once the current has been chosen we can apply Kirchhoff's voltage rule, namely that the sum of the voltage drops in any loop must be zero. We'll start at point *a* and move around the loop in the direction of the current *I*. As we move from point *a* to point *b* through the top resistor, we encounter a voltage drop that is equal to the product of the current and the resistance (i.e., $\Delta V_{ab} = +IR_1$). Note that since we traversed the resistor in the direction of the assumed positive current flow, we define ΔV_{ab} to be positive. Continuing from point *b* to point *c* through the bottom resistor, we encounter another voltage drop that is equal to the product of the current and the resistance (i.e., $\Delta V_{bc} = +IR_2$). Once again, since we are moving in the direction of assumed positive current flow, we define ΔV_{bc} to be positive.

Following *I* in the loop we encounter the negative terminal of the battery. As we move through the battery to the positive terminal, we have a voltage gain equal to the *emf* of the battery. Consequently we define ΔV_{ca} to be negative. Since we are now back at point *a* where we started, we can evaluate Kirchhoff's voltage rule, that the sum of the voltage drops must be zero.

$$\sum \Delta V_{ij} = \Delta V_{ab} + \Delta V_{bc} + \Delta V_{ca} = 0$$

$$IR_1 + IR_2 - V_b = 0$$

Here we have two voltage drops and one voltage gain combining to give zero, which we solve to get the expected result that the current is equal to the battery voltage divided by the sum of the resistances.

$$I = \frac{V_b}{R_1 + R_2}$$

In this example, the direction of the net positive charge flow was clear. In more complicated circuits, it may not be so easy to determine the correct directions of the currents. The good news is that we don't need to know these directions ahead of time. We can simply choose directions arbitrarily; if we happen to choose an incorrect direction for a specific current, then its value will turn out to be negative, indicating the direction to be opposite to what we assumed.

10.6 Simple Circuit Example

We'll now solve a slightly more complicated example that involves more than one current. Consider the circuit shown in Figure 10.2 consisting of a battery and three resistors.

We have two nodes in this circuit (points *b* and *c*). Our first job is to choose the directions for the currents. We'll assume I_1 passes through R_1 from *a* to *b*. We'll also assume that I_2 passes through R_2 from *b* to *c* and that I_3 passes through R_3 also from *b* to *c*.

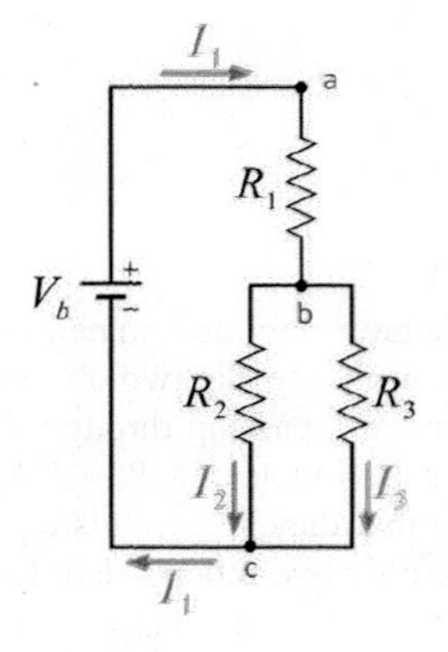

FIGURE 10.2 A multiloop circuit containing three resistors and an ideal battery. The current from the battery splits at node *b*.

Now that we have chosen directions for the currents, we can apply Kirchhoff's current rule. At node *b*, current I_1 enters the node and currents I_2 and I_3 leave the node. Consequently, Kirchhoff's current rule gives us one equation:

$$I_1 = I_2 + I_3$$

What happens at node *c*? Currents I_2 and I_3 enter and a new current leaves. However, this new current is not new at all; it must be I_1. We can see this by noting the result either from applying Kirchhoff's current rule to node *b*, or from noting that the current that enters the battery must be identical to the current that leaves the battery (I_1).

So far, we have one equation and three unknowns (I_1, I_2, and I_3). We need to generate two more equations in order to determine all of the unknowns. These two equations must come from applying Kirchhoff's voltage rules. We first consider the loop consisting of the battery, R_1 and R_2. Starting at point *a*, Kirchhoff's voltage rule becomes

$$+I_1R_1 + I_2R_2 - V_b = 0$$

If we apply Kirchhoff's voltage rule to the loop consisting of the battery, R_1 and R_3, we obtain

$$+I_1R_1 + I_3R_3 - V_b = 0$$

We now have three equations and three unknowns. We can solve these equations in any number of ways. For example, we can first rewrite this last equation by replacing I_3 by its value in the first equation (i.e., $I_3 = I_1 - I_2$). We are then left with two equations in two unknowns (I_1 and I_2). We can solve for I_2 in the second equation (i.e., $I_2 = (V - I_1R_1)/R_2$) and plug it into the new third equation to get an expression for I_1, the current through the battery, as shown. We can then use this value for I_1 in the original second equation to solve for I_2 and then again in the original third equation to solve for I_3. The final results for the currents are

$$I_1 = \frac{V_b(R_2 + R_3)}{R_1R_2 + R_1R_3 + R_2R_3}$$

$$I_2 = \frac{V_bR_3}{R_1R_2 + R_1R_3 + R_2R_3}$$

$$I_3 = \frac{V_bR_2}{R_1R_2 + R_1R_3 + R_2R_3}$$

We have now determined all of the currents in the circuit. Before we leave this example, we want to make two observations: First, what about the other loop in this circuit, namely leaving *b* passing through R_2 and R_3 to return to *b*? Kirchhoff's voltage rule applied to this loop yields: $I_2R_2 - I_3R_3 = 0$. If we look at our existing solutions for I_2 and I_3, we see that they satisfy this equation, as they must. There is no new information here. There are only two independent loops in this circuit; you can use any two you like and you will get the same results.

Second, we can check our calculation of I_1 by noting that R_2 and R_3 are in parallel and could be replaced by an equivalent resistance $R_2R_3/(R_2 + R_3)$. This equivalent resistance is then in series with R_1; consequently, we know I_1 must just be equal to V_b divided by $R_1 + R_2R_3/(R_2 + R_3)$. Simplifying this expression, we see these two expressions for I_1 are identical, as they must be!

10.7 Two Loop Example

The two previous examples could have been done by the appropriate use of series and parallel combinations of resistors. We'll now do an example that cannot be done in this manner. Consider the circuit shown in Figure 10.3 consisting of two batteries and three resistors.

We start, as always, by choosing directions for the currents in the circuit. Here, we'll call I_1 the current leaving the positive terminal of the battery 1, I_2 the current passing through R_2 that enters node a, and I_3 the current that leaves node a and passes through R_3. As before, there are two nodes in the circuit that will yield one independent equation from Kirchhoff's current rule. From node a, we obtain the equation

$$I_1 + I_2 = I_3$$

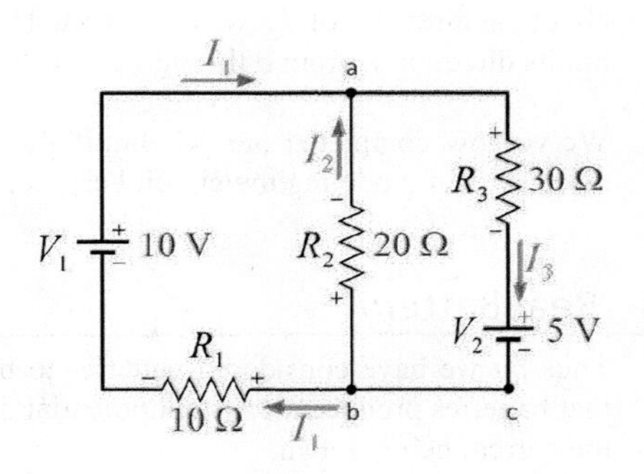

FIGURE 10.3 A two loop circuit composed of three resistors and two ideal batteries.

Note that the node c equation is $I_3 = I_2 + I_1$, containing no new information, since I_1 is the current that passes through the battery V_1.

We now need to use Kirchhoff's voltage rule to generate two more equations. There are three loops; any two will do. Let's start at node a and go clockwise around the outer loop. We have a voltage drop across R_3 $(+I_3R_3)$, a voltage drop $(+V_2)$ across battery 2, a voltage drop across R_1 $(+I_1R_1)$, and a voltage gain $(-V_1)$ through battery 1. Therefore, our second equation is

$$I_3R_3 + V_2 + I_1R_1 - V_1 = 0$$

To obtain our third equation, let's start from node a and go clockwise around the left loop. We have a voltage gain across R_2 $(-I_2R_2)$, a voltage drop across R_1 $(+I_1R_1)$, and a voltage gain $(-V_1)$ through battery 1. Therefore, our last equation becomes

$$-I_2R_2 + I_1R_1 - V_1 = 0$$

We now have three equations and three unknowns. We can solve these equations in any number of ways and we obtain

$$I_1 = \frac{V_1(R_2 + R_3) - V_2R_2}{R_1R_2 + R_1R_3 + R_2R_3}$$

$$I_2 = \frac{-V_2R_1 - V_1R_3}{R_1R_2 + R_1R_3 + R_2R_3}$$

$$I_3 = \frac{V_1R_2 - V_2(R_1 + R_2)}{R_1R_2 + R_1R_3 + R_2R_3}$$

Substituting in the values for the components, we obtain $I_1 = +0.364$ A, $I_2 = -0.318$ A, and $I_3 = +0.045$ A. Note that the value of I_2 is negative, indicating that our assumption about the direction of I_2 was incorrect. The magnitude of the current is certainly 0.318 A, but its direction is from *a* through R_2 to *b*.

We've now completed our Kirchhoff Rules examples; we'll close this unit with a brief discussion of modeling the circuit behavior of real batteries.

10.8 Real Battery

Thus far we have considered batteries to be ideal voltage sources, i.e., we have assumed that batteries produce a constant potential difference across their terminals, independent of the current being drawn.

In a real battery, the voltage across the terminals can decrease if a large enough current is being drawn from the battery. A good model for a real battery is to include an internal resistance *r* in series with a voltage source V_o, as shown in Figure 10.4(a).

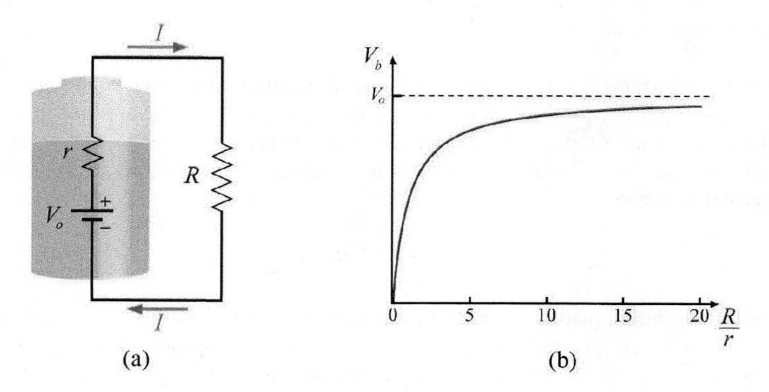

FIGURE 10.4 (a) A model for a real battery consists of an internal resistance *r* in series with a voltage source V_o. (b) A plot of V_b, the voltage that appears across the terminals of the battery shown in (a) as a function of R/r.

If no load resistance is connected to the battery, there is no complete circuit, no current will be drawn, and the potential difference across the terminals of the battery will be V_o. If we connect a load, consisting of a single resistor of resistance *R*, current will be drawn and the potential difference across the terminals of the battery will be less than V_o. In particular, the current *I* will be equal to $V_o/(R+r)$. Therefore, the voltage that appears at the terminals will be equal to the *emf* of the battery minus the voltage drop across the internal resistance ($V_o - Ir$). We can rewrite V_b in terms of the ratio of the load resistance *R* to the internal resistance *r* to obtain

$$V_b = V_o \frac{\frac{R}{r}}{1+\frac{R}{r}}$$

V_b quickly approaches V_o as the ratio R/r becomes large, as shown in Figure 10.4(b).

In practice we want the effective internal resistance of the battery to be small compared to the load, so that the voltage across the terminals remains close to the specified voltage, the *emf* of the battery. This is one of the reasons a 12-volt car battery is so much larger than a 9-volt radio battery.

10.9 Summary

We began by introducing Kirchhoff's voltage and current rules, the only two rules needed to perform circuit analysis. Kirchhoff's voltage rule is essentially a statement of the conservation of energy: namely, if you start at any point on a circuit and follow that circuit around a complete loop, returning to where you started, the net change in the electric potential is zero. Kirchhoff's current rule is essentially a statement of charge conservation: namely, the sum of the currents flowing into any node of the circuit is equal to the sum of the currents flowing out of that node.

We then addressed some sign conventions for the direction of current flow and changes in voltage that we choose to use in the application of these rules. Namely, we define the direction of the current (indicated by $I > 0$) to be the direction of the flow of net positive charge, and we define a positive voltage change to be a voltage drop when the element is traversed in the direction of the assumed positive charge flow. We applied these rules to several circuits, including some with multiple loops.

Finally, we introduced a model for a real battery as a voltage source in series with an internal resistance. When a load resistance is connected to the terminals of the battery, a current is drawn, resulting in a reduced voltage at the terminals. This reduction becomes small as the ratio of the load resistance to the internal resistance becomes large.

MAIN POINTS

Kirchhoff's Voltage and Current Rules

KVR: The sum of the voltage drops across circuit elements in any closed loop is zero.

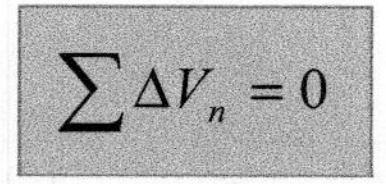

$$\sum \Delta V_n = 0$$

Examples:

Outer Loop (cw):

$$I_3R_3 + V_2 + I_1R_1 - V_1 = 0$$

Left Loop (cw):

$$-I_2R_2 + I_1R_1 - V_1 = 0$$

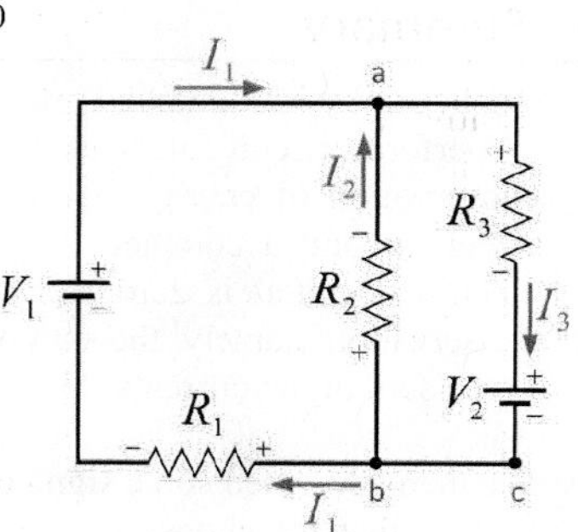

KCR: The sum of the currents flowing into any node in the circuit is equal to the sum of the currents flowing out of that node.

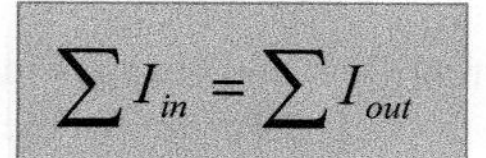

$$\sum I_{in} = \sum I_{out}$$

Example:

$$I_1 + I_2 = I_3$$

Model of a Real Battery

A real battery can be modeled as a voltage source in series with an internal resistance.

$$V_b = V_o \frac{\frac{R}{r}}{1 + \frac{R}{r}}$$

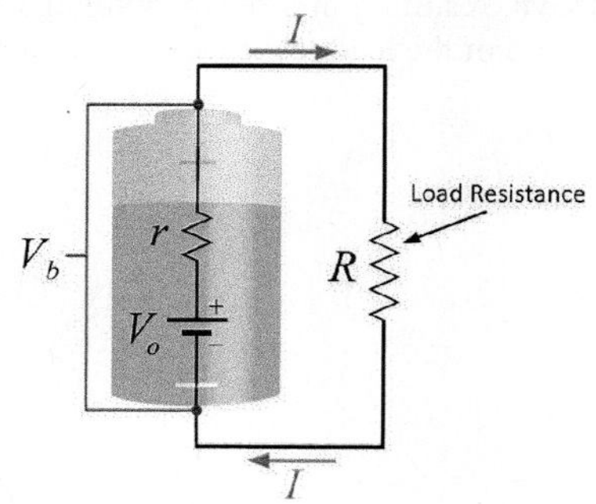

PROBLEMS

1. Circuit with Two Batteries and Six Resistors: A circuit is constructed with six resistors and two batteries, as shown in Figure 10.5. The battery voltages are $V_1 = 18\ \text{V}$ and $V_2 = 12\ \text{V}$. The positive terminals are indicated with a + sign. The values for the resistors are $R_1 = R_5 = 50\ \Omega$, $R_2 = R_6 = 81\ \Omega$, $R_3 = 80\ \Omega$, and $R_4 = 121\ \Omega$. The positive directions for the currents I_1, I_2, and I_3 are indicated by the directions of the arrows. (a) What is V_4, the magnitude of the voltage across the resistor R_4? (b) What is I_3, the current that flows through the resistor R_3? A positive value for the current is defined to be in the direction of the arrow. (c) What is I_2, the current that flows through the resistor R_2? (d) What is I_1, the current that flows through the resistor R_1? (e) What is $V(a) - V(b)$, the potential difference between the points a and b?

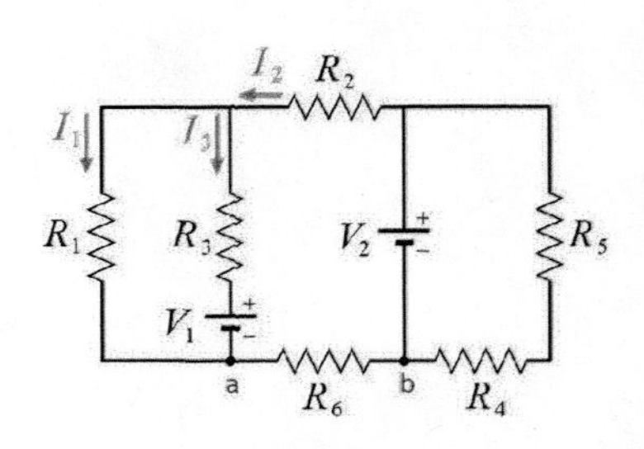

FIGURE 10.5 Problem 1

2. Non-Ideal Battery: A circuit is constructed with five resistors and one real battery, as shown in Figure 10.6. We model the real battery as an ideal *emf* $V_o = 12\ \text{V}$ in series with an internal resistance r as shown. The values for the resistors are $R_1 = R_3 = 55\ \Omega$, $R_4 = R_5 = 71\ \Omega$, and $R_2 = 112\ \Omega$. The measured voltage across the terminals of the battery is $V_b = 11.52\ \text{V}$. (a) What is I_1, the current that flows through the resistor R_1? (b) What is r, the internal resistance of the battery? (c) What is I_3, the current through resistor R_3? (d) What is P_2, the power dissipated in resistor R_2? (e) What is V_2, the magnitude of the voltage across the resistor R_2? (f) Resistor R_2 is now shorted out. How does the magnitude of the voltage across the battery, V_b, change: V_b *decreases*, V_b *increases*, or V_b *remains the same*?

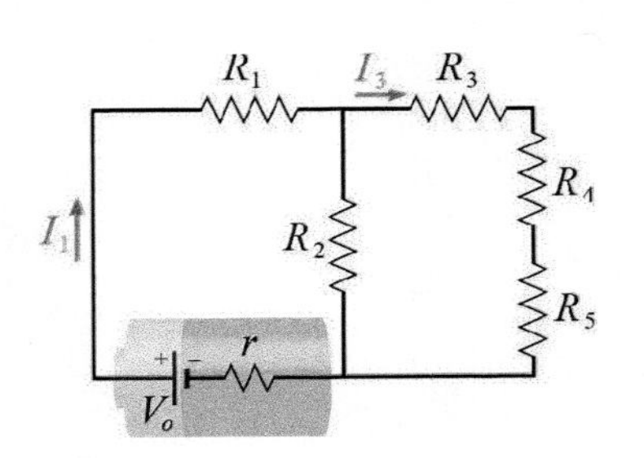

FIGURE 10.6 Problem 2

3. Current Divider (Interactive Example): The circuit in Figure 10.7 contains five resistors (R_1, R_2, R_A, R_B, and R_C) with equal resistance R. Write an equation for the current I_2 in terms of the current I_1.

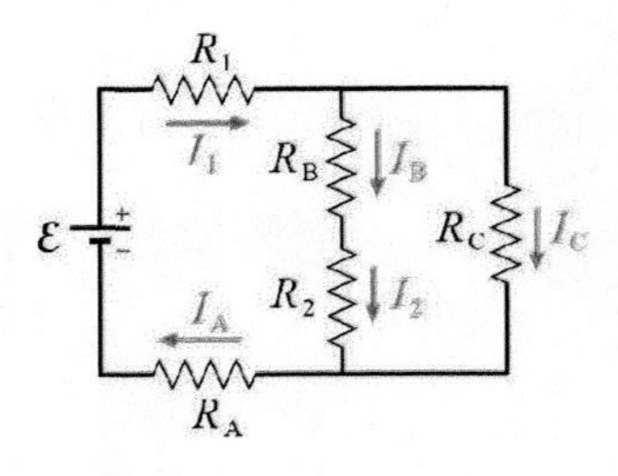

FIGURE 10.7 Problem 3

4. Multiloop (Interactive Example): Resistors R_1, R_2, R_3, and R_4 are arranged in a circuit, as shown in Figure 10.8. The direction of positive currents I_1, I_2, I_3, and I_4 through the resistors are shown (i.e., negative current values correspond to currents in the opposite direction to that shown in the figure). The values for the resistors and the batteries in the circuit are $R_1 = 100\ \Omega$, $R_2 = 200\ \Omega$, $R_3 = 30\ \Omega$, $R_4 = 400\ \Omega$, $V_1 = 4$ V, and $V_2 = 12$ V. What is the value of I_2, the current through R_2?

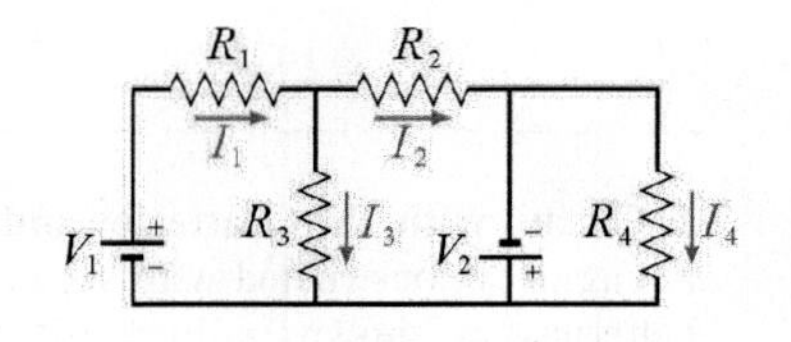

FIGURE 10.8 Problem 4

UNIT

11

RC CIRCUITS

11.1 Overview

In the last unit we analyzed circuits containing batteries and resistors using Kirchhoff's rules. We will now analyze circuits that include capacitors and will discover qualitatively different behaviors for the currents and voltages.

We'll start by considering a circuit that uses a battery to charge a capacitor through a resistor. We'll then calculate what happens when the battery is removed and the capacitor is discharged. We'll then discuss the energy flow in these circuits. We will close by considering a more complicated circuit that contains two capacitors and two resistors.

11.2 Qualitative Description of Charging a Capacitor

We will begin with a qualitative discussion of the behavior of currents and voltages in the circuit shown in Figure 11.1. In Unit 8 we considered circuits containing just a battery and a capacitor and saw that the voltage across the capacitor was always the same as the battery voltage, and that the charge on the capacitor was just equal to the product of this voltage with the capacitance. Implicit in this picture was the idea that the capacitor acquired its charge essentially the instant the battery was connected to it. Adding a resistor to the circuit will limit the rate at which charge will flow to the capacitor, and hence the rate at which the voltage across the capacitor will approach the battery voltage.

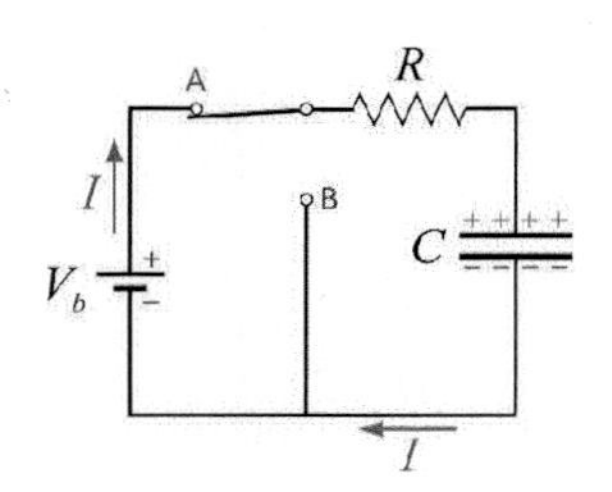

FIGURE 11.1 The capacitor is initially uncharged when the switch is closed. Current flows from the ideal battery through the resistor to charge the capacitor.

In the circuit shown in Figure 11.1, initially the switch is open and the capacitor is uncharged. In this configuration, there is no complete circuit, so no current flows. The potential difference between points A and B is V_b, the *emf* of the battery. At time $t = 0$, the switch is closed (i.e., moved to position A) and a circuit is completed so that current will start to flow. Our goal here is to understand the magnitude and direction of this current as a function of time.

The new wrinkle here is the addition of the capacitor. Recall that the voltage drop across a capacitor is equal to its charge divided by its capacitance. Initially, the capacitor is uncharged so that the voltage across it is zero (i.e., initially the capacitor acts as a wire). Therefore, initially this circuit looks like it contains only a battery and a resistor, resulting in an initial current equal to the battery voltage divided by the resistance ($I = V / R$).

So far there's nothing different about this circuit. However, as time goes on, new behaviors emerge. In particular, as the current flows through the capacitor, the top plate becomes positively charged and the bottom plate becomes negatively charged. Therefore, a non-zero voltage develops across the capacitor, resulting in a reduced voltage across the resistor since by Kirchhoff's voltage rule, the sum of these voltages must equal the battery voltage. Since the voltage across the resistor is equal to the product of the current and the resistance ($V = IR$), we see the current I must decrease. Asymptotically, the current goes to zero as the charge on the capacitor approaches its maximum value corresponding to the capacitor's voltage becoming equal to the battery voltage ($Q_{max} / C = V_b$).

11.3 Quantitative Description of Charging a Capacitor

We will now determine the time dependence of the current and the voltages in this circuit. The good news is that we start just as we have always started with Kirchhoff's rules. We choose the current to be clockwise as shown, and apply Kirchhoff's voltage rule, beginning at point A. We first encounter a voltage drop across the resistor ($+IR$), and then another voltage drop across the capacitor ($+q / C$), and finally a voltage gain through the battery ($-V_b$).

The resulting equation then is the only equation we can write since there is only one loop.

$$IR + \frac{q}{C} - V_b = 0$$

At first glance, it looks like we've got a problem since it appears we have two unknowns, I and q. In fact, there is no problem here because I and q are not independent; we know $I \equiv dq/dt$. Therefore, we can replace I with its equivalent expression in terms of q to arrive at a single equation with only one unknown, q:

$$R\frac{dq}{dt} + \frac{q}{C} - V_b = 0$$

This equation has only one variable, but it is a differential equation. We can solve this equation by inspection (that's the mathematician's word for guessing). Namely, we want a function $q(t)$, which when differentiated results in something proportional to the original function. The one function we know that does that is the exponential. We also know that $q(0) = 0$ and $q(\infty) = CV$.

The function

$$q(t) = CV_b(1 - e^{-t/RC})$$

satisfies the differential equation and the known constraints at $t = 0$ and $t = \infty$. A plot of the time dependence of this function $q(t)$ is shown in Figure 11.2(a). We see that as t increases, $q(t)$ increases exponentially from zero toward its asymptotic value of CV_b.

The current $I(t)$ can be obtained from $q(t)$ by differentiating with respect to time.

$$I(t) = \frac{V_b}{R}e^{-t/RC}$$

Note that the constraints $I(0) = V/R$ and $I(\infty) = 0$ are automatically satisfied in this expression. The plot in Figure 11.2(b) shows that as t increases, $I(t)$ decreases exponentially from its maximum value of V_b/R toward its asymptotic value of zero.

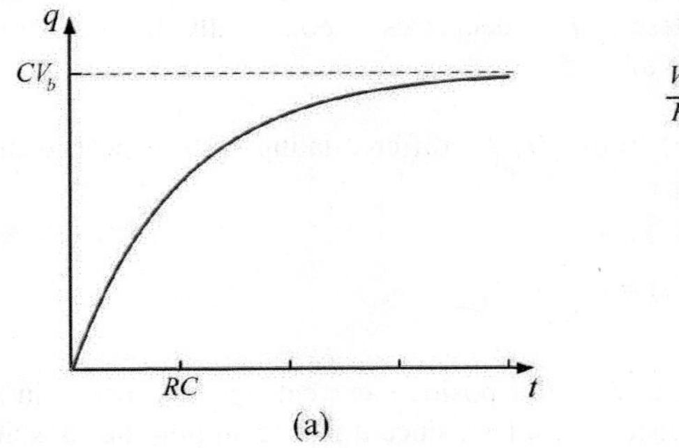

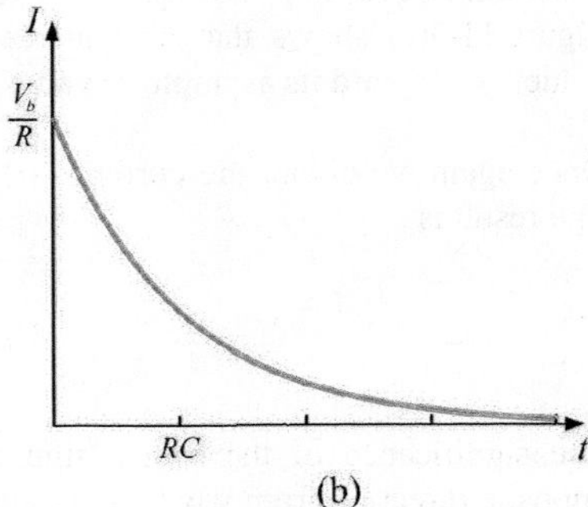

FIGURE 11.2 (a) The time dependence of q, the charge on the capacitor, in the discharging circuit shown in Figure 11.1. (b) The time dependence of I, the current, in the charging circuit shown in Figure 11.1.

11.4 Discharging a Capacitor

We now move the switch from position A to position B (Figure 11.1), removing the battery from the circuit. What will happen?

At the instant the switch is moved, there will be no time for charge to flow on or off the capacitor. Therefore, the charge on the capacitor immediately after the switch is moved will have the same value it had just before the switch was moved; let's call it q_o. Therefore, there will be a voltage ($V_o = q_o / C$) across the capacitor at this time. Consequently, a current will begin to flow. This current flows from the positively charged plate of the capacitor through the resistor and back to the negatively charged plate of the capacitor. Note that this current is in the opposite direction of the current that charged the capacitor. This current will discharge the capacitor.

Let's do the calculation. We apply Kirchhoff's voltage rule starting at point B and move clockwise in the sense of the initial current direction. We first encounter a voltage drop across the resistor (+*IR*) and then another voltage drop across the capacitor ($+q / C$). The resulting equation then states

$$IR + \frac{q}{C} = 0$$

Replacing I by dq / dt, we once again obtain a differential equation for q:

$$R\frac{dq}{dt} + \frac{q}{C} = 0$$

The solution to this equation is once again an exponential:

$$q(t) = q_o e^{-t/RC}$$

Note that we have redefined $t = 0$ to be the time the switch is moved to B and that the constraints ($q(0) = q_o$ and $q(\infty) = 0$) are satisfied by this expression for $q(t)$. The plot in Figure 11.3(a) shows that as t increases, $q(t)$ decreases exponentially from its initial value, q_o, toward its asymptotic value of zero.

Once again we obtain the current $I(t)$ from $q(t)$ by differentiating with respect to time. The result is

$$I(t) = -\frac{q_o}{RC} e^{-t/RC}$$

The significance of the minus sign is that the positive current actually flows in the opposite direction from what we assumed, as it must, since it is discharging the capacitor. The plot in Figure 11.3(b) shows that as t increases, the *magnitude* of $I(t)$ also decreases exponentially from its maximum value of q_o / RC toward its asymptotic value of zero.

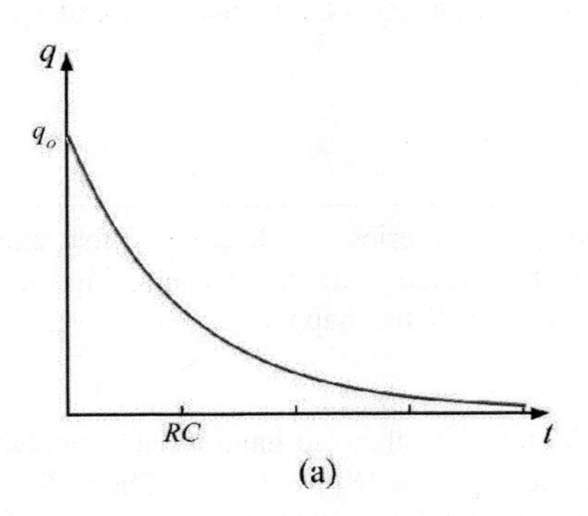

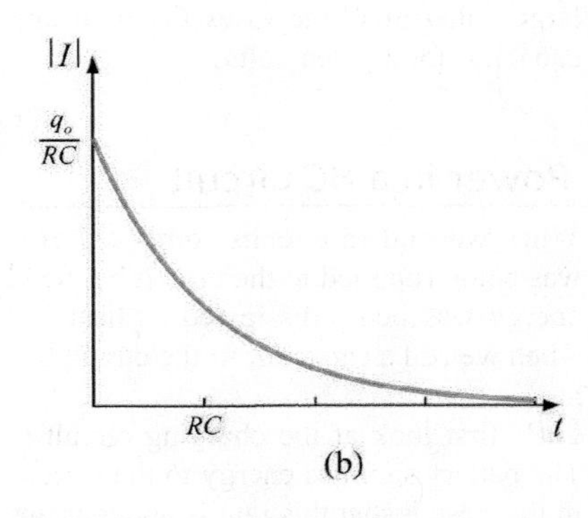

FIGURE 11.3 The time dependence of q, the charge on the capacitor, as it is discharged (the switch in Figure 11.1 is moved to B at $t = 0$). The time dependence of the *magnitude* of the current as the capacitor is discharged (the switch in Figure 11.1 is moved to B at $t = 0$). The sign of the current is actually *negative*, indicating that the *direction* of the discharging current is opposite that of the charging current.

11.5 Time Constants

We have just calculated the currents and voltages in circuits containing a resistor and a capacitor. We've seen that for both charging and discharging circuits, the time dependence of the charge on the capacitor and the current through the circuit is exponential. The rate of the exponential is determined by the product of the resistance and the capacitance. We call this product, RC, the **time constant** τ of the circuit. The time constant should have the units of time, and it does since $1\ \Omega = 1$ V/A and 1 F $= 1$ C/V, resulting in $1\ \Omega$F $= 1$ C/A $= 1$ second.

For any quantity that changes exponentially, the fractional change in a given interval is always the same, no matter when you start the interval. For example, consider the current in an RC circuit. We have shown that the current at any time t is just the initial current I_o times the exponential factor $e^{-t/\tau}$. When $t = \tau$, the exponential factor is e^{-1}, which is about 0.37. In other words, when the elapsed time is one time constant, the current is about 37% of its initial value. When $t = 2\tau$, the exponential factor is e^{-2}. Since exponents add, we can write e^{-2} as e^{-1} times e^{-1}. In other words, the current at $t = 2\tau$ is equal to 37% of 37% (or about 14%) of its initial value. Thus, we see that the current at $t = 2\tau$ is 37% of what it was at $t = 1\tau$. In exactly the same way, the current at $t = 3\tau$ will be 37% of what it was at $t = 2\tau$, etc. This is the signature behavior of an exponential function: for a fixed change in the exponent (e.g., 1 time constant), the value of the function changes by a fixed factor (e.g., 0.37).

Thus we see that, for a given RC circuit, it is the *time constant* that characterizes the time scale of the charging or discharging process. This time constant is determined by the values of R and C. For example, larger values of R and C lead to larger time constants. This makes sense since, for example, a large value of R reduces the current in the circuit; thereby reducing the rate at which charge is being stored on the capacitor. Similarly, a large value of C increases the total amount of charge ($Q = CV$) that is stored on the capacitor for a given voltage.

11.6 Power in a *RC* Circuit

When we studied circuits composed of resistors and batteries, we discovered that energy was being supplied to the circuit by the battery at a constant rate that was equal to the rate energy was being dissipated as heat in the resistors. What happens to the energy flow when we add a capacitor to the circuit?

Let's first look at the charging circuit (see Figure 11.1) that we have already analyzed. The battery supplies energy to the circuit at a rate $P_{Battery} = IV_b$, as before. The difference in this case is that this rate is not constant as the current I is decreasing exponentially with time as $e^{-t/RC}$, as shown in Figure 11.4. Energy is being dissipated in the resistor at a rate $P_R = I^2 R$, as before. Once again, we see that this rate is not constant; in fact it decreases exponentially with time as $e^{-2t/RC}$, since it is proportional to the square of the current. From Figure 11.4, we can see that all of the energy that is being supplied by the battery is not being dissipated as heat in the resistor. What is happening to the rest of this energy?

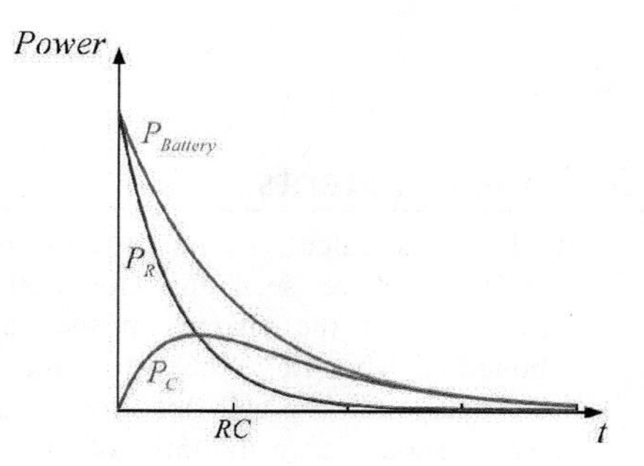

FIGURE 11.4 The time dependence of the power supplied by the battery ($P_{Battery}$), the power dissipated in the resistor (P_R), and the power stored in the capacitor (P_C) during the charging process in the circuit shown in Figure 11.1.

The capacitor is gaining charge as time increases. As the capacitor gains charge, it gains energy (i.e., $U_C = q^2/2C$). The capacitor is storing the energy being supplied by the battery that is not being dissipated as heat in the resistors. Indeed, this storage of energy is the main reason capacitors are used in circuits. For example, the flash unit in your camera requires a release of a large amount of energy in a short time, which is accomplished by discharging a capacitor through a small resistance.

The rate at which energy is stored in this circuit is equal to the product of the voltage across the capacitor and the charging current.

$$P_C = V_C I = \frac{q}{C} I$$

Consequently, the time dependence of this rate is the product of two exponential terms $e^{-t/RC}(1 - e^{-t/RC})$, which gives rise to the peaked time dependence shown (i.e., the rate at which energy is stored in the capacitor increases from zero at $t = 0$, reaches a maximum and then decreases toward its asymptotic value of zero).

Using these results, we can confirm that the rate at which energy is supplied by the battery is equal to the rate at which energy is dissipated in the resistor plus the rate at which it is stored in the capacitor.

11.7 Example: Charging Two Capacitors

It is possible to write down the appropriate Kirchhoff's rules for any *RC* circuit. If the *RC* circuit has at least one node, we will be left with coupled differential equations that are often difficult to solve. In this course we will restrict ourselves to circuits that can be reduced to the same form as the *RC* circuits we have studied up to this point. The key will be to determine the appropriate values for *R* and *C*.

In the example shown in Figure 11.5 we start with C_1 and C_2 uncharged, and at time $t = 0$ the switch is moved to position A. Here we have a single loop circuit so that the solution can be obtained by writing down Kirchhoff's voltage rule for this loop. We'll assume the current I leaves the positive terminal of the battery and moves clockwise as shown. We'll start at point A and first encounter a voltage drop across a resistor ($+IR_1$), then a voltage drop across a capacitor ($+q/C_1$), and another voltage drop across a capacitor ($+q/C_2$). Note that we have used the same q for each capacitor since they are in series. Continuing, we encounter a voltage drop across a resistor ($+IR_2$), and finally a voltage gain through the battery ($-V_b$). Putting this altogether we get:

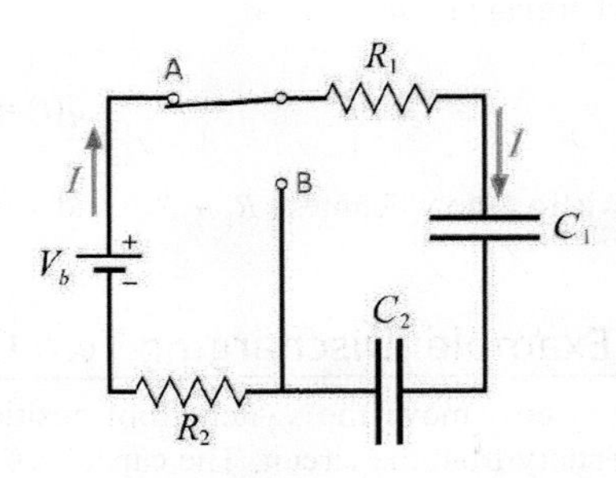

FIGURE 11.5 Capacitors C_1 and C_2 are initially uncharged when the switch is closed. Current flows from the ideal battery V_b through the resistors R_1 and R_2 to charge the capacitors.

$$I(R_1 + R_2) + q\left(\frac{1}{C_1} + \frac{1}{C_2}\right) - V_b = 0$$

Replacing I by dq/dt, we obtain the final differential equation for $q(t)$:

$$\frac{dq}{dt}(R_1 + R_2) + q\left(\frac{1}{C_1} + \frac{1}{C_2}\right) - V_b = 0$$

We note that this equation has exactly the same form as that for our initial charging circuit except that R has been replaced by $(R_1 + R_2)$ and $1/C$ has been replaced by $(1/C_1 + 1/C_2)$. Consequently, we expect the time dependence to be exponential with a time constant τ equaling the product of the series combination of the two resistors and the series combination of the two capacitors.

$$\tau = (R_1 + R_2)\frac{C_1C_2}{C_1+C_2}$$

In order to write down the exact form for $q(t)$, we need to consider the initial and asymptotic values for the charge. We know $q(0)$ is 0, since both capacitors are initially uncharged. To determine the asymptotic value for q, we first note that C_1 and C_2 are connected in series, which means that $q_1 = q_2 = q$. The asymptotic value for the voltage across the combination of C_1 and C_2 is just the battery voltage V_b. Therefore, the asymptotic value for q is just the charge on the equivalent capacitance C for the series combination of C_1 and C_2 multiplied by the battery voltage V_b ($q(\infty) = CV_b$).

Our final expression for $q(t)$ then becomes absolutely identical to that of the original *RC* charging circuit,

$$q(t) = CV_b(1 - e^{-t/RC})$$

where R now denotes ($R_1 + R_2$) and $1/C$ is equal to ($1/C_1 + 1/C_2$).

11.8 Example: Discharging Two Capacitors

We now move the switch from position A to position B in Figure 11.5, removing the battery from the circuit. The capacitors will now discharge through the single resistor R_1. To determine $q(t)$, we apply Kirchhoff's voltage rule to this loop. We will leave the assumed direction of current the same as before, but we expect to find that it will turn out to be negative, indicating the discharging current moves in the opposite direction to that of the charging current. Starting at point B and moving clockwise, we obtain

$$IR_1 + q\left(\frac{1}{C_1} + \frac{1}{C_2}\right) = 0$$

Once again we see that this equation is identical to that of the original *RC* discharging circuit except that R has been replaced by R_1 and $1/C$ has been replaced by ($1/C_1 + 1/C_2$). Consequently, we expect the time dependence to be exponential with a time constant τ given by

$$\tau = R_1 \frac{C_1C_2}{C_1+C_2}$$

Furthermore, the initial value of q is just q_o, the value of the charge on either capacitor at $t = 0$, the time the switch is moved to position B. The asymptotic value of q is 0; therefore, we arrive at a solution for $q(t)$, which is identical to that of the original *RC* discharging circuit,

$$q(t) = q_o e^{-t/RC}$$

where R now denotes R_1 and $1/C$ is equal to ($1/C_1 + 1/C_2$).

11.9 Summary

In this unit, we analyzed circuits composed of capacitors, resistors, and batteries. These circuits have a qualitatively different behavior from those we had studied before. In these circuits the current is not constant but has an exponential dependence on time.

We began by considering a circuit composed of a battery, a capacitor, a resistor, and a switch. Closing the switch results in the capacitor acquiring charge. Applying Kirchhoff's voltage rule to this circuit, we obtained a differential equation for this charge q. The solution to this equation has q increasing exponentially with a time constant τ equal to the product of the resistance and the capacitance toward the asymptotic value of the battery *emf*, as shown in Figure 11.2(a). The current was found by differentiating the expression for q, resulting in an exponentially decreasing current toward zero with the same time constant.

Opening the switch causes the capacitor to discharge through the resistor. We determined that both the charge on the capacitor and the magnitude of the current through the circuit decreased exponentially to zero with the same time constant τ. The direction of the current was opposite to that when the capacitor was being charged.

We then considered the energy flow in the *RC* charging circuit and determined that the rate at which energy was being supplied to the circuit by the battery decreased exponentially over time. Some of this energy was dissipated as heat in the resistor, with the remainder being stored in the capacitor.

Finally, we analyzed a charging/discharging circuit that contained two capacitors and two resistors. Applying Kirchhoff's rules, we found differential equations for the charge on the capacitors that were identical in form to those in the simpler circuits but having different time constants.

MAIN POINTS

KVR Applied to RC Circuits Generate Differential Equations

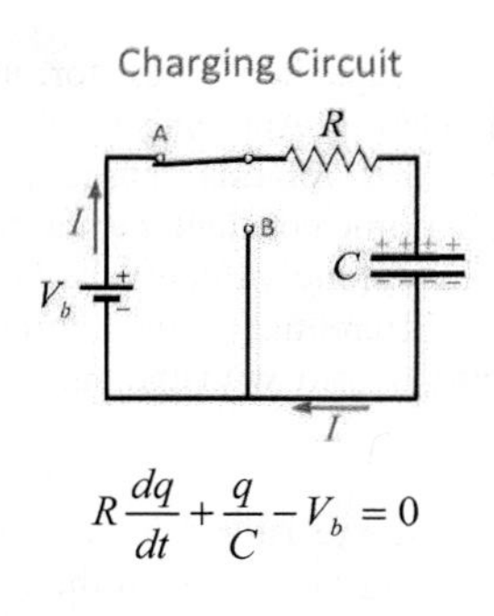

$$R\frac{dq}{dt}+\frac{q}{C}-V_b=0$$

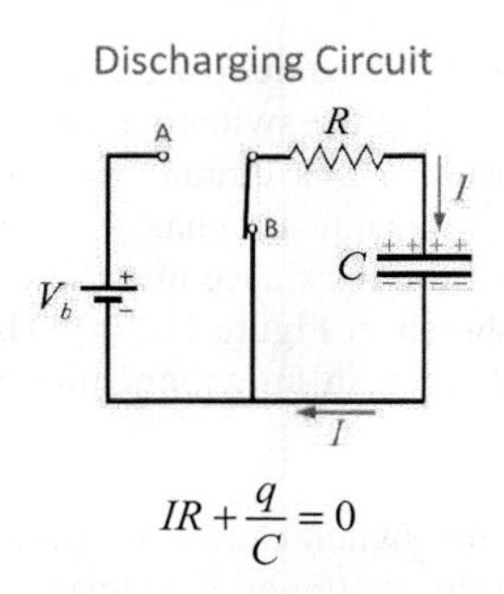

$$IR+\frac{q}{C}=0$$

Solutions are Exponentials with a Time Constant $\tau = RC$

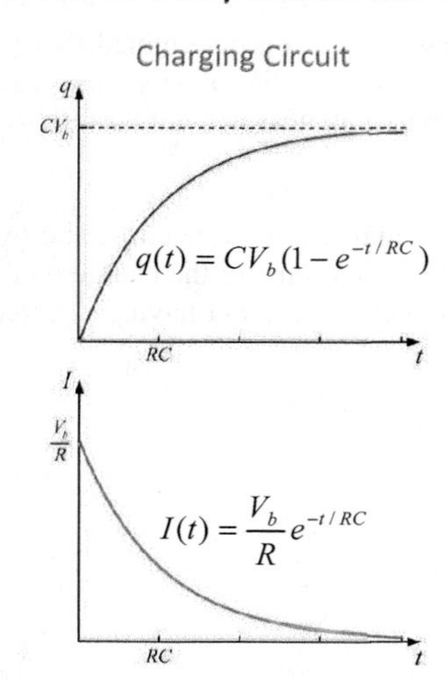

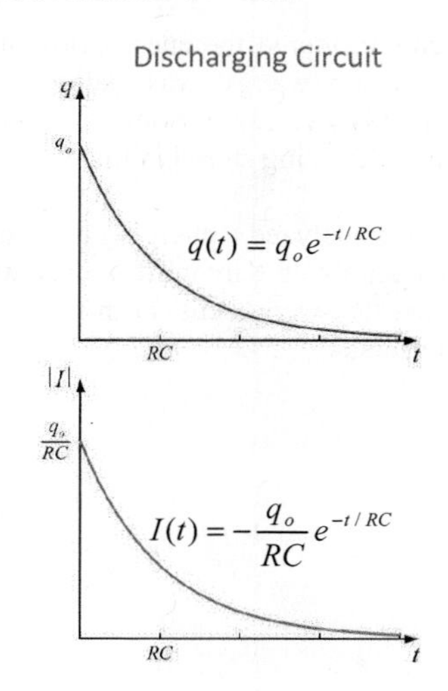

Energy Flow in an RC Circuit

Rate at which energy is supplied to the circuit by the battery equals the rate at which it is dissipated in the resistor plus the rate at which it is stored in the capacitor.

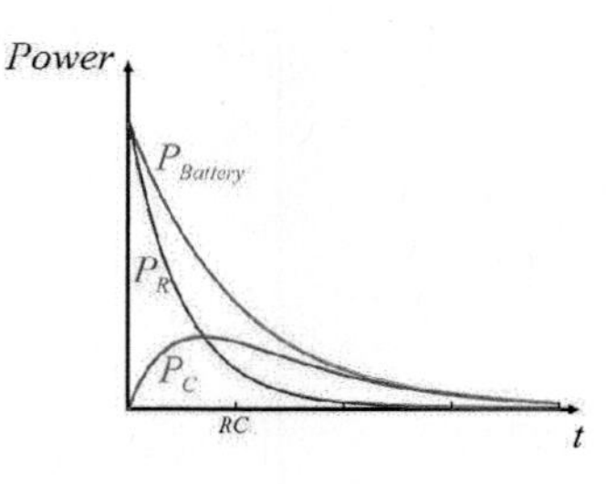

PROBLEMS

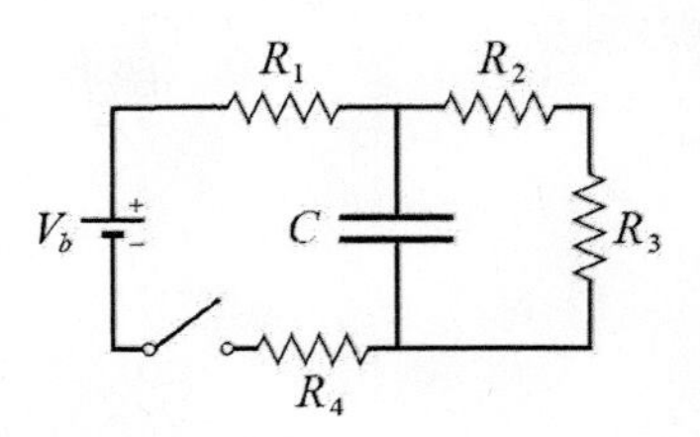

FIGURE 11.6 Problem 1

1. Two Loop *RC* Circuit 1: A circuit is constructed with four resistors, one capacitor, one battery, and a switch, as shown in Figure 11.6. The values for the resistors are: $R_1 = R_2 = 60\ \Omega$, $R_3 = 53\ \Omega$, and $R_4 = 103\ \Omega$. The capacitance is $C = 53\ \mu\text{F}$ and the battery voltage is $V_b = 24$ V. (a) The switch has been open for a long time when, at time $t = 0$, the switch is closed. What is $I_1(0)$, the magnitude of the current through the resistor R_1 just after the switch is closed? (b) What is $I_1(\infty)$, the magnitude of the current that flows through the resistor R_1 after the switch has been closed for a long time? (c) What is $Q(\infty)$, the charge on the capacitor after the switch has been closed for a very long time? (d) What is $I_2(\infty)$, the magnitude of the current through resistor R_2 after the switch has been closed for a very long time? (e) What is $I_C(0)$, the magnitude of the current through the capacitor just after the switch is closed?

2. Two Loop *RC* Circuit 2: A circuit is constructed with four resistors, one capacitor, one battery, and a switch, as shown in Figure 11.7. The values for the resistors are: $R_1 = R_2 = 58\ \Omega$, $R_3 = 117\ \Omega$, and $R_4 = 94\ \Omega$. The capacitance is $C = 83\ \mu\text{F}$ and the battery voltage is $V_b = 12$ V. The positive terminal of the battery is indicated with a + sign. (a) The switch has been open for a long time when, at time $t = 0$, the switch is closed. What is $I_4(0)$, the magnitude of the current through the resistor R_4 just after the switch is closed? (b) What is $Q(\infty)$, the charge on the capacitor after the switch has been closed for a very long time? (c) After the switch has been closed for a very long time, it is then opened. What is $Q(t_{open})$, the charge on the capacitor at a time $t_{open} = 636\ \mu\text{s}$ after the switch was opened? (d) What is $I_{C,\max}(closed)$, the current that flows through the capacitor whose magnitude is maximum during the time when the switch is closed? A positive value for the current is defined to be in the direction of the arrow shown. (e) What is $I_{C,\max}(open)$, the current that flows through the capacitor whose magnitude is maximum during the time when the switch is open?

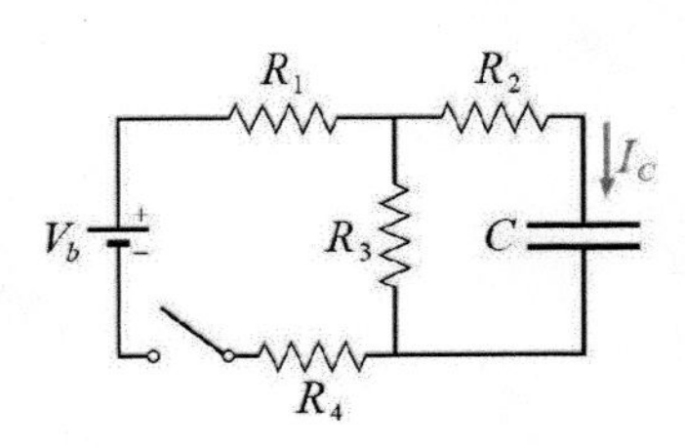

FIGURE 11.7 Problem 2

3. Three *R*, Two *C* (INTERACTIVE EXAMPLE): The switch in the circuit shown in Figure 11.8 has been open for a long time when, at time $t = 0$, it is closed. The values of the circuit elements are: $V_b = 12$ V, $R_1 = 110\ \Omega$, $R_2 = 220\ \Omega$, $R_3 = 330\ \Omega$, $C_1 = 40\ \mu\text{F}$, and $C_2 = 80\ \mu\text{F}$. What is $Q_{2,final}$, the charge on C_2 a long time after the switch is closed?

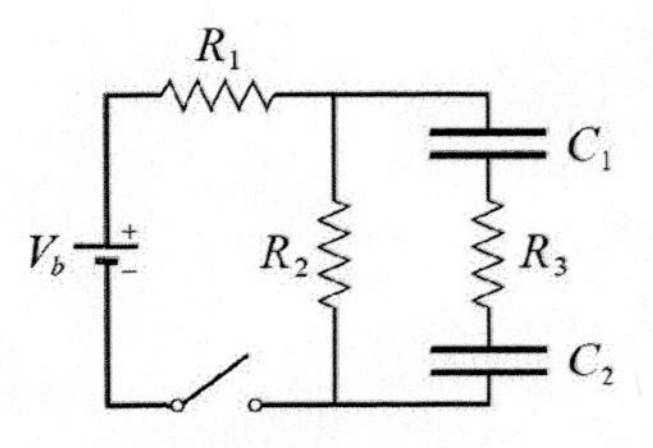

FIGURE 11.8 Problem 3

Part III: Magnetism

Unit 12

Magnetism

12.1 Overview

We will begin our study of magnetism with some observations about bar magnets, compass needles, and current-carrying wires. We'll see that these magnetic effects can be described by the existence of another vector field, the magnetic field, which is created by electric charges in motion.

Before we learn how to calculate these magnetic fields, we will first study the forces they produce on charged particles. We will begin by introducing the Lorentz force, the general equation that describes the force on a charged particle as it moves through regions containing electric and magnetic fields. We will conclude by working a few examples.

12.2 Magnetic Observations

Magnetic effects from natural magnets have been known for over 2000 years. Until the nineteenth century, magnetism was taught as a subject that was totally separate from electricity. We now see electricity and magnetism as parts of a completely unified theory. How did this change come about?

We'll start by describing some experiments that illustrate this shift in understanding. The existence of magnetic forces is often demonstrated with the use of bar magnets, as shown in Figure 12.1. The ends of a bar magnet are usually called the north and south poles. If the north pole of one bar magnet is brought close to the south pole of another bar magnet, we observe a strong attraction between the magnets. On the other hand, if the north pole of one bar magnet is brought close to the north pole of another bar magnet, we observe a strong repulsion between the magnets.

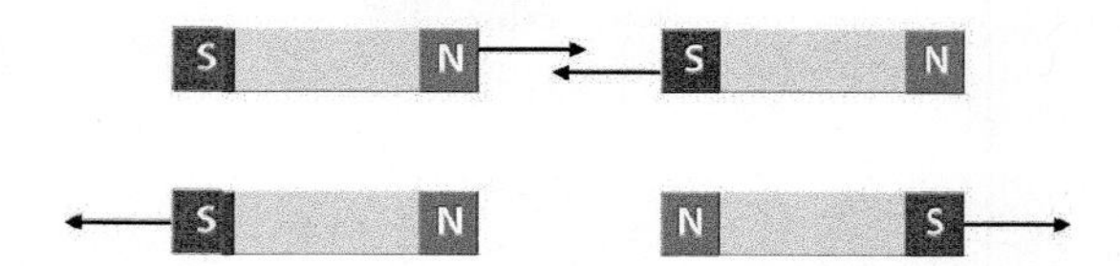

FIGURE 12.1 Bar magnets: Attractive forces exist between unlike poles and repulsive forces exist between like poles.

Compass needles are light bar magnets that can pivot about their center. When not in close proximity to other magnets, the north pole of a compass needle aligns itself with the geographic direction of north. When the compass needle is in close proximity to a bar magnet, however, its alignment is more complicated, as shown in Figure 12.2. The alignment of the north pole of the compass needle traces out a pattern that is very reminiscent of that produced by the electric field of a dipole.

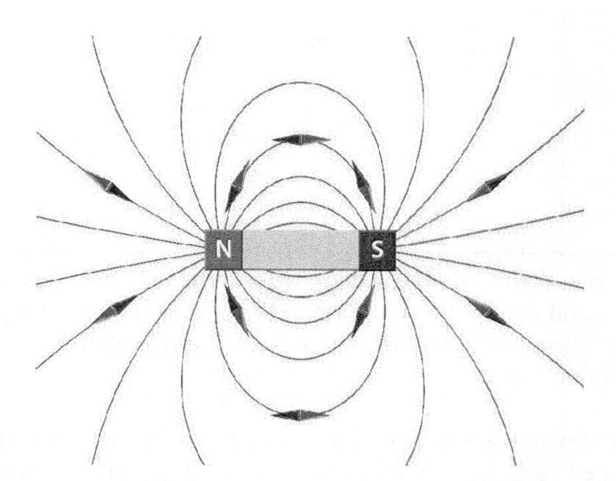

FIGURE 12.2 The alignment of compass needles in the presence of a bar magnet. The result is reminiscent of the field lines produced by an electric dipole.

One possible explanation for this similarity is that there exists another new kind of entity in the world called **magnetic charge**, in which like magnetic charges repel and unlike magnetic charges attract. This explanation fails, however, once we try to isolate these charges. For example, if we cut a bar magnet in half, we do not end up with a north charge and a south charge, but instead we get two bar magnets, each with a north and a south pole. Many attempts have been made to isolate such individual magnetic charges (called monopoles) because within the framework of quantum mechanics, the existence of magnetic monopoles can lead to an explanation for the quantization of electric charge. To date, however, no such experiment has produced any evidence for the existence of magnetic monopoles.

12.3 The Source of the Magnetic Field

Since magnetic charges do not account for our observations with bar magnets, we need to look elsewhere for an explanation. We will now offer two more observations that will lead to our current framework for understanding magnetic effects.

The first observation is that when a compass is brought near a wire carrying a large electric current, the needle's alignment changes. If the current is stopped, the needle once again points north. A natural interpretation of these results is that electric currents can create magnetic effects, much as a natural bar magnet can. Indeed, if we wish to describe magnetic effects in terms of the interaction of the compass needle with a **magnetic field**, then it seems that the origin of such a magnetic field would be electric currents. Since we know electric charges are always in motion inside materials, perhaps magnetic materials are just those materials that support internal currents that can produce magnetic fields.

In fact, our current understanding is that all magnetic fields are created by electric currents, charges in motion. At this point, we could move directly to a discussion of exactly how these magnetic fields are produced by currents. Instead, however, we choose to simply take the existence of magnetic fields as a given and first explore the nature of magnetic forces. We will see they are somewhat more complicated than electric forces, as illustrated by our second observation.

Consider two parallel wires, as shown in Figure 12.3. We know these wires do not exert electric forces on each other because they are electrically neutral. Connecting wire 1 to a battery produces a current in wire 1, which in turn produces a magnetic field in the vicinity of wire 2. In this case, we still observe no forces between the wires. If we now connect wire 2 to a battery so that current flows through it, we observe something different; namely the two wires now exert forces on each other. If the currents are in the same direction, the forces are attractive; if the currents are in opposite directions, the forces are repulsive.

These forces cannot be electric forces since the wires are still electrically neutral, even when they carry current. In fact, these forces are magnetic forces. It appears that magnetic fields can produce forces on electric charges as well as on magnets. Since there was no force on wire 2 when it carried no current, we conclude that magnetic fields exert forces on charges *only* when they are moving.

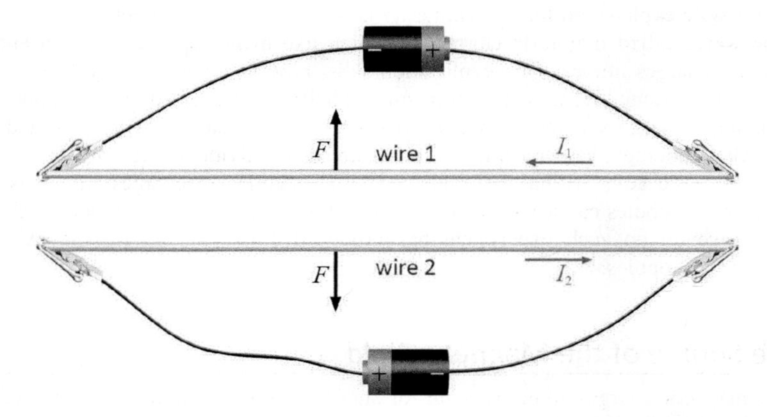

FIGURE 12.3 Two current-carrying wires exert magnetic forces on each other. The forces are attractive if the currents are in the same direction and are repulsive if the currents are in opposite directions.

12.4 Magnetic Force

We have just seen that two wires, each carrying current, will exert forces on each other. Our interpretation of this situation is that wire 1 creates a magnetic field in the vicinity of wire 2 and the charges that are moving in wire 2 experience a force due to this magnetic field.

Using a compass to map the direction of the magnetic field produced by a long straight current-carrying wire, we find that the magnetic field is always in a plane perpendicular to the wire and rotates around in either a clockwise or counterclockwise direction, depending upon the sense of the current.

For example, if the current shown is going into the page, as indicated by the x, the magnetic field, usually labeled B, rotates in a clockwise direction, as shown in Figure 12.4(a).

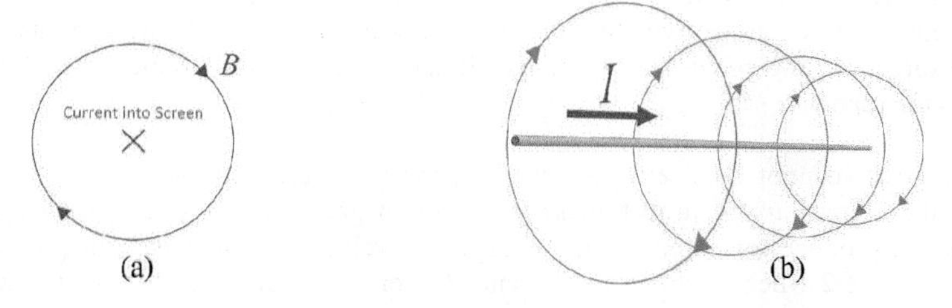

FIGURE 12.4 (a) An electric current directed into the page (indicated by x) produces a clockwise magnetic field B. (b) A perspective view of the direction of the magnetic field produced by a current-carrying wire.

Therefore, in the case where I_1 and I_2 are parallel, I_2 creates a magnetic field at the location of wire 1 that points out of the page, as shown in Figure 12.5. The force on the positive current in wire 1 points up. I_1, on the other hand, creates a magnet field at the location of wire 2 that points into the page, resulting in a force on the positive current in wire 2 that points down.

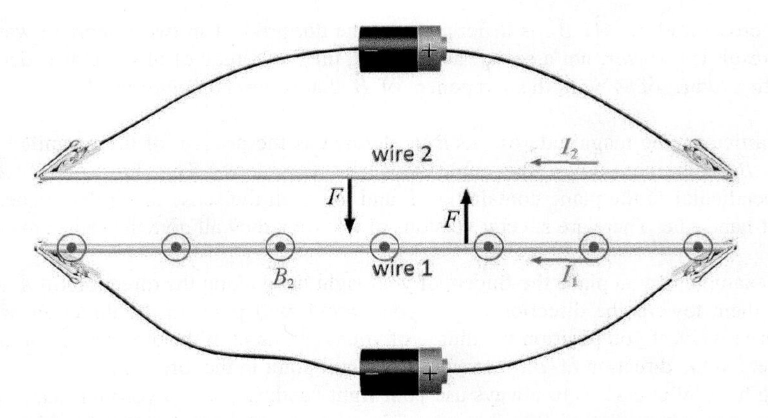

FIGURE 12.5 The current I_2 produces a magnetic field at wire 1 that is directed out of the page (as indicated by •). This field exerts an upward force on wire 1 since I_1 is in the same direction as I_2.

If we switch the direction of I_2, the magnetic field at wire 1 flips to point into the page, which results in a force on wire 1 that points down. At wire 2, the field created by I_1 remains the same, but the direction of the positive current is reversed so that the force on wire 2 now points up.

What do these results tell us about the magnetic force? First, it's clear that the force is perpendicular to both the direction of the current and the direction of the magnetic field. Second, the direction of the force reverses when either the direction of the current or the direction of the magnetic field is reversed.

There is a mathematical structure that naturally describes these observations and it is called the **cross product**. The general expression for the magnetic force on charge q moving with velocity $\vec{v}$ at a point in space where the magnetic field $\vec{B}$ is given by

$$\vec{F} = q\vec{v} \times \vec{B}$$

The SI unit for the magnetic field is the tesla (T), where 1 tesla = 1 N/A·m. We will define the cross product and investigate some of its properties in the next section.

12.5 Cross Products

To this point we have pretty much exclusively used the dot product to combine two vectors; for example, the flux was calculated using $\vec{E}\cdot d\vec{A}$ and the potential was calculated using $\vec{E}\cdot d\vec{l}$. The result of $\vec{A}\cdot\vec{B}$ is a scalar that measures the product of $\vec{A}$ with the component of $\vec{B}$ that is parallel to $\vec{A}$.

The cross product, $\vec{A}\times\vec{B}$, is different from the dot product in two important ways: first, the result is a *vector*, not a scalar, and second, the magnitude of this vector is determined by the product of $\vec{A}$ with the component of $\vec{B}$ that is *perpendicular* to $\vec{A}$.

In particular, the magnitude of $\vec{A}\times\vec{B}$ is defined as the product of the magnitudes of $\vec{A}$ and $\vec{B}$ with the sine of the angle between $\vec{A}$ and $\vec{B}$. The direction of $\vec{A}\times\vec{B}$ is perpendicular to the plane containing $\vec{A}$ and $\vec{B}$, with the sense being determined by the right-hand rule. There are several versions of this rule; they all give the same answer.

For example, if you place the fingers of your right hand along the direction of $\vec{A}$ and then curl them toward the direction of $\vec{B}$, your thumb will point in the direction of $\vec{A}\times\vec{B}$. Alternatively, if you position the thumb of your right hand in the direction of $\vec{A}$ and your fingers in the direction of $\vec{B}$, then your palm will point in the direction of $\vec{A}\times\vec{B}$. The key thing here, of course, is to always use your right hand; if you use your left hand, you will get the opposite direction. The features of the cross product are illustrated in Figure 12.6.

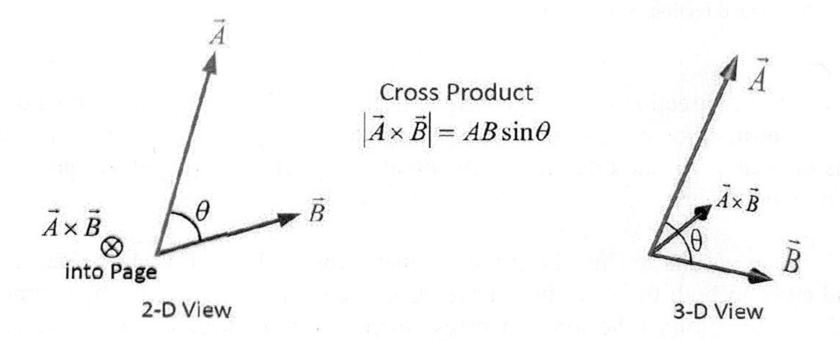

FIGURE 12.6 The definition of the cross product.

Let's do an example to illustrate the use of the cross product in our magnetic force context. Consider a region of space containing a uniform magnetic field $B = 0.75\text{ T}$ pointing to the right, as shown in Figure 12.7. A proton ($q = 1.6\times10^{-19}\text{ C}$) moves with speed v = 300 m/s, traveling at an angle of 30° with respect to the magnetic field as shown. The magnitude of the force on the proton is just the product $qvB\sin(30°) = 1.8\times10^{-17}\text{ N}$. The direction is given by the right hand rule to be into the page.

Note that if the particle had been an electron instead of a proton, the magnitude of the force would have been the same, but the direction would be just opposite (out of the page).

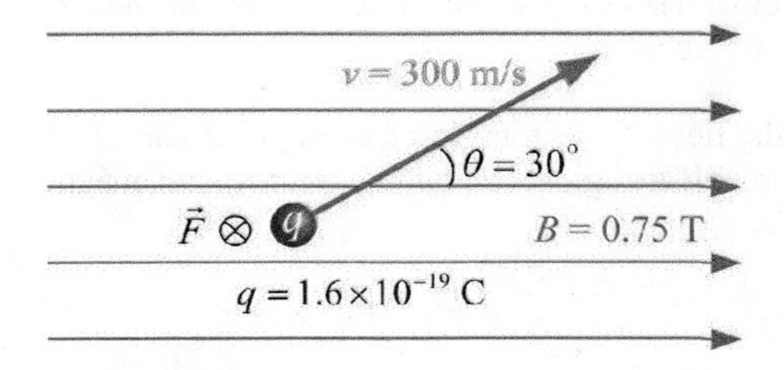

FIGURE 12.7 An illustration of the force law: $\vec{F} = q\vec{v}\times\vec{B}$

12.6 Velocity Selector

We will now consider a few examples that illustrate the magnetic force and its use in creating charged particle beams. The first example is the velocity selector.

Here, the idea is to produce a collimated beam of charged particles that all have pretty much the same speed and the same direction. The direction is defined by a channel in an absorber, as shown in Figure 12.8. If charged particles with a variety of speeds and directions are incident on this absorber, only those particles initially entering the channel and having a direction parallel to the channel will exit at the opposite end.

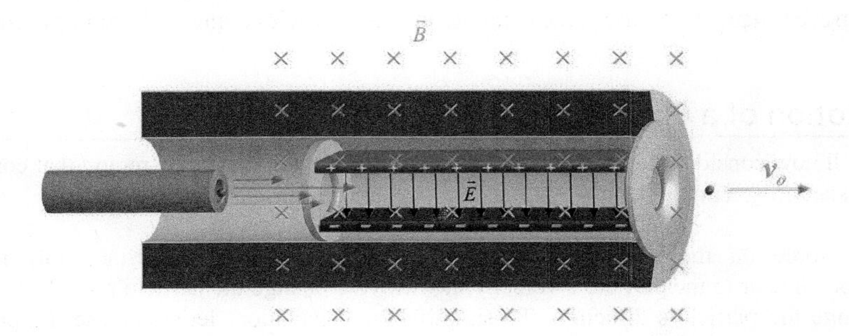

FIGURE 12.8 The velocity selector. Only those particles with velocity $v_o = E/B$ emerge from the end of the collimator since the electric and magnetic forces cancel for particles with speed v_o.

So far, we're halfway there. The exiting particles are all traveling in the same direction, but can have a wide variety of speeds. In order to select just those particles with a particular speed, we will create electric and magnetic fields throughout the channel. The idea is that we first orient these fields so that the force from the electric field is in the opposite direction of the force from the magnetic field. Since the magnetic force depends on the speed of the particle, there will be a unique speed such that the magnitudes of the electric and magnetic forces on the particle are equal, thereby producing a zero net force on the particle. Only these particles with that unique speed will now exit the channel;

particles with all other speeds will be deflected by the non-zero combination of the electric and magnetic forces.

How do we orient the fields? The total force acting on a particle having charge q, moving with velocity $\vec{v}$ through a region containing electric and magnetic fields, is called the **Lorentz force** and is given by

$$\vec{F} = q\vec{E} + q\vec{v} \times \vec{B}$$

The electric force is parallel to the electric field. Let's choose the direction of the electric field to be down. We then need the direction of the magnetic force to be up. Since the magnetic force is given by $\vec{F}_{magnetic} = q\vec{v} \times \vec{B}$, we choose to direct the magnetic field into the page as shown. A positively charged particle will then experience an electric force that is down and a magnetic force that is up. Since $\vec{B}$ is perpendicular to $\vec{v}$, the magnitude of the magnetic force is just qvB. The magnitude of the electric force is just qE. Setting these two magnitudes equal, we determine v_o, the speed of the particles that pass through undeflected:

$$v_o = \frac{E}{B}$$

There is no net force on particles with this velocity, allowing them to traverse the channel unhindered. Note that the magnetic force on positive particles with speed less than E / B will be smaller than the electric force resulting in these particles being deflected downward, while the magnetic force on positive particles with speed less than E / B will be greater than the electric force resulting in these particles being deflected upward.

12.7 Motion of a Charge in a Uniform Magnetic Field

We'll now consider the case of a charged particle moving through a region that contains a constant magnetic field. What will be the trajectory of this particle?

The force on the particle is the magnetic force $\vec{F} = q\vec{v} \times \vec{B}$. Since this force is perpendicular to the particle's velocity, it will NOT change the particle's speed, but it will change the particle's direction. To simplify the calculation, let's suppose the particle's initial velocity is perpendicular to the magnetic field, as shown in Figure 12.9. In this case, the motion will be confined to a plane perpendicular to the magnetic field since any change in velocity must be in the direction of the magnetic force, which is always perpendicular to the magnetic field.

The magnitude of the magnetic force in this case, qvB, is constant throughout the motion since the particle's speed does not change. This constant force, then, gives rise to a centripetal acceleration of constant magnitude; the particle undergoes uniform circular motion.

The radius of this motion can be determined from Newton's second law. In particular, the acceleration (v^2 / R) is equal to the force (qvB) divided by the mass (m), yielding

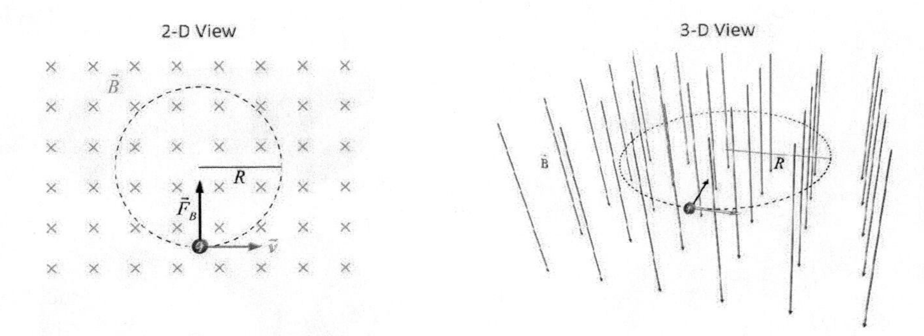

FIGURE 12.9 The force $\vec{F}_B$ on a particle of charge q, mass m, and velocity v produced by a magnetic field $\vec{B}$ produces uniform circular motion with radius $R = mv / qB$.

$$R = \frac{mv}{qB}$$

We can see from this expression that for a fixed magnetic field, the radius of curvature is directly proportional to the momentum (mv) of the particle. In fact, this radius of curvature is the quantity that is measured in modern magnetic particle detectors in order to determine the momentum of the particle.

12.8 The Cyclotron

We'll close with a final example: the invention of the cyclotron, an early particle accelerator. In 1929, E. O. Lawrence noticed an important feature of the equation we just derived for the radius of curvature of a charged particle's motion in a constant magnetic field. Namely, if we rewrite this equation in terms of the angular velocity $\omega = v / R$, we see that the R's cancel and that we end up with an expression for the angular velocity of the motion that depends only on the magnetic field and the charge to mass ratio of the particle.

$$R = \frac{mv}{qB} = \frac{m(\omega R)}{qB} \qquad \Rightarrow \qquad \omega = \frac{qB}{m}$$

In other words, if the initial velocity were to increase, the radius would increase, but the time it takes to make one revolution stays the same.

Lawrence exploited this fact to create the first cyclotron, a magnetic resonance accelerator. In particular, he placed two conducting cavities in the shape of two semicircles inside a constant magnetic field, as shown in Figure 12.10. He then applied an alternating voltage to these conductors (called Dees) at a frequency that matched the orbital frequency $f = \omega / 2\pi = (q / m)(B / 2\pi)$. Consequently, particles at rest near the center would be accelerated across the gap, travel in a small semicircle and then be

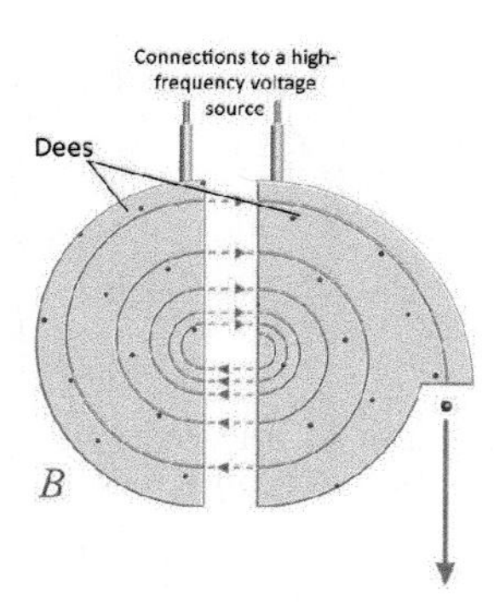

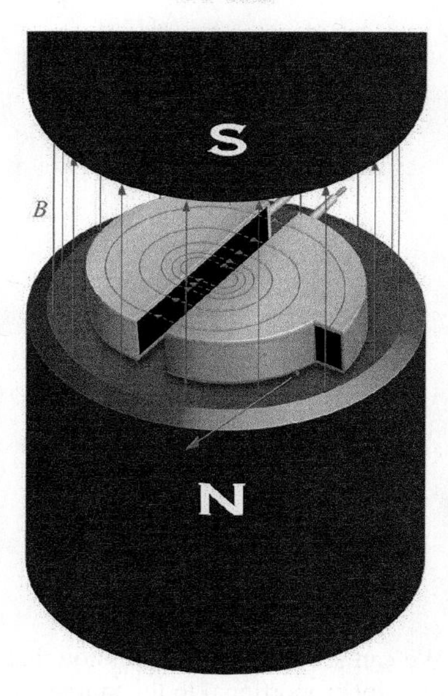

FIGURE 12.10 The Cyclotron: Alternating voltage applied to conductors (Dees) at frequency $f = (q/m)(B/2\pi)$ causing particles to continue to accelerate until they leave the magnetic field.

accelerated across the Dees, gaining energy, and then would travel in a larger semicircle until it would once again be accelerated across the Dees, increasing its energy and radius. This process would continue until the particle was extracted from the cyclotron at a significantly higher energy.

Early cyclotrons accelerated particles up to a few million electron volts. One electron volt is the energy an electron has after being accelerated through one volt. Millions of volts were not needed here, though, since the particles made many orbits before leaving the cyclotron, achieving their final energy in increments each time they went from one Dee to the other.

These machines typically used magnetic fields on the order of $1-2$ T and accelerating voltages of $50-100$ kV. There is a natural limitation to the energies that can be attained with the cyclotron since the key feature (the period being independent of the velocity) only holds for particles with non-relativistic speeds.

12.9 Summary

We began by making observations of the behavior of bar magnets. The magnets have two poles, usually called north and south, with the property that unlike poles attract and like poles repel. We considered a natural explanation in terms of a magnetic charge, but had to abandon this explanation since our attempt to isolate this charge by cutting a bar magnet in half actually produces two new bar magnets.

In fact we describe all magnetic effects in terms of a magnetic field that is created by the motion of electric charges, i.e., electric currents. Moreover, magnetic fields exert forces on electric charges that are in motion. We introduced the mathematical form for this force in terms of a cross product of vectors. In particular, the magnetic force $\vec{F}$ on charge q moving with velocity $\vec{v}$ through a region containing a magnetic field B is given by $\vec{F} = q\vec{v} \times \vec{B}$.

We then considered a few examples that illustrate this magnetic force and its use in creating charged particle beams. We began by noticing that if we orient electric and magnetic fields such that they are perpendicular to each other, then the forces they produce on a charged particle moving in a direction perpendicular to both of these fields will be in opposite directions. Therefore, for particles traveling at a unique velocity given by the ratio of E to B, the net force will be zero. We can use this fact to create a device that selects only those particles having this particular velocity.

We then noticed that a charged particle moving in a uniform magnetic field will experience a constant force that is perpendicular to both the magnetic field and the velocity, and will therefore undergo uniform circular motion. The radius of this motion is proportional to the speed of the particle and inversely proportional to the strength of the magnetic field.

Finally, we discussed an important feature of this motion that was exploited to create the cyclotron, a particle accelerator. Namely since the angular velocity of the particle depends only on the magnetic field and the charge to mass ratio of the particle, an alternating voltage can be applied at the corresponding orbit frequency to incrementally increase the particle's energy until it reaches a radius of the physical size of the cyclotron.

MAIN POINTS

Empirical Results Indicate Something New, the Magnetic Field

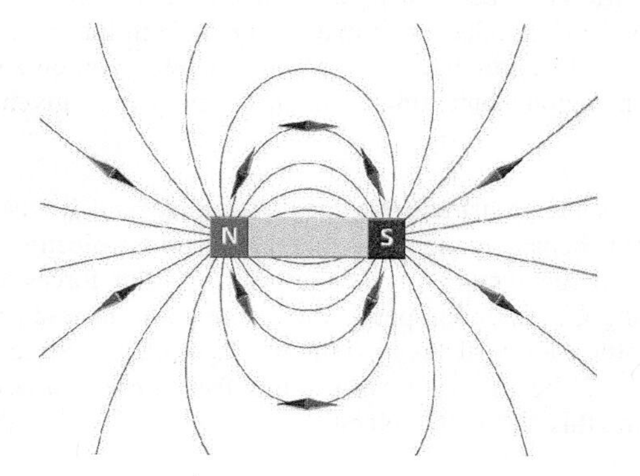

Compass needles surrounding a bar magnet align themselves with the magnetic field produced by the bar magnet.

Magnetic Fields are Produced by Currents, Charges in Motion

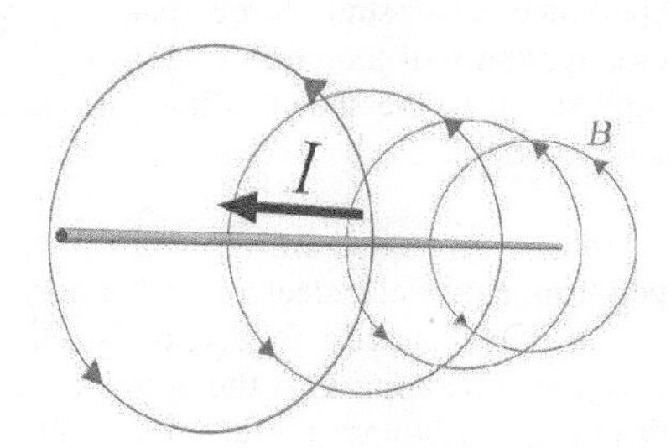

The Magnetic Force

Magnetic fields exert forces on electrical charges in motion.

$$\vec{F}_B = q\vec{v} \times \vec{B}$$

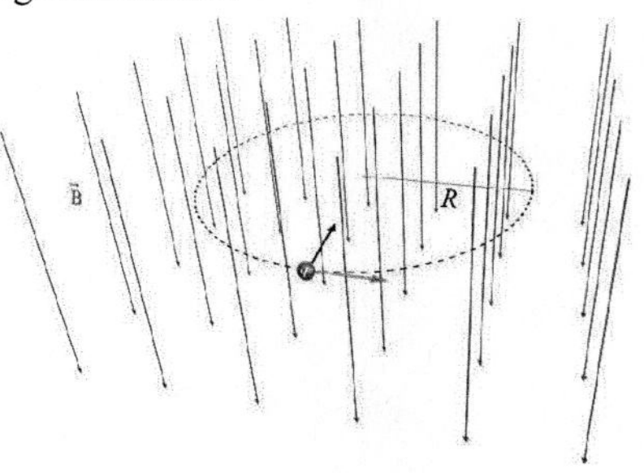

PROBLEMS

1. Motion in a Magnetic Field 1: A charged particle of mass $m = 6.3\times10^{-8}$ kg, moving with constant velocity in the y-direction enters a region containing a constant magnetic field $B = 3.8$ T aligned with the positive z-axis as shown. The particle enters the region at $(x, y) = (0.9\text{ m}, 0)$ and leaves the region at $(x, y) = (0, 0.9\text{ m})$ a time $t = 560$ μs after it entered the region. (a) With what speed v did the particle enter the region containing the magnetic field? (b) What is F_x, the x-component of the force on the particle at a time $t_1 = 186.7$ μs after it entered the region containing the magnetic field? (c) What is F_y, the y-component of the force on the particle at a time $t_1 = 186.7$ μs after it entered the region containing the magnetic field? (d) What is q, the charge of the particle? Be sure to include the correct sign. (e) If the velocity of the incident charged particle were doubled, how would B have to change (keeping all other parameters constant) to keep the trajectory of the particle the same?

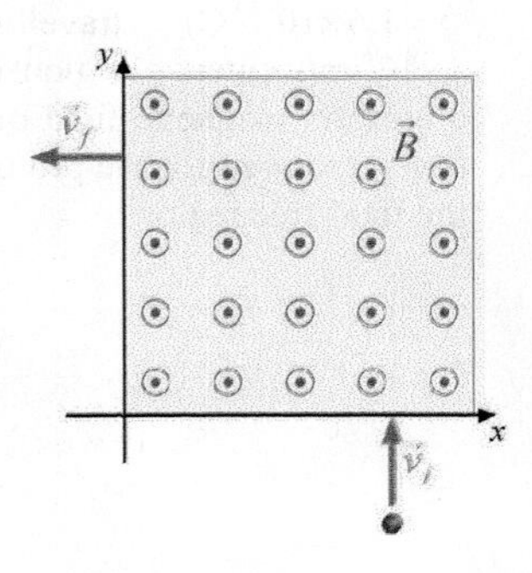

FIGURE 12.11 Problem 1

(i) Increase B by a factor of 2.
(ii) Increase B by less than a factor of 2.
(iii) Decrease B by less than a factor of 2.
(iv) Decrease B by a factor of 2.
(v) There is no change that can be made to B to keep the trajectory the same.

2. Motion in a Magnetic Field 2: A proton ($q = 1.6\times10^{-19}$ C, $m = 1.67\times10^{-27}$ kg) moving with constant velocity in the $+x$ direction enters a region containing a constant magnetic field that is directed along the z-axis as shown. The proton enters the region at $(x, y) = (0, 0)$. The magnetic field extends for a distance $D = 0.57$ m in the x-direction. The proton leaves the field having a velocity vector $(v_x, v_y) = (2\times10^5\text{ m/s}, 1.3\times10^5\text{ m/s})$. (a) What is v, the magnitude of the velocity of the proton as it entered the region containing the magnetic field? (b) What is R, the radius of curvature of the motion of the proton while it is in the region containing the magnetic field? (c) What is h, the y-coordinate of the proton as it leaves the region containing the magnetic field? (d) What is B_z, the z-component of the magnetic field? Note that B_z is a signed number. (e) If the incident velocity v were increased, how would h and θ change, if at all?

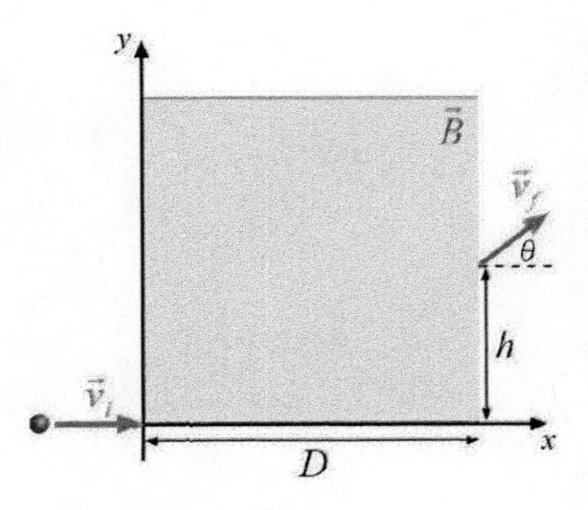

FIGURE 12.12 Problem 2

(i) h and θ would both increase.
(ii) h and θ would both decrease.
(iii) h would increase and θ would decrease.
(iv) h would decrease and θ would increase.
(v) Neither h nor θ would change.

3. Motion in a B Field (INTERACTIVE EXAMPLE): A proton ($m = 1.67 \times 10^{-27}$ kg, $q = 1.6 \times 10^{-19}$ C) travelling with speed 1×10^{6} m/s enters a region of space containing a uniform magnetic field of 1.2 T. What is the time t required for the proton to re-emerge into the field-free region?

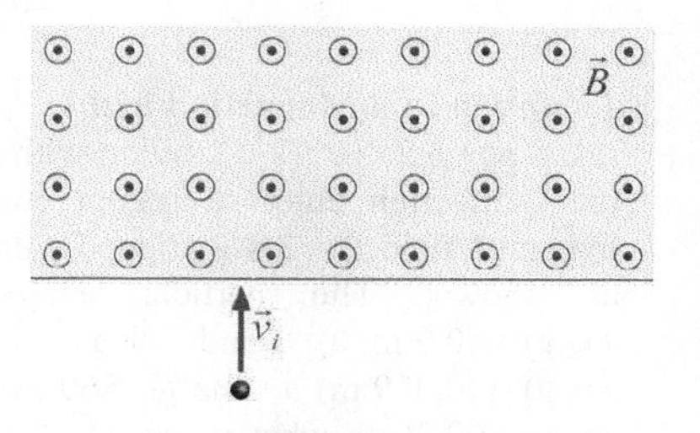

FIGURE 12.13 Problem 3

UNIT

13

FORCES AND TORQUES ON CURRENTS

13.1 Overview

We will continue our study of magnetism by examining the forces and torques exerted on current-carrying wires in magnetic fields. We'll start by calculating the forces on straight and curved wires in constant magnetic fields. We'll then calculate the torque on a closed loop of current in a uniform magnetic field. Next we introduce the useful concept of the dipole moment associated with a current loop. Finally, we will use this dipole moment to define the potential energy that is associated with a current loop located in a region of space that contains a uniform magnetic field.

13.2 Force on a Straight Current Segment

Last time we learned that charges moving in a uniform magnetic field experience a force that is proportional to the charge, the magnetic field, and the component of the velocity perpendicular to the magnetic field. The direction of the force is perpendicular to both the magnetic field and the velocity. All of this information can be captured in a single mathematical expression, namely that the force $\vec{F}$ is equal to the charge times the cross product of the velocity with the magnetic field ($\vec{F} = q\vec{v} \times \vec{B}$).

We'll now discuss what happens when a current-carrying wire is placed in a region of space that contains a uniform magnetic field. Since the current is composed of charges in motion, we expect that there will be some sort of force exerted on the wire. How do we go about calculating that force?

We'll start by considering a long straight wire that carries current I to the right, as shown in Figure 13.1. This wire passes through a region that contains a uniform magnetic field $\vec{B}$ directed into the page as indicated by the x's. Since, by convention, the direction of the current indicates the direction of the positive charge flow, we use the right hand rule to evaluate the cross product $\vec{v} \times \vec{B}$ and determine the net force on the charge carriers to be in the upward direction.

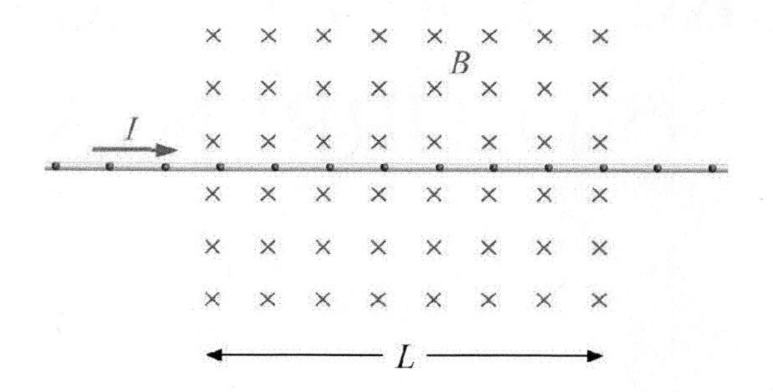

FIGURE 13.1 Straight wire segment carrying current I through a region of constant magnetic field $\vec{B}$.

We'll now calculate the magnitude of this force. Our first step is to realize that the net force $\vec{F}$ acting on the current will just be the vector sum of the forces acting on each of the individual charges, i.e.,

$$\vec{F} = \sum \vec{F}_i = q \sum \vec{v}_i \times \vec{B}$$

Consequently, we see that this net force $\vec{F}$ is proportional to the vector sum of the velocities of all the individual charges. This term, the vector sum of the velocities of the individual charges, is exactly the expression we need to determine v_{drift}, the average velocity of the individual charges. Namely, v_{drift} is just the vector sum of the individual velocities divided by N, the total number of charge carriers. We can therefore rewrite $\vec{F}$ as

$$\vec{F} = qN\vec{v}_{drift} \times \vec{B}$$

At any instant, the total number of charge carriers that will feel the magnetic force will just be the number of charge carriers in a length L of the wire. Therefore, $N = nAL$, where n is the number of charge carriers per unit volume in the wire, A is the cross-sectional area of the wire, and L is the length of the wire in the magnetic field. Consequently, we can write

$$\vec{F} = qnAL\vec{v}_{drift} \times \vec{B}$$

Now, the expression $nAqv_{drift}$ is just the current I, the amount of charge that flows through A per unit time. Putting this altogether, taking the current I to be a scalar and letting L become a vector whose direction is that of v_{drift}, we obtain our final simple result

$$\vec{F} = I\vec{L} \times \vec{B}$$

We can say that this force $\vec{F}$ is in fact exerted on the wire since the wire contains the charge carriers and these charge carriers are constrained to remain in the wire.

13.3 Force on a Curved Current Segment

We will now extend our discussion for the force on a straight wire to the more general case of a curved wire. We start with the picture of the curved wire carrying current I in a region containing a uniform magnetic field $\vec{B}$ directed to the right, as shown in Figure 13.2(a).

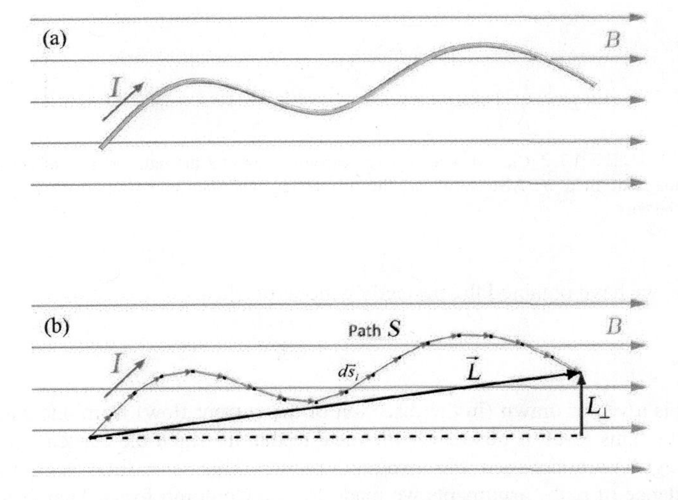

FIGURE 13.2 (a) Curved wire segment carrying current I through a region of constant magnetic field $\vec{B}$. (b) The wire is replaced with smaller current segments each with length ds.

Where do we start? Once again we can profitably use the ideas of calculus. Namely, we can imagine the curved wire to be made up of a very large number of segments, $d\vec{s}$, each of which is sufficiently small so that we can consider them to be straight, as shown in Figure 13.2(b). We can then apply our previous result for the force on a straight wire to each segment and then take the vector sum of the forces on all of these small segments to get the net force on the wire. Mathematically, this sum becomes an integral:

$$\vec{F} = \int I \, d\vec{s} \times \vec{B}$$

Now, we can decompose each segment $d\vec{s}$ into components that are parallel and perpendicular to the magnetic field. The component of $d\vec{s}$ that is parallel to the magnetic field will *not* contribute to the cross product since the cross product of two parallel vectors is zero. Consequently, in the equation for the total force $\vec{F}$, we can replace the vector $d\vec{s}$ with its perpendicular component $d\vec{s}$. Now, to actually do the integral, we would need the mathematical form that describes the curved wire. Rather than carrying out this procedure for some specific curve, we can get an important and totally general result by thinking about the integral as the limit of a sum. In particular, we see that $\vec{F}$ is proportional to the vector sum of the $d\vec{s}$'s. Shown in Figure 13.3 are the $d\vec{s}$ vectors for each segment of the wire, as well as their sum, which in the limit as $d\vec{s}$ goes to zero becomes the integral over the wire. This sum is in fact just equal to the projection of the vector $\vec{L}$ drawn from one end of the wire to the other that is perpendicular to the direction of $\vec{B}$.

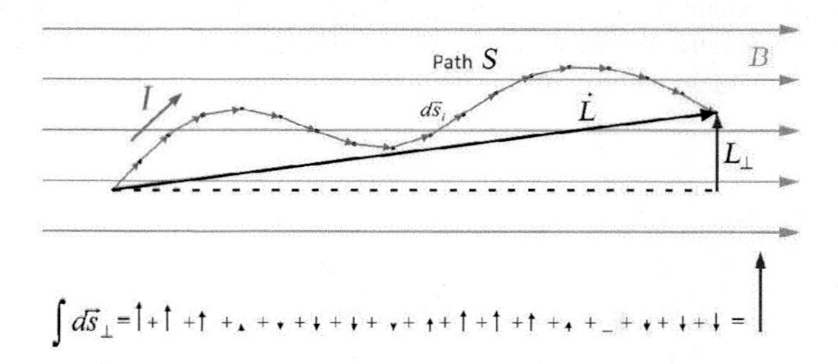

FIGURE 13.3 Curved wire segment carrying current I through a region of constant magnetic field $\vec{B}$. Also shown are the vectors $d\vec{s}_\perp$ and their integral over the length of the wire.

Therefore, we have obtained the perfectly general result:

$$\vec{F} = I\vec{L} \times \vec{B}$$

where $\vec{L}$ is a vector drawn (in the direction of the current flow) from one end of the wire to the other. This result is absolutely identical to that in which the curved wire is replaced by a straight wire between the two end points. This is a little reminiscent of the "independence of path" arguments we made for the Coulomb force. That argument led to the introduction of the electric potential; here we have no such direct analog. What we can say, though, is that the force on a wire in a uniform magnetic field does not depend on the exact shape of the wire, but rather only on the perpendicular distance to the field between the end points of the wire.

13.4 Force on a Current Loop

We now know how to calculate the forces that are exerted on individual segments of current-carrying wires located in a constant magnetic field. We now want to use this knowledge to determine what happens when we join these segments to make a current-carrying loop, as shown in Figure 13.4.

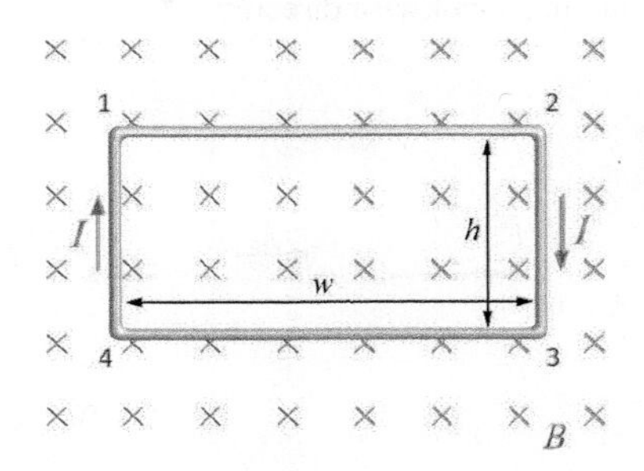

FIGURE 13.4 A rectangular loop of sides w and h carrying current I in a clockwise direction situated in a region of constant magnetic field $\vec{B}$.

The force on a current segment in a uniform magnetic field is just given by

$$\vec{F} = I\vec{L} \times \vec{B}$$

Since our segment is actually a closed loop, the vector $\vec{L}$ is zero. Consequently, the force on that curved segment (the closed loop) must be zero! We can verify this result by explicitly calculating the force on each of the four straight line segments and then adding them together to get the total force.

We'll start with the segment connecting the top left corner (1) to the top right corner (2). We calculate the force from the general expression: $\vec{F} = I\vec{L} \times \vec{B}$. Since the current is moving to the right and the magnetic field points into the page, the angle between the two is 90° and the force, $\vec{F}_{12}$, must be directed upward. The magnitude of the force is just equal to the product of the current, the length of the segment, and the magnetic field $(F_{12} = IwB)$.

We now note that the segment connecting the two bottom corners (3 and 4) carries the same current I, but its direction is to the left. Therefore, the force $\vec{F}_{34}$ must be equal in magnitude to $\vec{F}_{12}$, but its direction is just opposite, i.e., pointing downward. Therefore, the force on segment 12 exactly cancels the force on segment 34 (i.e., $F_{12} + F_{34} = 0$)!

Since the magnitudes of the currents in the two remaining segments (23 and 41) are equal, but their directions are opposite, we see that $\vec{F}_{23} + \vec{F}_{41}$ must also be zero. We have explicitly confirmed our original result. Although the force on an individual current segment may be non-zero, the net force on a closed loop is zero!

13.5 Torque on a Current Loop

We have just determined that the force on any current-carrying loop in a uniform magnetic field is zero. Nevertheless, current loops in uniform magnetic fields are often used to convert electrical energy into mechanical energy and vice versa (e.g., motors and generators). The explanation for this seeming contradiction is that while the force on a loop is always zero, the torque on that loop is usually not zero!

In Figure 13.5(a) we show the loop from the last section as viewed from the side. In this orientation, not only the force, but also the torque on the loop is zero. However, if we rotate the loop, as shown in Figure 13.5(b), we see the force remains zero, but now the torque becomes non-zero. In particular, the forces $\vec{F}_{12}$ and $\vec{F}_{34}$ will cause the loop to rotate in a clockwise direction.

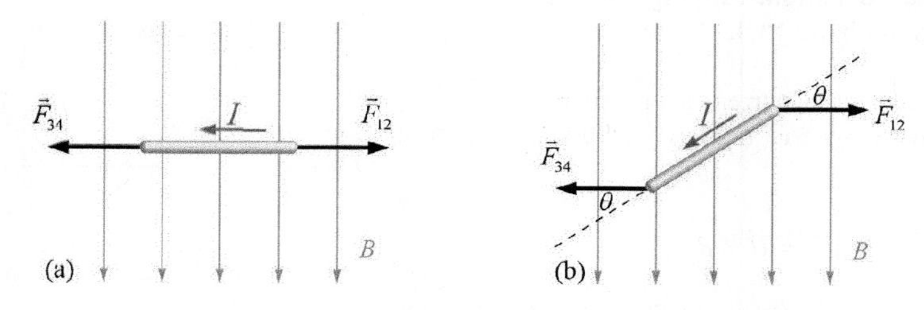

FIGURE 13.5 (a) The current loop for Figure 13.3 viewed from the side. Both the net force and the torque on the loop is zero. (b) The current *loop* rotated by an angle θ. In this case, the net force is zero, but the torque is not.

Recall that the **torque** on an object free to rotate about a certain axis is defined to be

$$\vec{\tau} = \vec{r} \times \vec{F}$$

where $\vec{F}$ is the applied force and $\vec{r}$ is the vector that defines the perpendicular distance from the axis to $\vec{F}$.

In Figure 13.5(b) we take the torque about an axis perpendicular to the page and passing through the middle of the loop. $\vec{F}_{12}$ gives rise to a torque of magnitude $\tau_{12} = (h/2)\vec{F}_{12}\sin\theta$. The direction of this torque is into the page; this torque tends to rotate the loop clockwise. Similarly, $\vec{F}_{34}$ gives rise to a torque of magnitude $\tau_{34} = (h/2)\vec{F}_{34}\sin\theta$. The direction of this torque is also into the page. Now the magnitudes of $\vec{F}_{12}$ and $\vec{F}_{34}$ are the same and equal to IwB. Consequently, τ, the magnitude of the total torque on the loop is just given by

$$\tau = IwhB\sin\theta$$

The direction of the torque is into the page. You can determine the sense of rotation from this direction by placing the thumb of your right hand in the direction of the torque vector and then your fingers curl in the direction of the rotation (which is clockwise in this case).

Note that the angular dependence of the torque (i.e., proportional to $\sin\theta$) makes sense in that at $\theta = 0°$, the plane of the loop is perpendicular to the field and the torque is zero; at $\theta = 90°$, the plane of the loop is parallel to the field and the torque is maximum. Note also that the torque is proportional to the current and the area of the loop. This feature is quite general; we will see it appear in all current loops we encounter.

Finally, we'll note that this torque we have just calculated is the basis for the operation of all electric motors. Namely, it is just this torque that is sustained into complete rotations by manipulating the current that is carried by the coil in a motor.

13.6 Dipole Moment of Current Loop

We've just seen that a current loop can experience a torque when placed in a uniform magnetic field. This torque depends on three things: the current in the loop, the area defined by the loop, and the orientation of the loop with respect to the magnetic field.

We will now combine all three of these things into one new quantity, the **magnetic dipole moment**, which we associate with the current loop. We'll start by defining an area vector $\vec{A}$ that will allow us to specify the orientation of the loop with respect to the magnetic field. The magnitude of $\vec{A}$ will just be the area defined by the loop and the direction of $\vec{A}$ will be perpendicular to the plane defined by the loop. You may recall that for a closed surface, we defined the area vector $\vec{A}$ to be perpendicular to the surface, with the positive direction defined to be outward. Here, for a current loop, we will define the positive direction of $\vec{A}$ using the direction of the current. In particular, if you curl the fingers of your right hand in the direction of positive current flow, then your thumb points in the direction of $\vec{A}$, as illustrated in Figure 13.6.

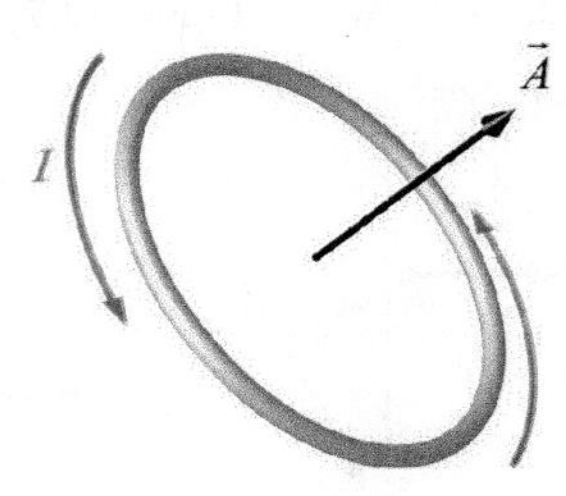

FIGURE 13.6 The direction of the area vector $\vec{A}$ is defined to be in the direction the thumb of your right hand points if your fingers curl in the direction of the current.

We define the magnetic dipole moment $\vec{\mu}$ to be the product of the current flowing around the loop with the area vector associated with the loop ($\vec{\mu} \equiv I\vec{A}$). If a current loop contains many turns (e.g., a coil), the magnetic moment $\vec{\mu}$ is the vector sum of the magnetic moments created by each turn separately, i.e., $\vec{\mu} = NI\vec{A}$, where N is the number of turns.

Once we have this definition, we see that we can rewrite our expression for the torque on a current loop in terms of this magnetic moment vector in a very compact form. Since the angle θ between $\vec{r}$ and $\vec{F}_{12}$ is identical to the angle between the magnetic moment vector and the $\vec{B}$ field, we can recast the form for the torque as

$$\vec{\tau} = \vec{\mu} \times \vec{B}$$

It may appear that our development in this section has been purely cosmetic, i.e., we defined a new quantity, the magnetic moment, so that we could obtain a simple equation for the torque. In fact, however, magnetic moments turn out to be quite important at all levels of physics. Interactions of all sorts of objects with magnetic fields are generally quantified in terms of magnetic moments. On a macroscopic scale, the behavior of a current loop in a magnetic field is very similar to that of a bar magnet. In each case, the magnetic moment vector (for a bar magnet, $\vec{\mu}$ is a vector aligned with the north pole of the magnet) rotates to align with the direction of the external magnetic field. At the microscopic scale, fundamental particles, such as the electron, possess intrinsic magnetic moments. We know about these magnetic moments through the particle's interaction with a magnetic field, but we can in no way identify a real current loop inside the electron that is responsible for creating this magnetic moment.

13.7 Potential Energy of Dipole in Magnetic Field

We have just seen that the torque on a current loop in a magnetic field depends on the orientation of the loop's magnetic moment vector with respect to the direction of the magnetic field. This torque may be used to rotate the loop and do work.

Let's consider the rotation of the loop as its magnetic moment vector moves from being perpendicular to the field to being parallel to the field, as shown in Figure 13.7. Recall from mechanics that the work done in a rotation is found by integrating the torque over the angular displacement:

$$W = \int \tau \, d\theta$$

Therefore, to calculate the work done in this case we need to integrate the torque from $\theta = 90°$ to $\theta = 0°$. To do this integral, we need to write the torque as $-\mu B \sin\theta$. The minus sign indicates that this torque gives rise to a rotation in the direction of *decreasing* θ. Since

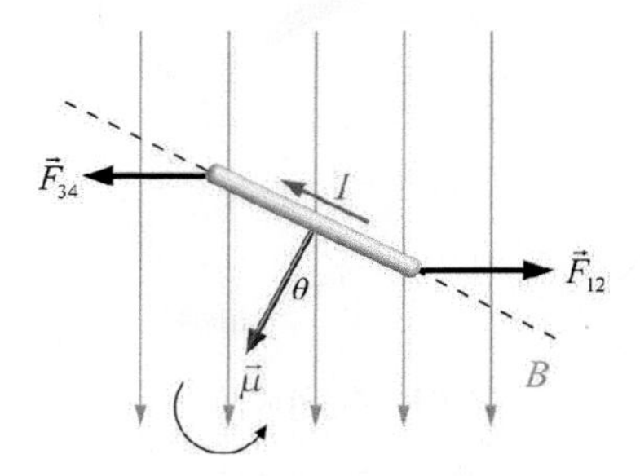

FIGURE 13.7 A current loop with magnetic moment vector $\vec{\mu}$ rotates in a counterclockwise direction from being perpendicular to the magnetic field $\vec{B}$ to parallel to $\vec{B}$.

$$\int \sin\theta \, d\theta = -\cos\theta$$

we can evaluate this integral between $\theta = 90°$ and $\theta = 0°$ to obtain W, the work done by the field:

$$W = \left[-\mu B(-\cos\theta)\right]_{90°}^{0°} = +\mu B\cos(0°) = +\mu B$$

This result makes sense; the field does positive work as the magnetic moment vector rotates from $\theta = 90°$ to $\theta = 0°$. Had we missed the mathematical subtlety of inserting the minus sign in the expression for the torque, we would have obtained a negative value for the work done by the field. Understanding the physics of this situation tells us that the work has to be positive; therefore, the physics tells us what the sign has to be. If we end up with the opposite sign, then we know that we must have made a sign error in the math somewhere along the line. In the matter of signs, always trust the answer you get from physics alone.

The general result, then, for the work done by the field when a current loop moves such that the angle between its magnetic moment vector and the $\vec{B}$ field changes from θ_1 to θ_2 is just equal to $\mu B(\cos\theta_2 - \cos\theta_1)$. Since this result only depends on the end points of the rotation, θ_1 and θ_2, we can now associate a change in potential energy with this work done in the usual way, i.e., we define the change in potential energy to be minus the work done by the field. Furthermore, we can assign a potential energy to every orientation of the dipole by choosing one such orientation to be the zero of potential energy. It is usual to choose the position of maximum torque ($\theta = 90°$) to be the zero. We then obtain an expression for the potential energy U at any angle θ by integrating minus the work done by the field from $\theta = 90°$ to $\theta = \theta$.

$$U \equiv -[-\mu B(-\cos\theta)]_{90°}^{\theta} = -\mu B\cos\theta = -\vec{\mu}\cdot\vec{B}$$

13.8 Summary

We began by calculating the force on a straight section of a current-carrying wire that is located in a region of uniform magnetic field. To make this calculation, we simply took the sum of the magnetic forces acting on all of the individual charge carriers. We arrived at the result that the total force on the straight wire is equal to the current multiplied by the cross product of the wire segment's length vector and the magnetic field.

We then considered the more general case of a curved wire segment in a uniform magnetic field. Here we treated the curved wire as the sum of a large number of small segments, each of which could be considered straight so that we could simply apply our previous result to each such segment. To get the final result, we summed the forces on all of the small segments. We obtained the interesting result that the force on the curved wire is exactly equal to the force that would be exerted on a straight wire joining the two end points of the curved wire and carrying the same current.

We then applied this knowledge of the forces exerted on current-carrying wire segments to determine the forces and torques on current loops. We first calculated the net force on a current loop in a uniform magnetic field and found that this force was exactly zero.

While the net force on a current loop in a uniform magnetic field is zero, the torque on that loop is usually not zero. We introduced an important new quantity $\vec{\mu}$, the magnetic dipole moment of the current loop, which we used to quantitatively describe the interactions of the loop with a uniform magnetic field. $\vec{\mu}$ is defined to be a vector whose direction is perpendicular to the plane of the loop in a sense determined by yet another right hand rule, and whose magnitude was equal to the product of the current with the area of the loop. We determined that the torque exerted on the loop could be written simply as the cross product of $\vec{\mu}$ with the uniform magnetic field $\vec{B}$. The interaction with $\vec{B}$ then causes $\vec{\mu}$ to rotate to align with $\vec{B}$, just like the poles of a bar magnet do.

Finally, we defined the potential energy of a current loop in a uniform $\vec{B}$ field to be the negative of the work done by the magnetic field to change the orientation of $\vec{\mu}$ with respect to $\vec{B}$. Choosing the zero of potential energy to be the position of maximum torque, we found we could write this potential energy U simply as minus the dot product of $\vec{\mu}$ with $\vec{B}$.

MAIN POINTS

Force on a Current-carrying Wire

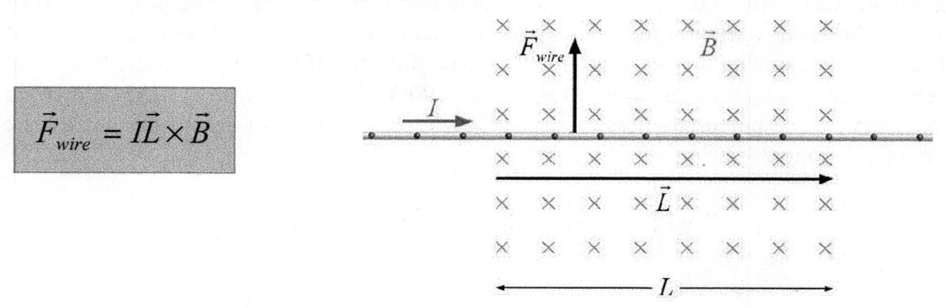

$$\vec{F}_{wire} = I\vec{L} \times \vec{B}$$

Magnetic Moments and Torques

Magnetic moment of a current-carrying loop

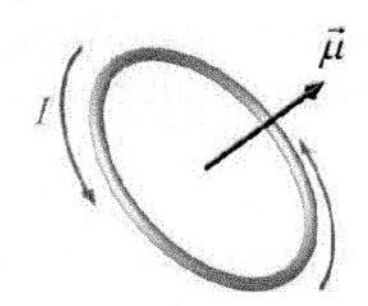

$$\vec{\mu} = NI\vec{A}$$

Torque on a loop

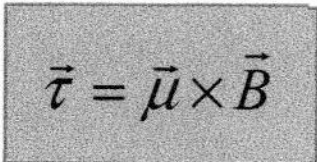

$$\vec{\tau} = \vec{\mu} \times \vec{B}$$

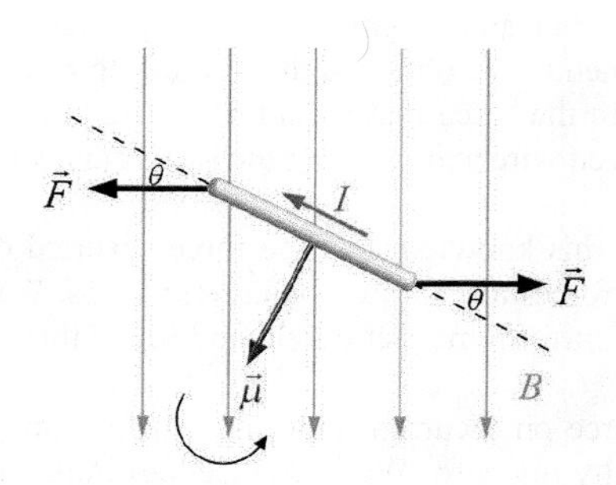

Potential Energy of a Magnetic Dipole

Rotating a magnetic moment in a magnetic field does work

$$U(\theta) = -\vec{\mu} \cdot \vec{B}$$

PROBLEMS

1. Rectangular Current Loop: A rectangular loop of wire with sides $h = 31$ cm and $w = 51$ cm is located in a region containing a constant magnetic field $B = 1.19$ T that is aligned with the positive y-axis as shown. The loop carries current $I = 281$ mA. The plane of the loop is inclined at an angle $\theta = 40°$ with respect to the x-axis.

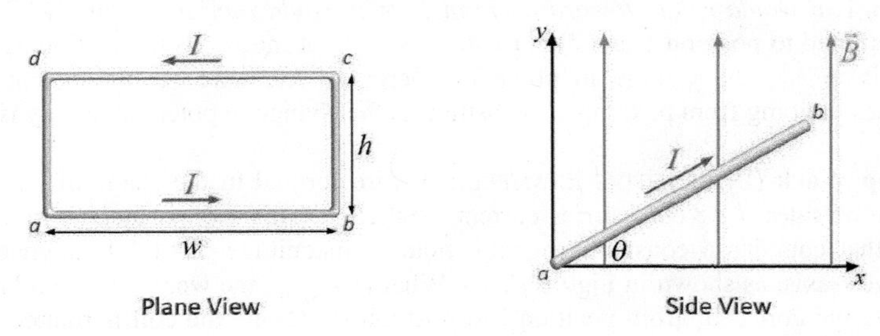

FIGURE 13.8 Problem 1

(a) What is μ_x, the x-component of the magnetic moment vector of the loop? (b) What is μ_y, the y-component of the magnetic moment vector of the loop? (c) What is τ_z, the z-component of the torque exerted on the loop? (d) What is F_{bc}, the magnitude of the force exerted on segment bc of the loop? (e) What is the direction of the force that is exerted on the segment bc of the loop?

(i) Along negative x-direction.
(ii) Along negative y-direction.
(iii) Along positive x-direction.
(iv) Along positive y-direction.
(v) None of the above.

2. Right Triangular Current Loop: A wire formed in the shape of a right triangle with base $L_{ab} = 12$ cm and height $L_{bc} = 72$ cm carries current $I = 402$ mA, as shown in position 1. The wire is located in a region containing a constant magnetic field $B = 0.83$ T aligned with the positive z-axis. (a) What is $F_{ac,x}$, the x-component of the force on the

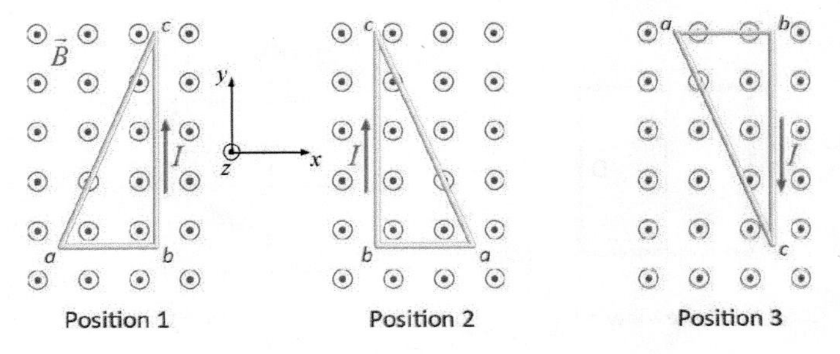

FIGURE 13.9 Problem 2

segment of the wire that connects points a and c in position 1? (b) What is $F_{ac,y}$, the y-component of the force on the segment of the wire that connects points a and c in position 1? (c) The wire is now rotated 180° about the y-axis to position 2, as shown. What is ΔU_{12}, the change in potential energy of the wire? Note that ΔU_{12} is a signed number. ΔU_{12} is positive if the potential energy in position 2 is higher than the potential energy in position 1. (d) The wire is now rotated back 90° about the y-axis toward position 1. If the wire is released from this position, how would it move: *it would rotate toward position* 1, *it would rotate toward position* 2, or *it would remain stationary*? The wire is now returned to position 1 and then rotated 180° about the x-axis to position 3, as shown. What is ΔU_{13}, the change in potential energy of the wire? If the potential energy increases in going from position 1 to position 3, the change in potential energy is positive.

3. Loop Work (INTERACTIVE EXAMPLE): A wire formed in the shape of an equilateral triangle of sides $d = 8$ cm, carries current $I = 0.25$ A. This loop is located in a region of space that contains a constant magnetic field of magnitude $B = 1.3$ T, aligned with the negative z-axis, as shown in Figure 13.10. What is $W_{\text{I to II}}$, the work you would have to do to rotate the coil 180° from position I to position II? (Note, the coil is rotated out of the plane of the page, not simply spun in the plane.)

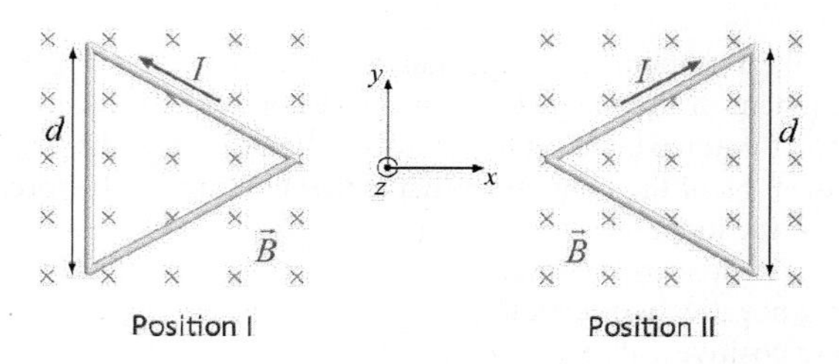

FIGURE 13.10 Problem 3

4. Loop Torque (INTERACTIVE EXAMPLE): A uniform magnetic field $B = 1.8$ T points in the $+x$ direction. A square loop with sides of length $d = 20$ cm, $N = 12$ turns, and current $I = 0.85$ A/turn pivots without friction around a pin about the z-axis, as shown in the figure below. (The z-axis points out of the page in the left-hand panel.) A mass M is hung

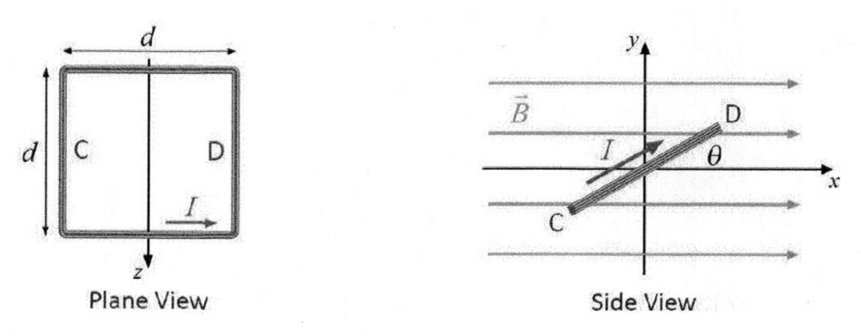

FIGURE 13.11 Problem 4

from one side of the loop (C or D), and the loop is in equilibrium when it makes an angle of 30° with respect to the x-z plane. Determine the size of the mass M and the side of the loop where the mass is hung.

UNIT

14

BIOT-SAVART LAW

14.1 Overview

Up until now, we have taken magnetic fields as a given and have investigated the resulting magnetic forces and torques. In this unit, we will learn how magnetic fields are created. Magnetic fields are produced by electric currents and we can calculate the magnetic fields produced by current distributions using the Biot-Savart law. We'll start by determining the magnetic field produced by an infinite straight current-carrying wire. We will then use this result to calculate the forces exerted by parallel current-carrying wires on each other. Finally, we will calculate the magnetic field along the axis of a circular current-carrying loop.

14.2 Biot-Savart Law

In Unit 12, we learned that magnetic fields are created by electric currents. The main evidence came from observations of the behavior of a compass needle in the vicinity of a wire. When there was no current in the wire, the compass needle was aligned with the Earth's magnetic field, but when a current began to flow, the needle moved to indicate the presence of a new magnetic field. This field was directed in a circular fashion around the wire, as shown in Figure 14.1.

How can we go about calculating this field that is produced by a current-carrying wire? Recall that when faced with the similar problem of calculating the electric field produced by a point charge, we needed to invoke a fundamental law, Coulomb's law, to make the calculation. Here, we will need to invoke a different fundamental law, the **Biot-Savart law**, to calculate the magnetic field produced by a current-carrying wire.

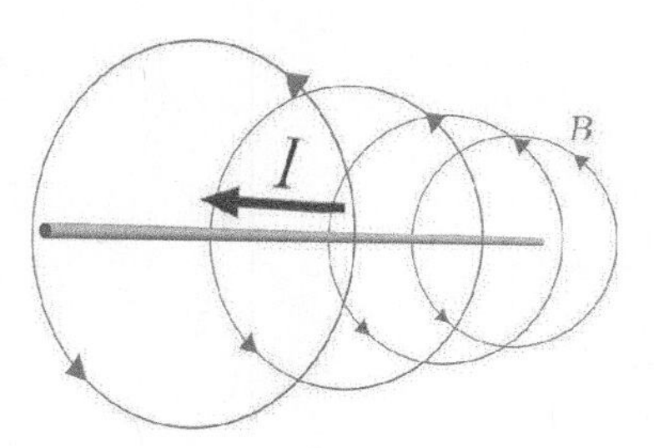

FIGURE 14.1 A current directed out of the page creates a magnetic field directed in the counterclockwise direction.

In particular, the magnetic field ($d\vec{B}$) that is produced by a current I in a segment of length $d\vec{s}$ at a point that is at a distance $\vec{r}$ from the segment is given by the expression:

$$d\vec{B} = \frac{\mu_o I}{4\pi} \frac{d\vec{s} \times \hat{r}}{r^2}$$

The constant μ_o is equal to $4\pi \times 10^{-7}$ T-m/A and clearly plays the role for magnetism that ε_o does for electricity, namely, to set the scale for the magnitude of the field.

Note that the vector $d\vec{s}$ takes the direction of I, the direction of the flow of positive charge, that $\vec{r}$ is a vector from the segment to the point at which we want to determine the magnetic field $d\vec{B}$, and that $\hat{r}$ is the unit vector that points along $\vec{r}$. In Figure 14.2, the direction of $d\vec{B}$ is determined by $d\vec{s} \times \vec{r}$, which is into the page. To see this, place the fingers of your right hand along $d\vec{s}$ and sweep them into the direction of $\vec{r}$; your thumb will then point into the page, giving the direction of $d\vec{B}$.

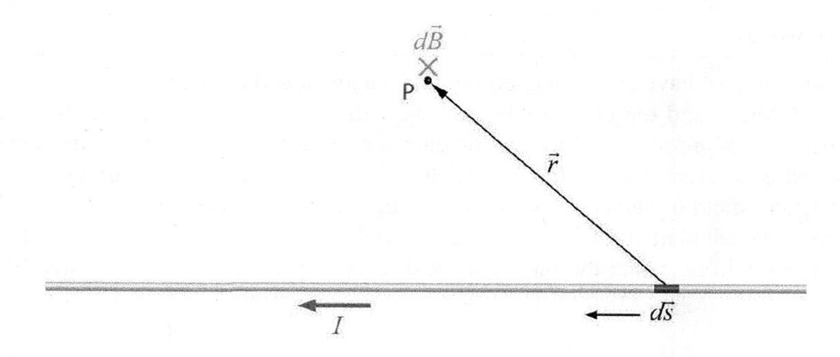

FIGURE 14.2 An illustration of the vectors used in the Biot-Savart law. The direction of the magnetic field is given by $d\vec{s} \times \hat{r}$, which in this case is directed into the page.

Note that the Biot-Savart law determines the magnetic field $d\vec{B}$ at point P produced by a current segment $I d\vec{s}$. To obtain the total magnetic field at P, we need to add up the contributions from all current segments, i.e., we will need to do an integral. We will do this calculation for an infinite straight wire in the next section.

14.3 Magnetic Field Produced by an Infinite Straight Wire

We will now use the Biot-Savart law to calculate the magnetic field $\vec{B}$ at point P created by an infinite straight wire carrying current I, as shown in Figure 14.3.

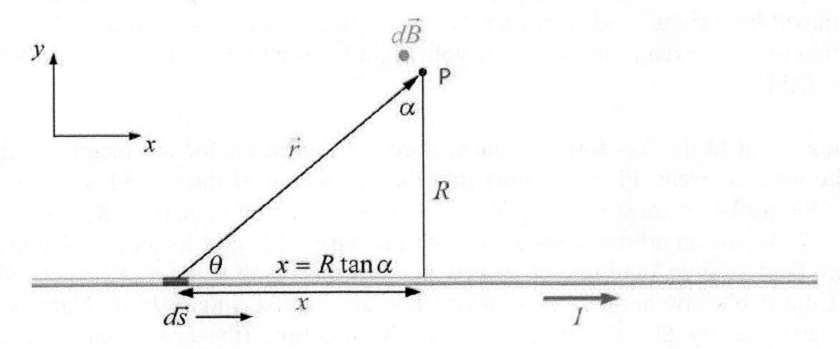

FIGURE 14.3 The geometry for calculating the magnetic field $\vec{B}$ at point P produced by current I in a long straight wire.

We'll start by investigating the direction of $\vec{B}$. Using the particular segment $d\vec{s}$ shown, we see that $I\,d\vec{s}\times\hat{r}$ yields a vector that points out of the page. As $d\vec{s}$ is moved along the wire, this result does not change! For all segments of the wire, $I\,d\vec{s}\times\hat{r}$ points out of the page. Therefore, we know the $\vec{B}$ field at point P will point out of the page. Now we just need to set up the integral to arithmetically sum up the dB contributions from all of the segments of the wire.

The most difficult part of evaluating the integral is the initial geometric setup. Our plan is to integrate over the angle α, as shown, from $-\pi/2$ to $+\pi/2$. The first step is to rewrite r in terms of R, the perpendicular distance to the wire.

$$r = \frac{R}{\cos\alpha}$$

The next step is to convert from ds to $d\alpha$.

$$ds = dx = \frac{R}{\cos^2\alpha}d\alpha$$

We then can determine the magnitude of the cross product used in the Biot-Savart law.

$$\left|ds\times\frac{\hat{r}}{r^2}\right| = \left(\frac{R}{\cos^2\alpha}d\alpha\right)\cos\alpha\left(\frac{\cos\alpha}{R}\right)^2 = \frac{1}{R}\cos\alpha\,d\alpha$$

Putting this altogether, we obtain

$$B = \frac{\mu_o I}{4\pi R} \int_{-\frac{\pi}{2}}^{+\frac{\pi}{2}} \cos\alpha \, d\alpha = \left(\frac{\mu_o I}{4\pi R}\right)(2) = \frac{\mu_o I}{2\pi R}$$

$\vec{B}$ is directed along the positive z-axis (out of the page) as discussed earlier. Our final result, then is that the magnitude of $\vec{B}$ is proportional to the current in the wire and inversely proportional to the perpendicular distance to the wire. The direction of $\vec{B}$ is determined by a right hand rule, namely if you place the thumb of your right hand in the direction of the current, the fingers of your right hand will curl around in the direction of the $\vec{B}$ field.

We now want to discuss some of the features of this result for the magnetic field of an infinite linear current. First, we note that the magnitude of the $\vec{B}$ field is proportional to I/R. Figure 14.4 shows the magnitude of $\vec{B}$ as a function of R from $R = 0$ to $R = 3R_o$, where R_o is just an arbitrary distance from the wire. This plot looks just like that for the electric field of an infinite line of charge. Perhaps this result is not all that surprising since both Coulomb's law and the Biot-Savart law are inverse-square laws. Namely, both of these laws specify that the field generated by a source (the charge element dq or the current segment $I\,ds$) is proportional to the magnitude of that source divided by the square of the distance from that source.

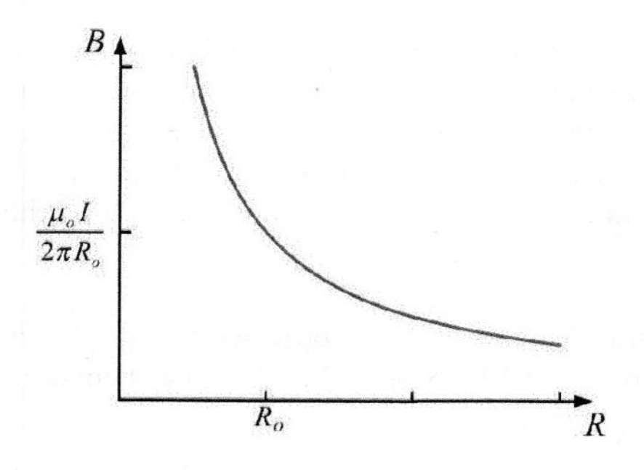

FIGURE 14.4 The magnitude of the magnetic field $\vec{B}$ produced by an infinite straight wire carrying current I as a function of R, the perpendicular distance from the wire.

For the point P in Figure 14.3, we determined that the direction of the field was in the positive z direction. We now want to make clear how that direction changes as the location of point P changes. Since the magnitude of the $\vec{B}$ field depends only on R, the perpendicular distance from the wire, we want to investigate the direction of the $\vec{B}$ field for all points a distance R from the wire, i.e., for all points on a circle of radius R centered on the wire, as shown in Figure 14.5.

The direction of the magnetic field is given by the direction of the cross product: $I\,d\vec{s} \times \hat{r}$. Since this direction is the same for all segments of the wire, we can simply evaluate the cross product for the segment that is in the plane of the circle that we are interested in, i.e., we can take $\vec{r}$ to be the same as $\vec{R}$.

We'll start with the point at the top of the circle to confirm our previous result: ($\vec{r}$ in the direction of $+y$), $I\,d\vec{s} \times \hat{r}$ points to the left (in the direction of $+z$). Next consider the point at the bottom of the circle. In this case, $I\,d\vec{s}\times\hat{r}$ points to the right (in the direction of $-z$). This result is completely general; any two points opposite each other on the circle will have $\vec{r}$ vectors, which are just the negative of each other; consequently, the direction of $\vec{B}$ (given by $I\,d\vec{s}\times\hat{r}$) at these points will also be exactly opposite each other. Furthermore, $\vec{B}$ must always be tangent to the circle since, from the definition of the cross product, it must be perpendicular to both the radius vector and the current.

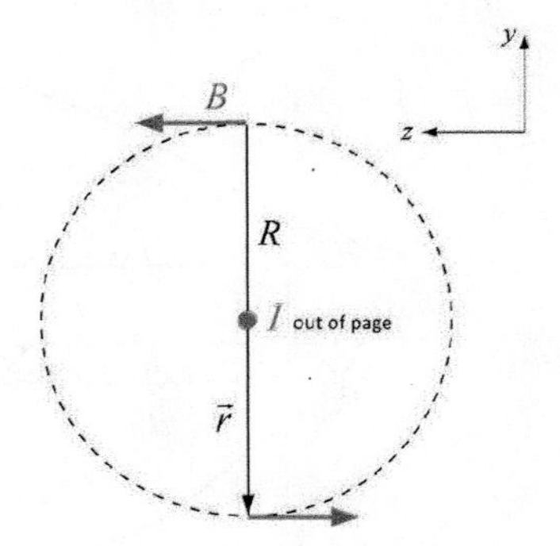

FIGURE 14.5 The direction of the magnetic field $\vec{B}$ produced by an infinite straight wire carrying current I out of the page.

As the points of consideration move around the circle, the pattern of the $\vec{B}$ field resembles that of a pinwheel, as shown in Figure 14.1. We can generate this pattern with the use of yet another right hand rule. Namely, if you place the thumb of your right hand in the direction of the current, your fingers will naturally curl in the direction of the magnetic field that is produced by that current. A current flowing out of the page generates a $\vec{B}$ field that rotates in a counterclockwise direction, while a current that flows into the page generates a $\vec{B}$ field that rotates in a clockwise direction.

14.4 Force between Two Parallel Current-Carrying Wires

Now that we know the form of the magnetic field produced by a long straight current-carrying wire, we can determine quantitatively an effect that was introduced in Unit 12, namely, the observation that two parallel current-carrying wires exert forces on each other.

Our interpretation of this effect is that each wire creates a magnetic field at the location of the other wire. This magnetic field then interacts with the current in the wire to produce a force on that wire.

In the case shown in Figure 14.6, the two parallel wires carry currents I_1 and I_2 in opposite directions. Let's first calculate the force on the bottom wire due to the current I_1 in the top wire. We must first determine the $\vec{B}$ field produced by I_1 at the location of the wire carrying I_2. This field is in fact the field we just calculated in the previous section: $B=\mu_o I_1/(2\pi d)$, where d is the separation of the two wires. The direction of the field is into the page as shown.

This field then gives rise to a force $\vec{F}_2$ on the bottom wire. This force on a length L of this wire is given by $\vec{F}_2 = I_2\vec{L}\times\vec{B}$. Since the $\vec{B}$ field is perpendicular to the wire, the magnitude of $\vec{F}_2$ is just given by

$$F_2 = I_2 L B_1 = \mu_o \frac{I_1 I_2 L}{2\pi d}$$

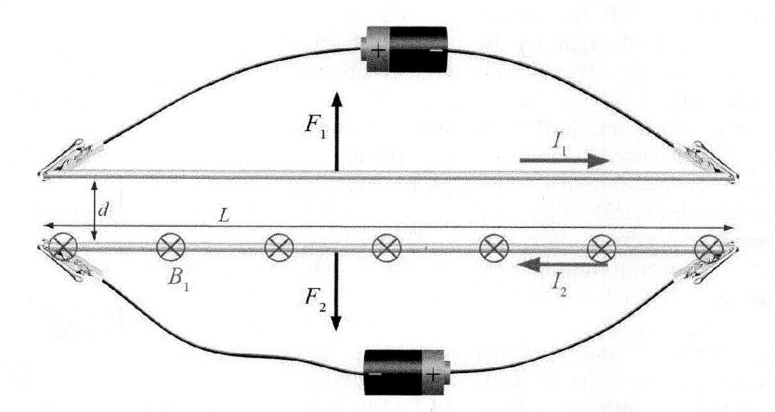

FIGURE 14.6 A force $\vec{F}_2$ is exerted on wire 2 due to the fact that its current (I_2) flows through a region containing a magnetic field ($\vec{B}_1$) created by the top wire carrying current I_1. Similarly, there is a force $\vec{F}_1$ exerted on the top wire due to the magnetic field produced by the bottom wire. These forces are of equal magnitude but have opposite directions.

The direction of this force is downward as determined by applying the right hand rule to evaluate $\vec{F}_2 = I_2\vec{L}\times\vec{B}$.

Similarly the current I_2 creates a magnetic field at the location of the top wire. This magnetic field then interacts with the current I_1 to create a force on the top wire. From Newton's third law, we know $\vec{F}_1 = -\vec{F}_2$. We leave it as an exercise for the reader to calculate the $\vec{B}$ field at I_1 and then evaluate $\vec{F}_1$ to demonstrate this fact.

14.5 Magnetic Field along the Axis of a Current Loop

We will now move on to another current example, a circular loop of radius *R* carrying current *I*, as shown in Figure 14.7. To simplify the calculation, we will restrict ourselves to finding the field $\vec{B}$ at all points along the axis of the loop.

Let's start by calculating the field at the center of the loop. All current segments in the loop will produce a $\vec{B}$ field at this point that points out of the page. This can be seen by either evaluating the cross product ($I\,d\vec{s}\times\hat{r}$) for each current segment or by simply applying the right hand rule for currents (place the thumb of your right hand in the direction of the current segment and your fingers curl in the direction of the *B* field produced by that segment). We can use the Biot-Savart law to determine that each current segment produces a $\vec{B}$ field at the center of the loop with magnitude:

$$dB = \frac{\mu_o I}{4\pi R^2}\,ds$$

Integrating over the loop (i.e., the integral of *ds* over the loop is just the circumference $2\pi R$), we obtain

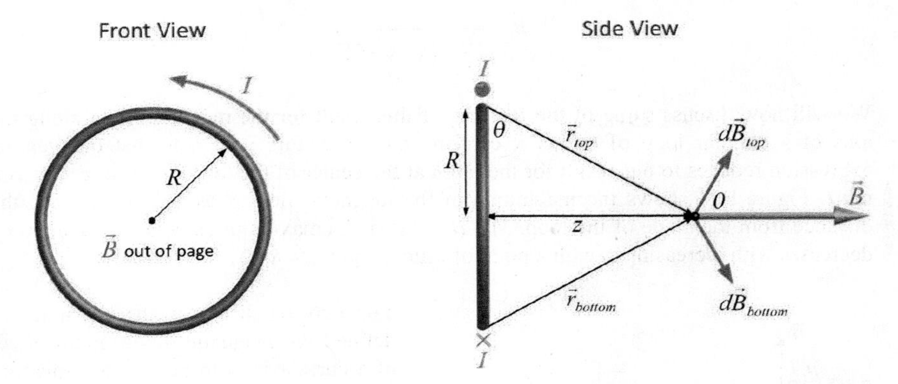

FIGURE 14.7 The magnetic field $\vec{B}$ produced by a loop of radius R carrying current I at all points along the axis of the loop is directed along that axis.

$$B = \frac{\mu_o I}{2R}$$

How will this $\vec{B}$ field change as we move along the axis away from the plane of the loop? We'll start by determining the direction of this field at all points along the z-axis, the axis of the loop. At any point on the axis, the field $d\vec{B}$ produced by the element $I\,ds$ will be perpendicular to the line connecting the current element with the point. As we move around the loop, these $d\vec{B}$ vectors trace out a cone around the z-axis. The total field is obtained by summing all these vectors. Consequently, we see that the direction of the field must be *parallel* to the axis of the current loop. Given the sense of the current shown in Figure 14.7, we can see that the magnetic field must point to the right for all points on the axis, even those on the axis to the left of the loop.

To determine the magnitude of the B field, we will again apply the Biot-Savart Law. The contributions to the z-component of the B field from all segments of the loop are all the same and are given by

$$B = \int dB_z = \int \left(\frac{\mu_o I}{4\pi} \frac{ds}{r^2} \right) \cos\theta$$

We now rewrite $\cos\theta$ and r^2 in terms of R and z:

$$B = \int \left(\frac{\mu_o I}{4\pi} \frac{ds}{(R^2 + z^2)^2} \right) \frac{R}{\sqrt{R^2 + z^2}}$$

The integral, therefore, simplifies to the integral of ds, which is just equal to the circumference of the circle ($2\pi R$). We therefore arrive at our final result for the magnitude of the magnetic field at a distance z from the loop:

$$B = \frac{\mu_o I}{2} \frac{R^2}{\left(R^2 + z^2\right)^{\frac{3}{2}}}$$

We will now discuss some of the features of this result for the magnetic field along the axis of a circular loop of radius R carrying current I. First, we note that our general expression reduces to our result for the point at the center of the loop if we set $z = 0$, as it must. Figure 14.8 shows the magnitude of the magnetic field B as a function of z, the distance from the plane of the loop. The B field has its maximum value at $z = 0$ and then decreases with increasing z, with a point of inflection at $z = R/2$.

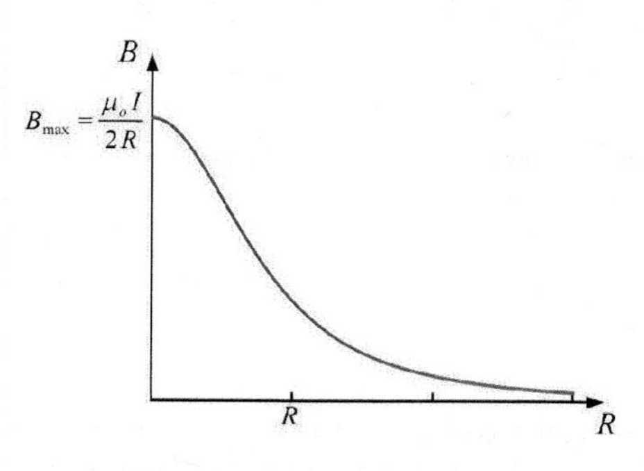

FIGURE 14.8 The magnitude of the magnetic field $\vec{B}$ along the axis of a loop of radius R carrying current I as a function of z, the distance from the plane of the loop.

You may recall that in the last unit, we defined the magnetic dipole moment $\vec{\mu}$ of a current loop to be a vector quantity whose magnitude is given by the product of the current I and the area $\vec{A}$ of the loop. The direction of $\vec{\mu}$ is given by a right hand rule that in this case would lead to $\vec{\mu}$ being aligned with the positive z-axis, the same direction as the magnetic field at all points along that axis.

The long-distance behavior of the $\vec{B}$ field produced along the axis can be found by taking the limit of the general expression when $z >> R$. If we replace R^2 by $\mu / \pi I$ in this expression, we obtain the result: $B_z \approx \mu_o \mu / (2\pi z^3)$.

At this point, we can start to see the motivation for calling $\vec{\mu}$ a dipole moment and for identifying a current loop as a dipole. If you look back at Unit 2, we defined an electric dipole as equal positive and negative charges separated by a distance a. It is usual to define the electric dipole moment $\vec{p}$ as a vector whose direction points from the negative charge to the positive charge and whose magnitude is the product of the charge Q and the distance a. The $\vec{E}$ field of an electric dipole along its axis is aligned with electric dipole moment $\vec{p}$, just as the $\vec{B}$ field of a magnetic dipole along its axis is aligned with its magnetic dipole moment $\vec{\mu}$. Further, if you look at the form for the electric dipole field along its axis in the long-distance limit, you will find that the magnitude of $\vec{E}$ is proportional to p/r^3, just as the magnitude of $\vec{B}$, the field of the current loop along its axis in the long-distance limit, is proportional to μ / z^3.

14.6 The Off-Axis B Field of a Current-Carrying Loop

We've just calculated the magnetic field produced by a circular current loop for all points along the central axis of the loop. What about all the other points that do not lie along the central axis of the loop? It's clear that, in principle, we can follow exactly the same procedure to determine these values. We just calculate the field produced by small segments of the loop at any point of interest and then just add them up. In practice, however, the integrals become difficult to do analytically for points not along the central axis. We can, however, use computers to calculate these integrals numerically to any desired precision.

We start by inspecting the symmetry and realizing that the magnetic field can only be a function of r and z. To display the results of the numerical calculation, we will define the x-axis such that the point of interest has $x = 0$. We then display our results in the y-z plane, as shown in Figure 14.9.

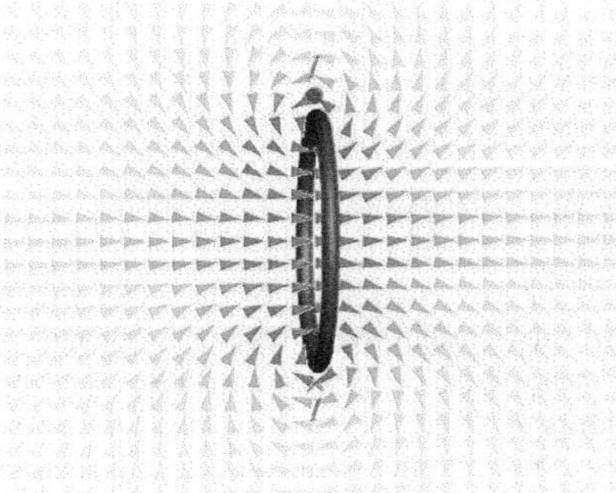

FIGURE 14.9 Arrows depicting the strength and direction of the magnetic field produced by a circular current loop in the x-y plane.

We can see that the field is aligned with the z-axis in the plane of the loop, but then curls around the loop as we move away from that plane. The magnitude of the field increases as we get near the current.

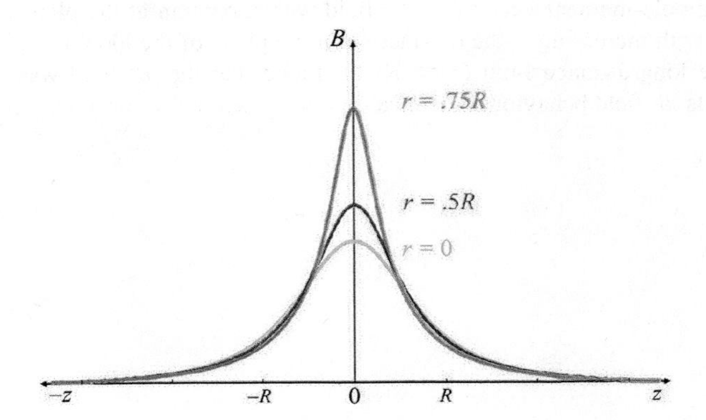

FIGURE 14.10 Arrows depicting the strength and direction of the magnetic field produced by a circular current loop in the x-y plane.

Figure 14.10 shows the magnitude of the magnetic field as a function of z for three values of r (0, 0.5R, and 0.75R). We choose to display these numerical results because the circular loop is an important example since it is the model for a magnetic dipole and because we will expand this numerical calculation in the next unit to model another important example, the solenoid, as a collection of closely-packed circular current loops.

14.7 Summary

We began by introducing the Biot-Savart law in which the magnetic field $d\vec{B}$ at a point displaced from a current segment $I\,d\vec{s}$ by an amount $\vec{r}$ is given by

$$d\vec{B} = \frac{\mu_o I}{4\pi}\frac{d\vec{s}\times\hat{r}}{r^2}$$

where the constant $\mu_o = 4\pi\times10^{-7}$ T-m/A.

We then applied this law to the calculation of the magnetic field $\vec{B}$ at a perpendicular distance R from an infinite straight wire carrying current I. We determined that the magnitude of this filed was given by $B = \mu_o I/(2\pi R)$. The direction of the field at any point on a circle of radius R from the wire is tangent to that circle with its direction determined by a right hand rule. Namely, if you place the thumb of your right hand in the direction of the current, your fingers will curl in the direction of the $\vec{B}$ field.

We then used this result to calculate the forces exerted by two parallel current-carrying wires. We determined that these forces were equal and opposite as they must be from Newton's third law, and that their magnitude was proportional to the product of the currents and inversely proportional to the separation between the wires.

Finally, we calculated the magnetic field along the axis of a circular current-carrying loop. We found that the field was directed along the axis of the loop, parallel to the loop's magnetic dipole moment vector $\vec{\mu}$. The field was maximum at the plane of the loop and decreased with increasing z, the distance from the plane of the loop, as $R^2/(R^2+z^2)^{3/2}$. Taking the long-distance limit ($z >> R$) we found that the $\vec{B}$ field was proportional to μ/z^3. This $\vec{B}$ field behavior is identical to the $\vec{E}$ field behavior of an electric dipole.

MAIN POINTS

Biot-Savart Law

A fundamental law that determines the magnetic field produced by a current distribution.

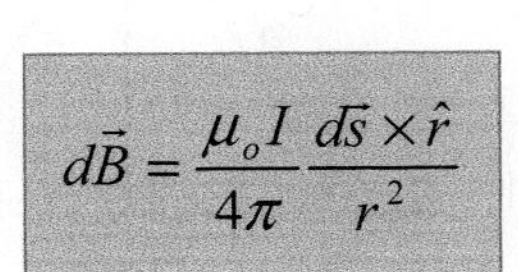

$$d\vec{B} = \frac{\mu_o I}{4\pi}\frac{d\vec{s} \times \hat{r}}{r^2}$$

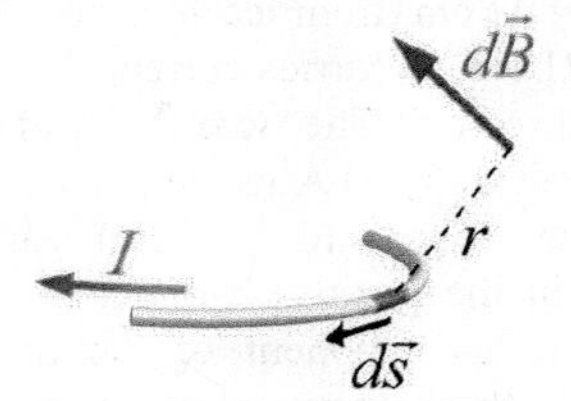

Forces between Current-Carrying Wires

Forces exist between current-carrying wires because the current in wire 1 produces a magnetic field in the region of the wire 2, which then exerts a force on the current in wire 2.

$$F_1 = F_2 = \frac{\mu_o}{2\pi d} I_1 I_2 L$$

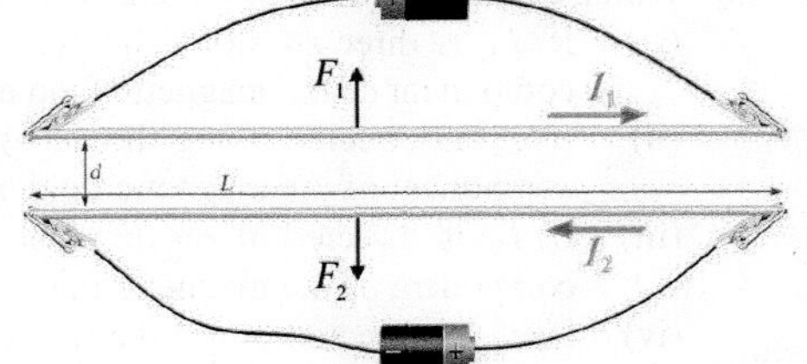

The Magnetic Field along the Axis of a Circular Current Loop

Biot-Savart law can be applied to calculate the magnitude of the magnetic field along the axis of a circular current-carrying loop.

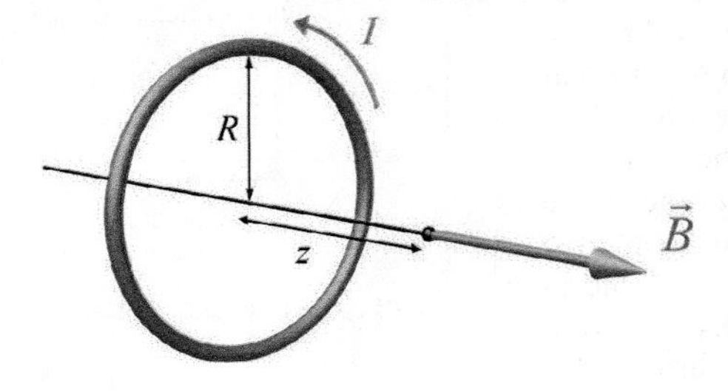

$$B = \frac{\mu_o I}{2}\frac{R^2}{\left(R^2 + z^2\right)^{\frac{3}{2}}}$$

PROBLEMS

1. Three Infinite Straight Wires: Three infinite straight wires are fixed in place and aligned parallel to the z-axis as shown. The three wires are each a distance $d = 42$ cm from each other. The wire at $(x, y) = (-21$ cm, $0)$ carries current $I_1 = 2.2$ A in the negative z-direction. The wire at $(x, y) = (21$ cm, $0)$ carries current $I_2 = 1.3$ A in the positive z-direction. The wire at $(x, y) = (0, 36.4$ cm$)$ carries current $I_3 = 9.0$ A in the positive z-direction. (a) What is $B_x(0,0)$, the x-component of the magnetic field produced by these three wires at the origin? (b) What is $B_y(0,0)$, the y-component of the magnetic field produced by these three wires at the origin? (c) What is $F_{1,x}$, the x-component of the force exerted on a one-meter length of the wire carrying current I_1? (d) What is $F_{1,y}$, the y-component of the force exerted on a one-meter length of the wire carrying current I_1? (e) What is $F_{2,x}$, the x-component of the force exerted on a one-meter length of the wire carrying current I_2? (f) Another wire is now added, also aligned with the z-axis at $(x, y) = (0, -36.4$ cm$)$. This wire carries current I_4. Which of the following statements is true?

(i) If I_4 is directed along the positive z-axis, then it is possible to make the y-component of the magnetic field equal to zero at the origin.

(ii) If I_4 is directed along the negative z-axis, then it is possible to make the y-component of the magnetic field equal to zero at the origin.

(iii) If I_4 is directed along the positive z-axis, then it is possible to make the x-component of the magnetic field equal to zero at the origin.

(iv) If I_4 is directed along the negative z-axis, then it is possible to make the x-component of the magnetic field equal to zero at the origin.

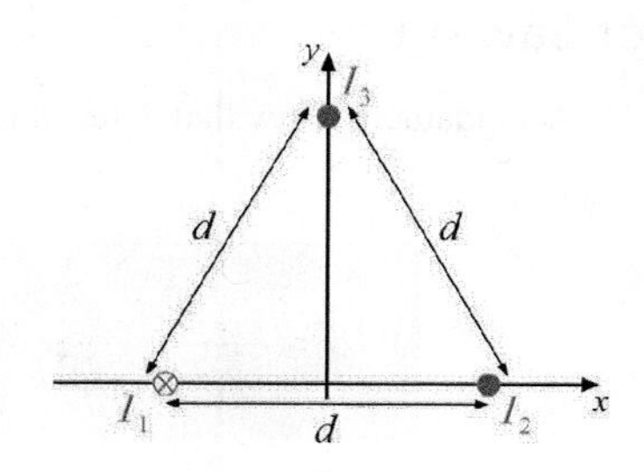

FIGURE 14.11 Problem 1

2. Wire and Rectangular Loop: A rectangular loop of wire with sides $h = 36$ cm and $w = 78$ cm carries current $I_2 = 0.304$ A. An infinite straight wire, located a distance $L = 24$ cm from segment ad of the loop as shown, carries current $I_1 = 0.552$ A in the positive y-direction. (a) What is $F_{ad,x}$, the x-component of the force exerted by the infinite wire on segment ad of the loop? (b) What is $F_{bc,x}$, the x-component of the force exerted by the infinite wire on segment bc of the loop? (c) What is $F_{Net,y}$, the y-component of the net force exerted by the infinite wire on the loop? (d) Another infinite straight wire, aligned with the y-axis, is now added at a distance $2L = 0.48$ cm from segment bc of the loop as shown. A current, I_3, flows in this wire. The loop now experiences a net force of zero. What is the direction of I_3: *along the positive y-direction*, or *along the negative y-direction*? (e) What is the magnitude of I_3?

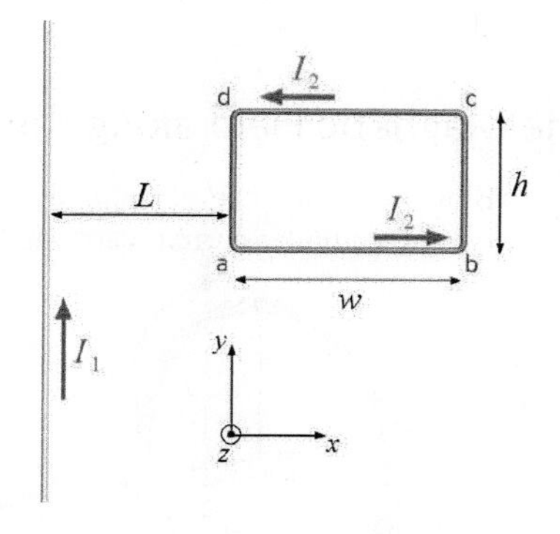

FIGURE 14.12 Problem 2

UNIT 15

AMPÈRE'S LAW

15.1 Overview

In the last unit, we introduced the Biot-Savart law, which allowed us to calculate the magnetic field produced by an arbitrary current distribution. We will now introduce another fundamental law, Ampère's law, which relates the integral of the magnetic field around a closed path to the current passing through the area defined by that path. In some sense, Ampère's law is the magnetic analog of Gauss' law; it is always true and in cases with a large amount of symmetry, can be used to simplify the calculation of the magnetic field.

We will begin by evaluating the line integral of the magnetic field produced by an infinite straight current-carrying wire around a variety of paths to illustrate the applicability of Ampère's law. We will then use Ampère's law to calculate the magnetic field produced by examples that illustrate the symmetries that can be used; namely, an infinite straight current-carrying wire and an infinite sheet of current.

15.2 Review of Gauss' Law

We'll begin with a brief review of Gauss' law in order to motivate our introduction of Ampère's law.

You'll recall that we represented the electric field produced by a point charge by a set of radial field lines emanating from the charge, with the number of these lines being proportional to the magnitude of the charge. The electric flux through any surface was defined to be

$$\Phi_E \equiv \int_{surface} \vec{E} \cdot d\vec{A}$$

and was found to be proportional to the number of these field lines leaving the surface. From this representation, we arrived at Gauss' law, the statement that the flux through any closed surface is proportional to the total charge enclosed.

$$\oint_{surface} \vec{E} \cdot d\vec{A} = \frac{Q_{enclosed}}{\varepsilon_o}$$

Gauss' law holds true for all charge distributions, but was found to be especially useful for actually determining the electric field created by charge distributions that possessed significant symmetries. We will find an analogous situation with Ampère's law and magnetic fields.

15.3 Motivation of Ampère's Law

Electric charges generate electric fields; we chose the simplest charge distribution, that of a single point charge, to develop Gauss' law. Since electric currents generate magnetic fields, we will choose the simplest current, that of an infinite straight wire, to develop Ampère's law.

In the last unit, we used the Biot-Savart law to calculate the magnetic field $\vec{B}$ at a perpendicular distance R from an infinite straight wire carrying current I. We determined that the magnitude of this field was proportional to the current, and inversely proportional to the perpendicular distance from the wire.

$$B = \frac{\mu_o I}{2\pi R}$$

The direction of the field at any point on a circle of radius R from the wire was tangent to that circle with its sense determined by a right hand rule. Namely, if you place the thumb of your right hand in the direction of the current, your fingers will curl in the direction of the $\vec{B}$ field.

We note two important features of this result. First, the direction of the $\vec{B}$ field is always tangent to any circle in the plane perpendicular to the wire with the wire at its center.

Second, the magnitude of the field decreases as $1/R$, while the circumference of the circle grows as R.

Consequently, if we consider the line integral of $\vec{B}$ around any circle concentric with the wire, we will always obtain the same value. In particular, we see that:

$$\oint_{circle} \vec{B}\cdot d\vec{l} = \frac{\mu_o I}{2\pi R}\oint_{circle} dl = \mu_o I$$

This simple equation we call the integral form of Ampère's law. The amazing thing is that this equation holds not only for this particular case, but for any path (it doesn't have to be a circle) and any current distribution (it doesn't have to be an infinite straight wire).

We will now do a couple of examples using this infinite straight current-carrying wire to demonstrate that this result holds for arbitrary paths.

15.4 Example: Integrating with Semicircles

We will once again consider the line integral of the $\vec{B}$ field produced by an infinite straight wire carrying current I. This time, however, we will change the path to one that is formed by connecting two semicircles (radius R_1 and R_2) with two straight segments (ab and cd), as shown in Figure 15.1.

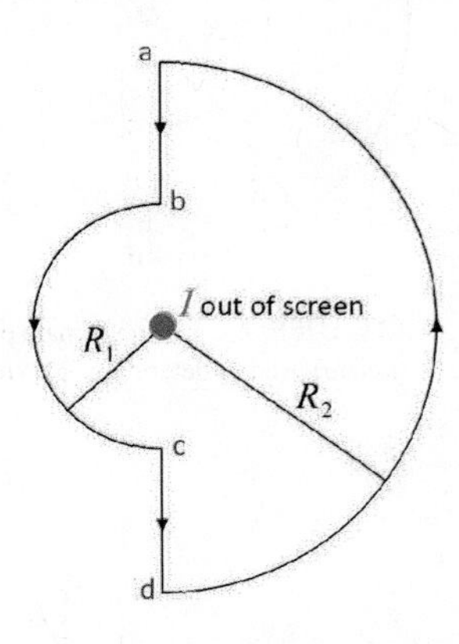

FIGURE 15.1 An alternate path ($abcda$) used to determine $\oint \vec{B}\cdot d\vec{l}$.

We will perform the integral by breaking it into four pieces (from a to b, from b to c, from c to d, and finally from d back to a).

The integral of $\vec{B}\cdot d\vec{l}$ is zero since $\vec{B}$ is perpendicular to $d\vec{l}$ at all points along the path. For the same reason, the same integral is also zero between c and d, the other radial segment. Therefore, the complete line integral will just be the sum of the contributions from the tangential segments connecting b to c and connecting d to a.

The magnitude of $\vec{B}$ at all points along the semicircle from b to c is a constant ($\mu_o I/2\pi R_1$). Further, the direction of $\vec{B}$ at all points along the semicircle from b to c is parallel to $d\vec{l}$ (i.e., tangent to the circle). Therefore, we can evaluate the integral from b to c:

$$\int_b^c \vec{B}\cdot d\vec{l} = B\int_b^c dl = \frac{\mu_o I}{2\pi R_1}\pi R_1 = \frac{\mu_o I}{2}$$

Similarly,

$$\int_d^a \vec{B} \cdot d\vec{l} = \frac{\mu_o I}{2}$$

Therefore, the integral around the complete path (*abcda*) is just equal to $\mu_o I$.

Note that this result is identical to what we obtained for the circular path in the preceding section. We'll now modify our path of integration one more time in order to make an important point.

In particular, we will change the path by flipping the semicircle of radius R_2 to the other side, as shown in Figure 15.2. Once again, we will perform the integral by breaking it into four pieces (from *a* to *b*, from *b* to *c*, from *c* to *d*, and finally from *d* back to *a*). Once again, we will perform the integral by breaking it into four pieces.

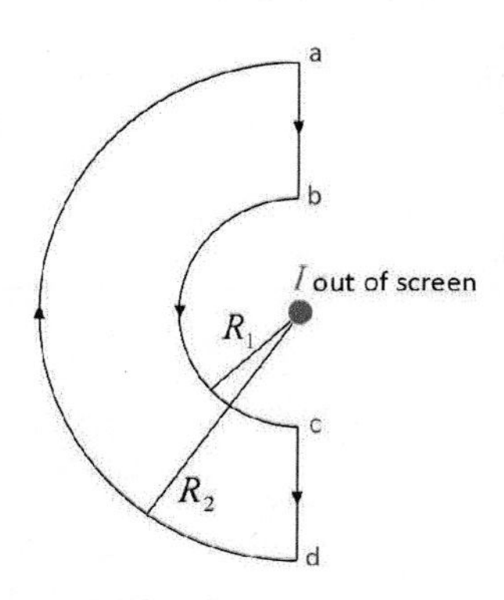

FIGURE 15.2 An alternate path (*abcda*) used to determine $\oint \vec{B} \cdot d\vec{l}$.

As in Figure 15.1, the contributions from the radial segments from *a* to *b* and from *c* to *d* are both zero since $\vec{B}$ is perpendicular to $d\vec{l}$ at all points along these paths. Once again, the complete line integral is just the sum of two tangential terms, the integral from *b* to *c* and the integral from *d* to *a*.

Looking at the integral from *b* to *c*, we see nothing has changed:

$$\int_b^c \vec{B} \cdot d\vec{l} = B\int_b^c dl = \frac{\mu_o I}{2\pi R_1} \pi R_1 = \frac{\mu_o I}{2}$$

The integral from *d* to *a*, however, is *not* the same as before! Flipping the outer semicircle has made an important difference. Namely, the path element ($d\vec{l}$) is now always *antiparallel* to the magnetic field! Consequently, the dot product, $\vec{B} \cdot d\vec{l}$, will be negative. Therefore, the integral from *d* to *a* is equal to *minus* $\mu_o I / 2$.

Putting this altogether, we see that the integral around the complete path (*abcda*) is now equal to zero!!

$$\oint_{abcda} \vec{B} \cdot d\vec{l} = \int_b^c \vec{B} \cdot d\vec{l} + \int_d^a \vec{B} \cdot d\vec{l} = 0$$

What has changed? Flipping the outer semicircle has produced a geometry in which the current is now *outside* of the closed path! In the previous geometry, the current was *inside* the closed path.

15.5 Ampère's Law

We have just demonstrated that for the $\vec{B}$ field produced by an infinite straight wire the integral of $\vec{B}\cdot d\vec{l}$ around some specific closed loops made up of semicircles centered on the wire were either equal to $\mu_o I$ or zero depending on whether the loop enclosed the current or not.

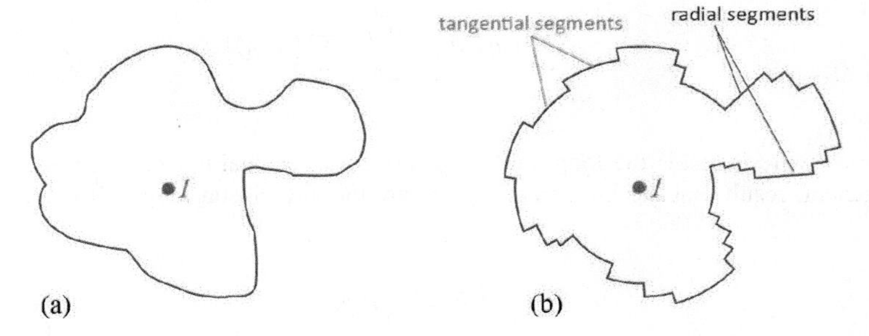

FIGURE 15.3 (a) An odd-shaped Amperian loop. (b) The approximation of the continuous loop by a succession of small segments that are alternately radial and tangential segments centered on the current-carrying wire (the • dot).

We now want to show that this result is in fact very general. First consider the closed loop shown in Figure 15.3(a). We can approximate this loop (or any other loop for that matter) as a succession of small segments that are alternately radial and tangential segments centered on the wire (Figure 15.3(b)). The more segments we use, the closer the approximation. The integral of $\vec{B}\cdot d\vec{l}$ around the closed loop is just the sum of the integral of $\vec{B}\cdot d\vec{l}$ over each segment.

$$\oint \vec{B}\cdot d\vec{l} = \sum_{\text{segments}}\left(\int \vec{B}\cdot d\vec{l}\right) = \sum_{\substack{\text{radial}\\ \text{segments}}}\left(\int \vec{B}\cdot d\vec{l}\right) + \sum_{\substack{\text{tangential}\\ \text{segments}}}\left(\int \vec{B}\cdot d\vec{l}\right)$$

This integral is zero for all the radial segments since $\vec{B}$ is always tangential, making the integral of $\vec{B}\cdot d\vec{l}$ equal to zero along all radial paths.

$$\sum_{\substack{\text{radial}\\ \text{segments}}}\left(\int \vec{B}\cdot d\vec{l}\right) = 0$$

The integral around the closed loop then is just the sum of the integrals along the tangential segments. Since $\vec{B}$ is tangential, the integral over any tangential segment is just equal to the integral of $B\,dl$.

$$\sum_{\substack{\text{tangential}\\\text{segments}}}\left(\int \vec{B}\cdot d\vec{l}\right)=\sum_{\substack{\text{tangential}\\\text{segments}}}\left(\int B\,dl\right)$$

Now, $B=\mu_o I/(2\pi R)$ and $dl = R\,d\theta$. Therefore, the integral over any tangential segment is just given by

$$\int_{\substack{\text{tangential}\\\text{segment}}} \vec{B}\cdot d\vec{l} = \int B\,dl = \int \frac{\mu_o I}{2\pi R} R\,d\theta = \frac{\mu_o I}{2\pi}\int d\theta$$

Therefore, the sum of all the tangential contributions will just be

$$\sum_{\substack{\text{tangential}\\\text{segments}}}\left(\int \vec{B}\cdot d\vec{l}\right)=\frac{\mu_o I}{2\pi}\sum_{\substack{\text{tangential}\\\text{segments}}}\left(\int d\theta\right)$$

Since the wire is inside the loop, the integral over $d\theta$ is equal to 2π. Therefore, we obtain the general result that the integral of $\vec{B}\cdot d\vec{l}$ around any closed loop will just be equal to $\mu_o I$.

$$\oint \vec{B}\cdot d\vec{l} = \mu_o I$$

If the wire had been outside of the loop, then the integral would have been zero since $\vec{B}\cdot d\vec{l}$ would be negative as much as positive as we integrate around the closed path.

Now suppose we had a collection of parallel infinite current-carrying wires passing through the loop. How would our result change? From superposition, we know that $\vec{B}$ at any point on the loop is just the vector sum of $\vec{B}$ produced by each wire separately. Consequently, the integral of $\vec{B}\cdot d\vec{l}$ around the loop is just the sum of the integrals needed for each wire separately.

Therefore, we have arrived at the general expression that we need; namely that the integral of $\vec{B}\cdot d\vec{l}$ around any closed loop is equal to μ_o multiplied by the current enclosed by the loop. This equation is called **Ampère's law**.

$$\oint_{loop} \vec{B}\cdot d\vec{l} = \mu_o I_{enclosed}$$

Actually, we should also worry about the case in which the currents are not perpendicular to the plane of the loop. In this case, the geometry is much more daunting, but the end result is the same. We'll now use Ampère's law to calculate the magnetic field produced by a few current distributions.

15.6 Magnetic Field inside a Wire

We begin by determining the magnetic field inside an infinite straight wire carrying current I, as shown in the wire segment of Figure 15.4.

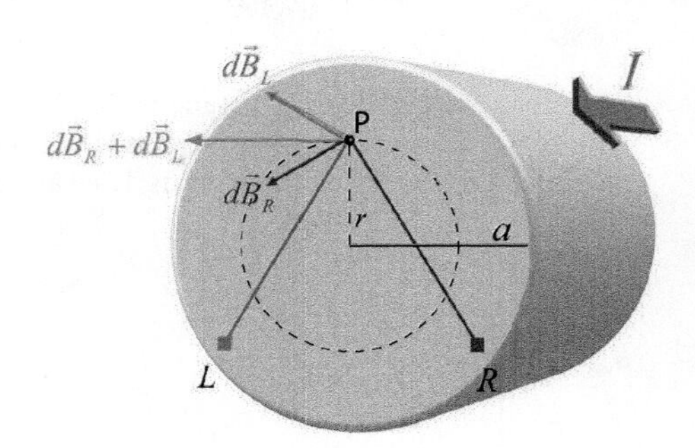

FIGURE 15.4 Calculation of the magnetic field inside a wire. Note that the radial components of the field produced by symmetric points (L and R) cancel, leaving the magnetic field to be totally tangential.

Our first job is to choose the closed path that will make our job of evaluating the closed integral of $\vec{B} \cdot d\vec{l}$ as simple as possible. A good choice would clearly be one for which the magnitude of the magnetic field is constant and the direction of the magnetic field relative to $d\vec{l}$ is always the same.

To choose such a path, we look to the symmetry of the situation. Here we have complete cylindrical symmetry about the center of the wire. Consequently, we know the magnitude of the $\vec{B}$ field can only depend on r, the distance from the center of the wire. The cylindrical symmetry also dictates that the direction of the $\vec{B}$ field can only be radial or tangential. If we choose a particular point P at a distance r from the center, as shown in Figure 15.4, we can see that the line connecting the center to that point bisects the circular cross section and that when we calculate the contributions to the magnetic field at point P from symmetric points in each half, as shown, the radial components cancel leaving only a tangential field.

Consequently, we know the field inside the wire must be only tangential and its magnitude depends only on r, the distance to the center. Therefore, to determine the field at point P, we will choose the path to be a circle of radius r centered on the wire. The magnitude of the $\vec{B}$ field at every point on this path is the same and the direction is always parallel to the path.

With this choice of path, $\vec{B} \cdot d\vec{l}$ becomes simply $B\,dl$. Since B is a constant, we only need to integrate dl around a circle. The result is simply the circumference of the circle. Therefore, the left-hand side of Ampère's law then becomes $B(2\pi r)$.

To calculate the right-hand side of the equation, we need to determine the current enclosed by the circle of radius r. Assuming the current is uniformly distributed throughout the wire, the fraction of the total current that is enclosed by this loop is given by the ratio of the area of the loop to the total area of the wire, i.e., the current enclosed is just equal to $I\,\pi r^2/(\pi a^2)$.

Equating the two sides of the Ampère's law equation allows us to determine the magnitude of B:

$$B(2\pi r) = \mu_o I \frac{r^2}{a^2} \qquad \Rightarrow \qquad B = \frac{\mu_o I}{2\pi a^2} r$$

Therefore, we see that the magnetic field increases linearly from zero at the center of the wire to a maximum value at the surface of the wire. Once outside the wire, we know the magnitude of the B field falls off as $1/r$, as shown in Figure 15.5.

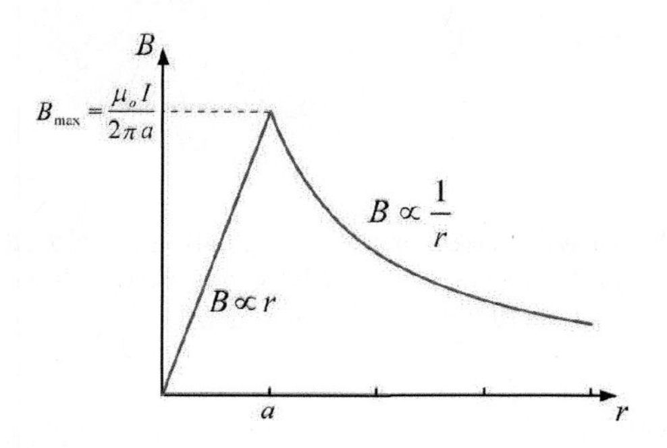

FIGURE 15.5 The magnitude of the magnetic field $\vec{B}$ produced by an infinite straight wire of radius a, carrying current I as a function of r, the perpendicular distance from the center of the wire.

15.7 Magnetic Field Produced by an Infinite Sheet of Current

You may recall that we could use Gauss' law to determine the electric field produced by charge distributions that exhibited spherical, cylindrical, or planar symmetries. The examples we chose were a single point charge for spherical symmetry, an infinite line of charge for cylindrical symmetry, and an infinite sheet of charge for planar symmetry. When we use Ampère's law to determine magnetic fields, there are only two symmetries, cylindrical and planar, that can be used since the source of the magnetic field is a current that cannot be a point, the generator of spherical symmetry.

We've just calculated the cylindrical case; we'll now turn to the planar symmetry exhibited by an infinite sheet of current. We'll model this infinite sheet as a set of closely packed wires (n wires per unit width), each carrying current I out of the page, as shown in Figure 15.6. If we consider the contributions to the $\vec{B}$ field at point P_t due to two wires equidistant from P_t ($\vec{B}_L$ and $\vec{B}_R$), we see that their y-components exactly cancel and their x-components add. Consequently, the $\vec{B}$ field points in the minus x direction at point P_t. Similarly, at the symmetric point P_b below the wires, the $\vec{B}$ field will point in the plus x direction.

We can then evaluate the $\vec{B}$ field at point P_t by choosing a rectangular path as shown. Integrating $\vec{B} \cdot d\vec{l}$ from b to c and from d to a yields zero, since $\vec{B}$ is perpendicular to $d\vec{l}$ for both segments. Integrating $\vec{B} \cdot d\vec{l}$ from a to b and from c to d each yields $+BL$, since $\vec{B}$ is parallel to $d\vec{l}$ for both segments.

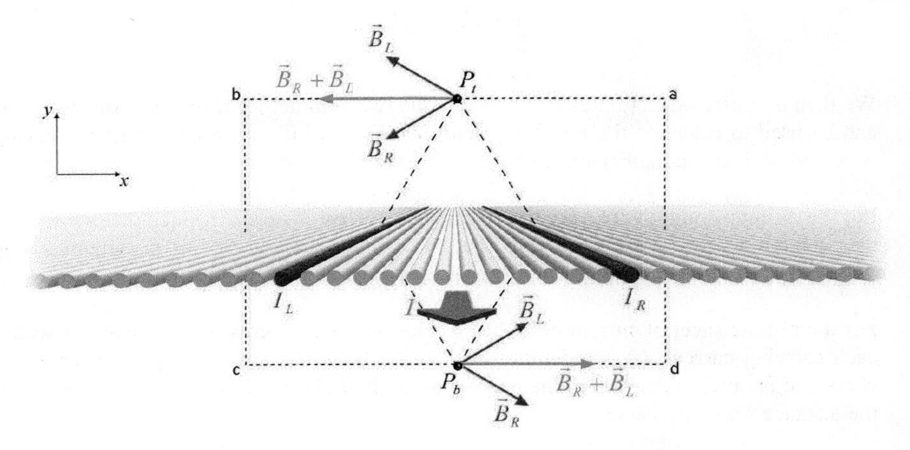

FIGURE 15.6 A model of an infinite current sheet as n wires per unit width, each carrying current I. The rectangle *abcd* can be used to calculate the $\vec{B}$ field at P_t using Ampère's law.

The current enclosed by the path is just equal to the number of wires (nL) times the current in each wire (I). Consequently, when we evaluate Ampère's law for this path, we obtain the result

$$B = \frac{1}{2}\mu_o nI$$

Note that this result for $\vec{B}$ is independent of the distance from the current sheet. The infinite current sheet produces a constant magnitude field everywhere in space, much like the infinite sheet of charge produced a constant electric field everywhere in space.

15.8 Summary

In this unit, we introduced Ampère's law, a fundamental law that relates the integral of the magnetic field around a closed path to the current passing through the area defined by that path.

We began by evaluating the line integral of the magnetic field produced by an infinite straight current-carrying wire around a variety of paths to illustrate the applicability of Ampère's law. In particular, we used a single circle centered on the wire and two sets of semicircles centered on the wire to demonstrate that the integral depended only on whether or not the wire was enclosed within the path.

We then pointed out how this result can be generalized to apply to any number of currents passing through any arbitrary closed loop. This final result is called Ampère's law:

$$\oint_{loop} \vec{B} \cdot d\vec{l} = \mu_o I_{enclosed}$$

We then investigated examples that illustrate the two symmetries for which Ampère's law can be used to calculate the magnetic field; namely, an infinite straight current-carrying wire (cylindrical symmetry) and an infinite sheet of current (planar symmetry).

For the infinite straight wire of radius a, carrying current I, we determined that the field at the center of the wire was zero and increased linearly to a maximum value $B_{max} = \mu_o I/(2\pi a)$ at the surface and then decreased as $1/r$ for all values of $r > a$.

For the infinite sheet of current described as closely packed wires (n wires per unit width, each carrying current I), we determined that the field was always perpendicular to the wires and parallel to the sheet. The magnitude of the field was a constant, independent of the distance from the sheet.

Main Points

Ampère's Law

A fundamental law that relates the path integral of the magnetic field around any closed loop to the total current enclosed by that loop.

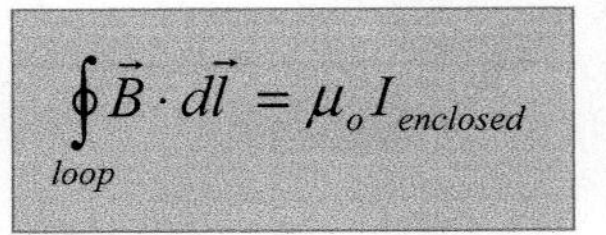

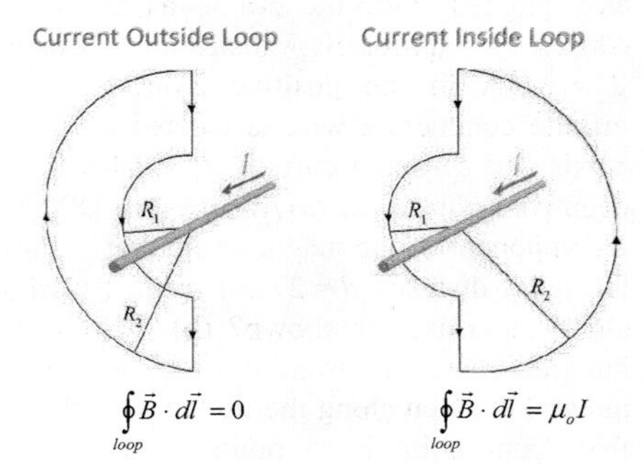

Ampère's Law Can Be Used to Calculate the Magnetic Fields Produced by Sufficiently Symmetric Current Distributions

A solid cylindrical wire of radius a carrying a uniform current I:

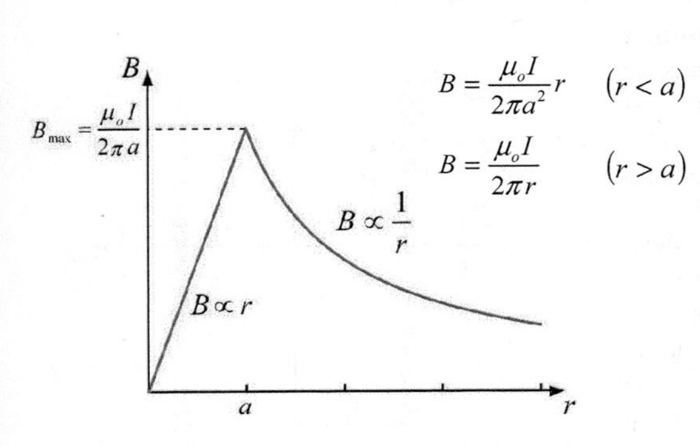

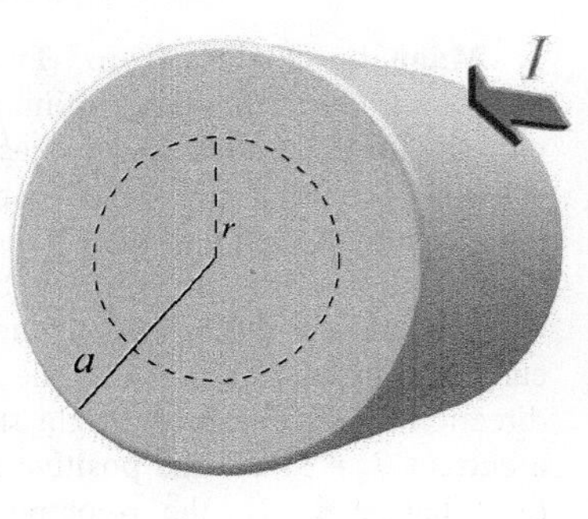

An infinite sheet of current (n wires/width each carrying current I):

$$B = \frac{1}{2}\mu_o nI$$

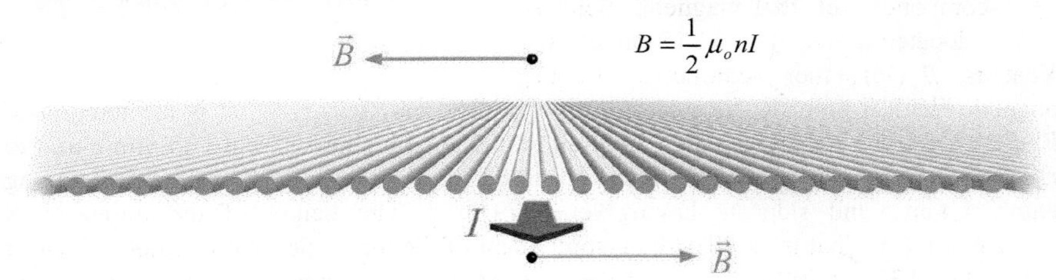

PROBLEMS

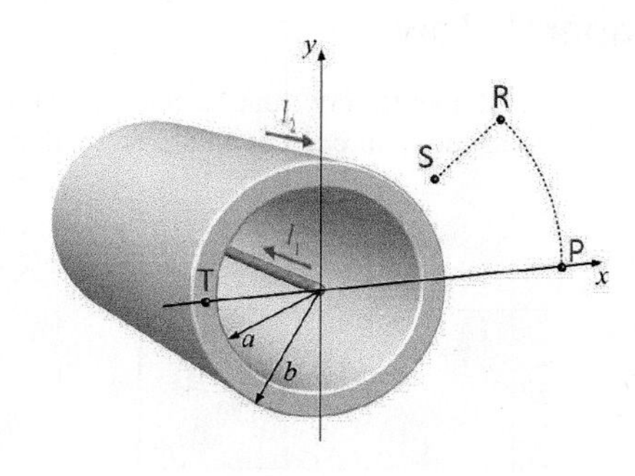

FIGURE 15.7 Problem 1

1. Magnetic Fields from Currents in a Wire and a Cylindrical Shell: A solid cylindrical conducting shell of inner radius $a = 5.6$ cm and outer radius $b = 7.9$ cm has its axis aligned with the z-axis as shown. It carries a uniformly distributed current $I_2 = 6.2$ A in the positive z-direction. An infinite conducting wire is located along the z-axis and carries a current $I_1 = 2.1$ A in the negative z-direction. (a) What is $B_y(\text{P})$, the y-component of the magnetic field at point P, located a distance $d = 27$ cm from the origin along the x-axis as shown? (b) What is the integral of $\vec{B} \cdot d\vec{l}$ from P to S, where the integral is taken along the dotted path shown, first from point P to point R at $(x, y) = (0.707d, 0.707d)$, and then to point S at $(x, y) = (0.6d, 0.6d)$? (c) What is $B_y(\text{T})$, the y-component of the magnetic field at point T, located at $(x, y) = (-6.5\text{ cm}, 0)$, as shown? (d) What is the integral of $\vec{B} \cdot d\vec{l}$ from S to P, where the integral is taken along a straight line path from point S to point P? (e) Suppose the magnitude of the current I_2 is now doubled. How does the magnitude of the magnetic field at $(x, y) = (2.8\text{ cm}, 0)$ change: $B(2.8\text{ cm}, 0)$ *increases*, $B(2.8\text{ cm}, 0)$ *decreases*, or $B(2.8\text{ cm}, 0)$ *remains the same*?

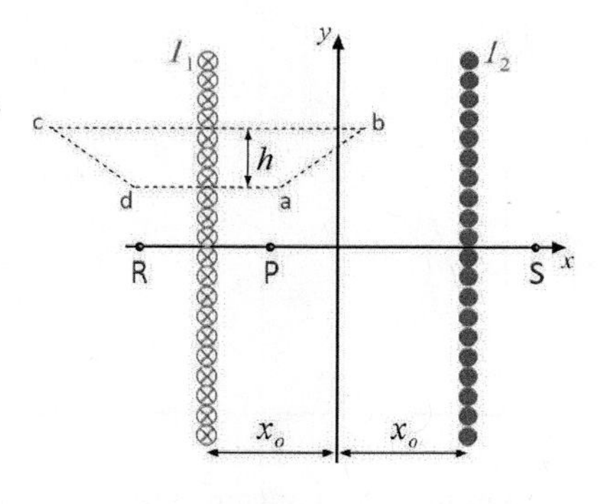

FIGURE 15.8 Problem 2

2. Magnetic Fields from Two Infinite Sheets of Current: Two infinite sheets of current flow parallel to the y-z plane as shown. The sheets are equally spaced from the origin by $x_o = 4.9$ cm. Each sheet consists of an infinite array of wires with a density $n = 14$ wires/cm. Each wire in the left sheet carries a current $I_1 = 2.2$ A in the negative z-direction. Each wire in the right sheet carries a current $I_2 = 3$ A in the positive z-direction. (a) What is $B_x(\text{P})$, the x-component of the magnetic field at point P, located at $(x, y) = (-2.45\text{ cm}, 0)$? (b) What is $B_y(\text{P})$, the y-component of the magnetic field at point P, located at $(x, y) = (-2.45\text{ cm}, 0)$? (c) What is $B_y(\text{R})$, the y-component of the magnetic field at point R, located at $(x, y) = (-7.35\text{ cm}, 0)$? (d) What is the integral of $\vec{B} \cdot d\vec{l}$, where the integral is taken around the dotted path shown, from a to b to c to d to a? The path is a trapezoid with sides ab and cd having length 12.8 cm, side ad having length 5.4 cm, and side bc having length 9.9 cm. The height of the trapezoid is $h = 12.6$ cm. (e) What is $B_y(\text{S})$, the y-component of the magnetic field at point S, located at $(x, y) = (7.35\text{ cm}, 0)$? (f) What is the integral of $\vec{B} \cdot d\vec{l}$, where the integral is taken along the dotted line shown, from a to b?

3. *B* Cylinders (INTERACTIVE EXAMPLE): Two very long coaxial cylindrical conductors are shown in Figure 15.9. The inner cylinder has radius $a = 2$ cm and caries a total current of $I_1 = 1.2$ A in the positive z-direction. The outer cylinder has an inner radius $b = 4$ cm, outer radius $c = 6$ cm, and carries a current of $I_2 = 2.4$ A in the negative z-direction. You may assume that the current is uniformly distributed over the cross-sectional area of the conductors. What is B_x, the x-component of the magnetic field at point P that is located at a distance $r = 5$ cm from the origin and makes an angle of 30° with the x-axis?

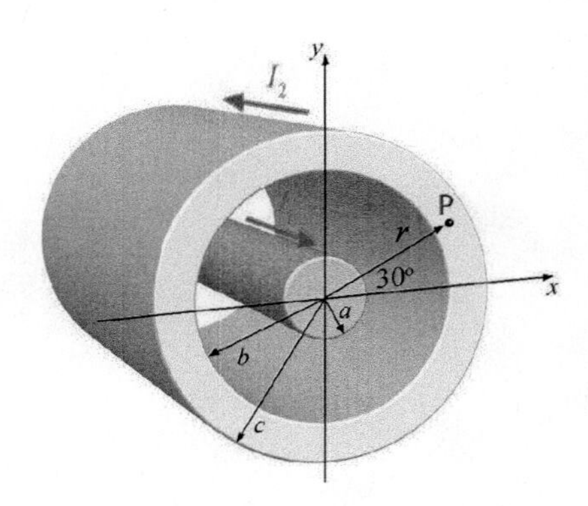

FIGURE 15.9 Problem 3

UNIT 16

MOTIONAL EMF

16.1 Overview

This unit marks the start of our study of **electrodynamics**. We will expand our study of electricity and magnetism to include situations in which $\vec{E}$ and $\vec{B}$ fields can change in time. We will discover qualitatively new phenomena that will require additions to be made to Maxwell's equations, the foundations of electricity and magnetism.

To begin, we will apply the Lorentz force law in the example of a conducting bar moving through a uniform magnetic field. We will discover that this force will cause a redistribution of charge in the bar, resulting in the creation of a potential difference, also called an *emf*, in the bar.

We will then consider two more examples, a conducting loop moving through the $\vec{B}$ field produced by an infinite straight current-carrying wire, and a conducting coil that rotates in a region of constant magnetic field. In each of these situations, we will find that *emf*'s will be generated in the moving conductors.

One reassuring comment is appropriate here. We will find that the geometry involved in these examples can sometimes get a bit involved. The good news is that the results for all of these examples can more easily be obtained using Faraday's law, which we will introduce next time. We choose to discuss motional *emf*'s first to place Faraday's law in context: that while *emf*'s generated when a conductor is in motion through a magnetic field (motional *emf*'s) can be understood either in terms of the Lorentz force causing a redistribution of the conduction electrons or Faraday's law, Faraday's law is more general; *emf*'s can be generated even when there is no motion of a conductor in a magnetic field, as long as the magnetic flux is changing in time.

16.2 Electrodynamics

To this point, the main connection we have seen between electricity and magnetism arises from the fact that magnetic fields are generated by electric charges in motion (i.e., electric currents). There is in fact a much deeper connection between electric and magnetic fields that we have not yet seen because we have restricted ourselves to electrostatics and **magnetostatics**, i.e., to cases in which the sources of the fields (charges for $\vec{E}$ fields and currents for $\vec{B}$ fields) do not change in time.

We will now remove that restriction by first looking at cases in which magnetic fields can change in time. We will discover new phenomena. In particular, we will learn that changing magnetic fields (or more properly a changing magnetic flux) can actually create electric fields. In the next two units we will develop Faraday's law, the last of a set of four equations called Maxwell's equations that completely describe electricity and magnetism. In the following four units, we will extend this study to inductors and then to oscillatory circuits and finally general alternating-current circuits.

Following that we will introduce Maxwell's displacement current, which will lead to a modification of Ampère's law so that a changing electric flux can generate a magnetic field. We then will have arrived at the culmination of our study of electricity and magnetism, an understanding of the deep connections between electric and magnetic fields, which can explain the existence of electromagnetic waves and the identification of light as an electromagnetic phenomenon. That's where we're going. Let's get started.

16.3 Motional Emf

We'll start by examining what happens when a conducting bar is moved with constant velocity $\vec{v}$ to the right through a region of constant magnetic field $\vec{B}$ pointing out of the page, as shown in Figure 16.1.

We know the conducting bar contains neutral atoms composed of electrons, protons, and neutrons. These particles have a net velocity to the right with the bar. The charged particles (protons and electrons) will feel a magnetic force ($\vec{F} = q\vec{v} \times \vec{B}$) due to this motion. In particular, an upward force will be exerted on the electrons and a downward force will be exerted on the protons. These forces will cause the conduction electrons to move upward. This redistribution of the charges within the conducting bar will establish an electric field $\vec{E}$ that points up. This electric field then will exert another force $\vec{F} = q\vec{E}$ on the charged particles in the bar. Since $\vec{E}$ points up, the force on the electrons due to the electric field opposes the force due to the magnetic field. Indeed, the redistribution of charges will stop once the electric field is strong enough that the *net force* on the electrons due to both the electric and magnetic fields is zero. Indeed, at equilibrium, the magnitudes of the magnetic force (qvB) and the electric force (qE) must be equal, which implies that an electric field with magnitude $E = vB$ must exist within the conducting bar.

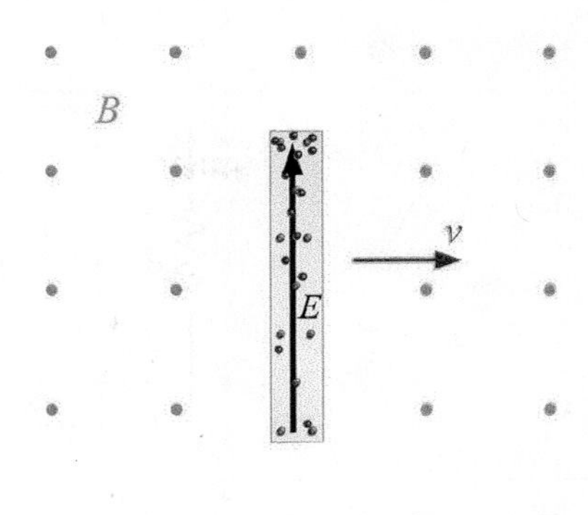

FIGURE 16.1 A conducting bar moves with speed v through a region of constant magnetic field $\vec{B}$, pointing out of the page. An induced electric field $\vec{E}$, directed upward, is created in the bar. This electric field is responsible for a potential difference (an *emf*) being generated between the ends of the bar.

You may recall that in all cases discussed so far in this book, the electric field inside a conductor has been equal to zero. Here, we have a non-zero $\vec{E}$ field in the conductor. What is different? Recall that the fundamental reason the electric field was zero inside a conductor was that the net force on the charges in the conductor had to be zero or else they would simply move until this was the case. We have exactly the same situation here—the charges in the conductor will move until the total force on them, electric *plus* magnetic in this case, is zero.

This induced electric field in the conductor creates a potential difference between the top and the bottom of the conductor given by

$$V = EL = vBL$$

This potential difference, or **motional *emf***, is real. For example, if we supply conducting rails and a resistor, as shown in Figure 16.2, this *emf* can drive a current in the completed circuit; the moving conductor effectively acts as a battery.

16.4 Power Considerations

We have just seen that if a conducting rod moves with constant velocity in a uniform magnetic field, as shown in Figure 16.2, a potential difference is produced between the ends of the rod. The magnitude of this potential difference is equal to the product of the magnetic field, the velocity, and the length of the rod. If this rod is placed on conducting rails connected by a resistor, this potential difference, or *motional emf*, will cause a current to flow in this circuit in the clockwise direction. We can determine the magnitude of this current from Kirchhoff's voltage rule. Namely, as we go around the loop in the clockwise direction, we see that the voltage drop across the resistor plus the motional *emf* must be zero. Consequently, the current in the circuit is just equal to the motional *emf* divided by the resistance.

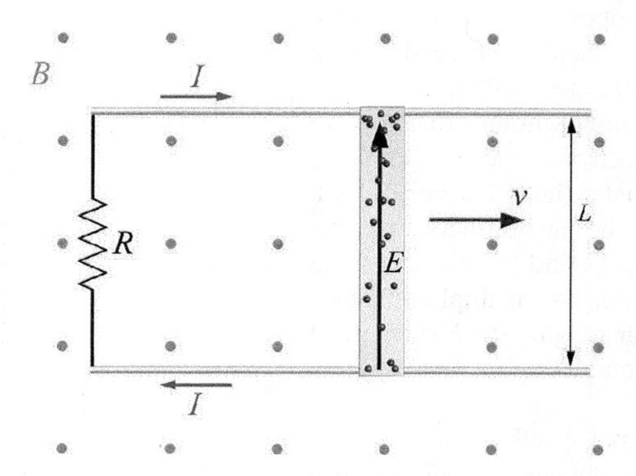

FIGURE 16.2 A conducting bar of length L moves on conducting rails with speed v through a region of constant magnetic field $\vec{B}$, pointing out of the page. The induced *emf* produces a clockwise current in the circuit.

$$I = \frac{V}{R} = \frac{vBL}{R}$$

Let's now look at the energy in this circuit. Energy is being provided by the *emf* and is being dissipated as heat in the resistor. Clearly the rate at which energy is being provided to the circuit (namely, the product of the current and the *emf*) is equal to the rate at which it is being dissipated in the resistor (namely, the product of the voltage across the resistor and the current), as it must be if energy is conserved.

$$P_{emf} = VI = (vBL)\left(\frac{vBL}{R}\right) = \frac{(vBL)^2}{R}$$

$$P_{dissipated} = I^2 R = \left(\frac{vBL}{R}\right)^2 R = \frac{(vBL)^2}{R}$$

The more interesting question is: Where is this energy that is being provided by the *emf* actually coming from? We can understand the source of this energy by realizing that the current that is flowing through the rod will also give rise to a Lorentz force on the rod since the charges in the rod are moving through a region containing a magnetic field. The direction of this force is to the left, as shown in Figure 16.3, in the opposite direction to the motion of the rod. In other words, this force will cause the rod to slow down. Consequently, if the rod is to maintain its constant velocity, an external agent must provide a force in the direction of motion of the rod that just cancels the Lorentz force. Indeed, the power provided by the external agent is just equal to the force exerted by the agent times the velocity of the rod. This power is exactly equal to the power provided to the circuit by the *emf*.

$$P_{external} = F_{external} v = (IBL)v = \left(\frac{vBL}{R}\right)BLv = \frac{(vBL)^2}{R}$$

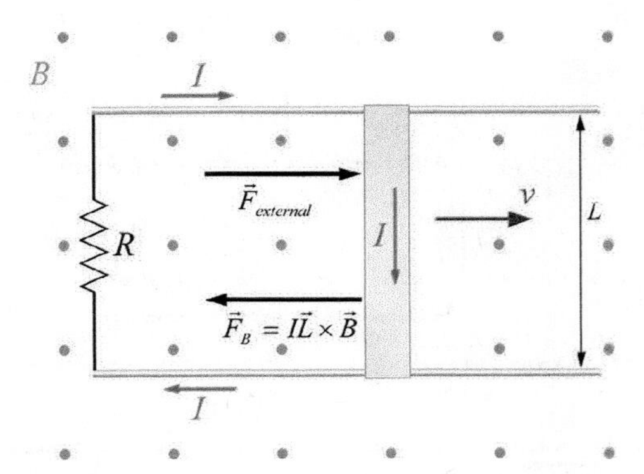

FIGURE 16.3 Since there is a current I in the conducting bar that is moving with constant speed through a region of constant magnetic field, there will be a magnetic force exerted on the bar. This force ($I\vec{L}\times\vec{B}$) is opposed to the direction of motion. Consequently, for this bar to move with constant velocity, there must be another force ($\vec{F}_{external}$) exerted in the direction of motion that just cancels the magnetic force. This external force provides the energy to the circuit that is ultimately dissipated as heat in the resistor.

16.5 Wire Loop Moving through a Non-Uniform Magnetic Field

In our last example we considered the motion of a conductor in a uniform magnetic field. We now want to look at what happens when we move a conductor through a non-uniform magnetic field.

Consider a rectangular wire loop of width w and length L moving away from an infinite straight wire carrying current I_o, as shown in Figure 16.4. The magnetic field at all points in the wire loop is directed out of the page. The corresponding magnetic force on all protons in the loop is to the right, while the force on all electrons in the loop is to the left. Since these protons and electrons are constrained to remain in the loop, the effect of this force is to create an electric field in the top and bottom segments of the loop in exactly the same way as the electric field in the conducting bar was created when it was moved through a magnetic field.

The important difference here is that the magnetic field produced by the current I_o falls off as $1/r$ so that the $\vec{B}$ field at the position of the top segment of the loop is smaller than the $\vec{B}$ field at the position of the bottom segment of the loop. Consequently, the $\vec{E}$ field induced in the top segment is less than the $\vec{E}$ field induced in the bottom segment. Since both $\vec{E}_{top}$ and $\vec{E}_{bottom}$ have the same direction (to the left), the *emf* induced in the loop is equal to the segment length L times the difference between $\vec{E}_{top}$ and $\vec{E}_{bottom}$.

$$\mathcal{E}_{loop} = L(E_{bottom} - E_{top}) = vL(B_{bottom} - B_{top})$$

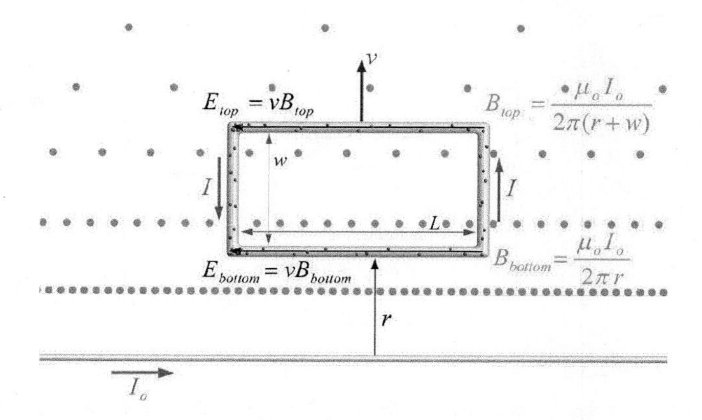

FIGURE 16.4 A rectangular loop moves with speed v away from a long straight wire carrying current I_o. The magnetic field on the bottom segment is bigger than that on the top segment, which leads to a larger electric field in the bottom segment than in the top segment. Consequently, an *emf* is induced in the loop, giving rise to a counterclockwise current.

If the loop has a finite resistance, a current will flow in the counterclockwise direction.

As opposed to the previous example, this induced *emf* is not constant in time. As the loop moves, the difference between the $\vec{B}$ field at the top segment and the bottom segment decreases, leading to a decrease in the induced *emf* as a function of time. Figure 16.5 shows this time dependence for the case that r_o, the initial position of the bottom segment, is equal to w.

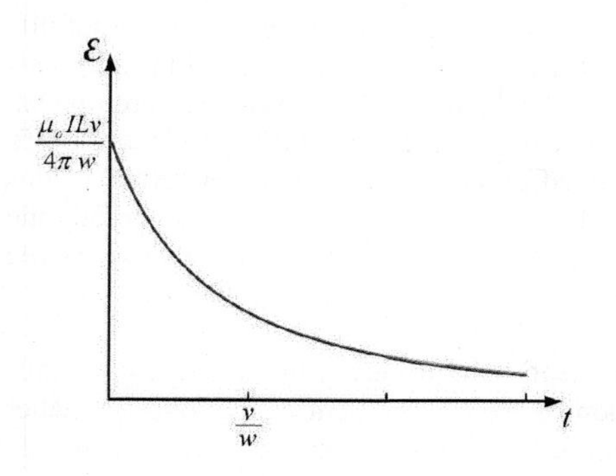

FIGURE 16.5 The *emf* induced in the loop in Figure 16.4 as a function of time.

16.6 The Generator

In the initial case of the conducting bar moving through a constant magnetic field, we demonstrated a conversion of mechanical energy (the force exerted by an external agent on the conducting bar over some distance) into electrical energy. The device usually used to make this conversion is a **generator** that is constructed by rotating a conducting loop (usually a coil made of many turns) in a constant magnetic field. We now want to show that this rotation can also create a motional *emf* that can be used to provide electrical power.

In Figure 16.6, we see a rectangular conducting loop rotating counterclockwise in a region of constant magnetic field. Let's now pause the rotation at an instant in which the loop makes an angle θ with respect to the magnetic field, as shown in Figure 16.7, and calculate the forces on the charge carriers due to their motion in this magnetic field. We see that the force on the electrons in the top segment is toward the front, while the force on the electrons in the bottom segment is toward the back. Consequently, we expect the electrons in the top segment to move to create an electric field that points to the front, while the electrons in the bottom segment will move to create an electric field that points to the back. The force on the electrons in the front and back segments, on the other hand, will not be directed along the segment so that we expect no redistribution of charge in these segments.

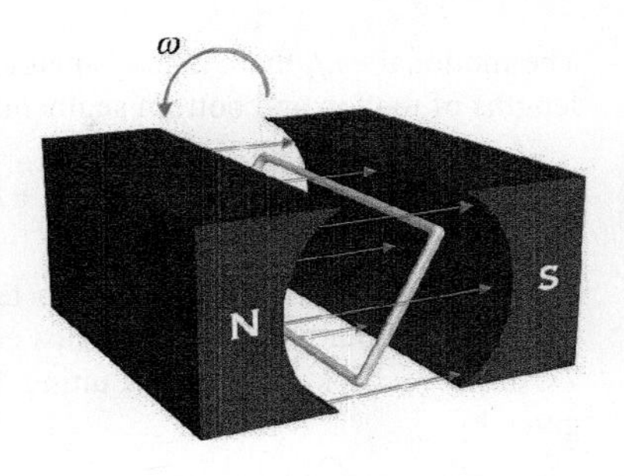

FIGURE 16.6 A rectangular loop rotating counterclockwise at angular velocity ω through a region of constant magnetic field $\vec{B}$.

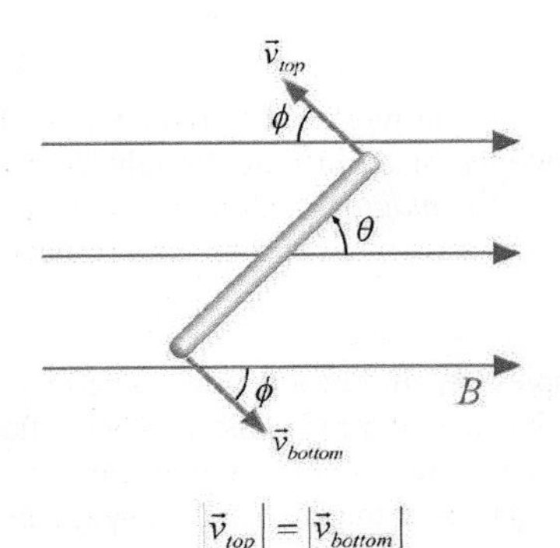

FIGURE 16.7 Side view of the rotating coil shown in Figure 16.6. The Lorentz force ($q\vec{v}\times\vec{B}$) on the charge carriers (the electrons) in the top segment is toward the front, while the force on the electrons in the bottom segment is toward the back. Consequently, the electrons in the top and bottom segments will redistribute themselves to create an *emf* in the loop.

The electric fields in the top and bottom segments point in the opposite directions so that when we go around the loop, the potential differences produced by these electric fields add together to produce a non-zero motional *emf*, as promised.

Let's now calculate the magnitude of this *emf*. The magnitude of the force on an electron in both the top and bottom segments is given by

$$F = \left|q\vec{v}\times\vec{B}\right| = qvB\sin\phi = qvB\cos\theta$$

We now determine the final electric field produced in each segment by setting the magnitude of the electric force (qE) equal to the magnitude of this magnetic force to obtain the result:

$$E = vB\cos\theta$$

The motional *emf*, then, is just equal to the product of this electric field and the sum of the lengths of the top and bottom segments.

$$\mathcal{E} = E(L_{top} + L_{bottom}) = 2EL = 2vBL\cos\theta$$

We can express our final result in terms of the rotational frequency ω, by noting that $\theta = \omega t$, and that the velocity is just equal to angular frequency times half of the length of the front or back segments. Putting this altogether, we obtain that the motional *emf* is given by

$$\mathcal{E} = \omega AB\cos(\omega t)$$

where A is equal to the area of the loop.

16.7 Connections

We've just demonstrated in three different examples how a conductor moving in a magnetic field can generate an *emf*. We now want to examine our results to find a connection between these three examples.

In the moving conducting bar example, we see that as the bar moves, the *area enclosed* by the circuit *increases*. In the moving loop in the vicinity of an infinite straight current-carrying wire example, we see that as the loop moves, the *magnetic field* through its area *decreases*. In the rotating loop example, the *relative orientation* of the loop with respect to the magnetic field *changes in time*.

What is the common link between these three examples? Well, in each case, something is changing in time; either the area enclosed by the bar and the rails, the magnetic field through the loop, or the orientation of the loop and the magnetic field. We will learn in the next unit that all of these changes are just different aspects of the change of a single new quantity, the magnetic flux. Indeed, we will be able to use this concept of the magnetic flux to formulate Faraday's law, one of the four fundamental laws of electricity and magnetism.

16.8 Summary

In this unit we investigated forces on the charge carriers in conductors in regions containing magnetic fields. In particular, we discussed (1) a conducting bar moving through a constant magnetic field, (2) a conducting loop moving away from an infinite straight current-carrying wire, and (3) a rotating conducting loop in a region containing a constant magnetic field. In all of these cases, we found that the Lorentz force on the charge carriers produced a redistribution of the conduction electrons so that a potential difference, a motional *emf*, was created in each if the conductors.

We stated, but did not prove, that the motional *emf* produced in each of these examples could be written simply as the time rate of change of a new quantity, the magnetic flux through the circuit. We will prove these claims in the next unit in which we introduce Faraday's law, one of the fundamental laws of electricity and magnetism. Faraday's law explains all the results in this unit and more! Indeed, Faraday's law represents the first important step in establishing the deep connections between electric and magnetic fields, which ultimately will explain the existence of electromagnetic waves and the identification of light as an electromagnetic phenomenon.

MAIN POINTS

Motional EMF

The Lorentz force on charge carriers in a conductor in a magnetic field can give rise to a redistribution of the conduction electrons so that a potential difference, a motional *emf*, can be created in that conductor.

Examples

Conducting Bar on Rails

Motional EMF: $\mathcal{E} = vLB$

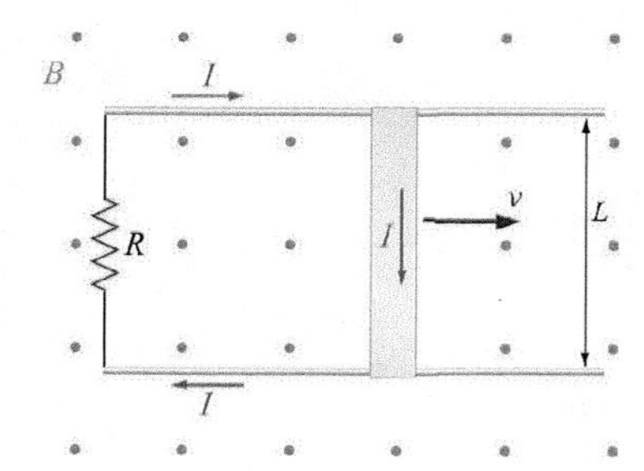

Wire Loop and a Straight Current

Motional EMF: $\mathcal{E} = vL(B_{bottom} - B_{top})$

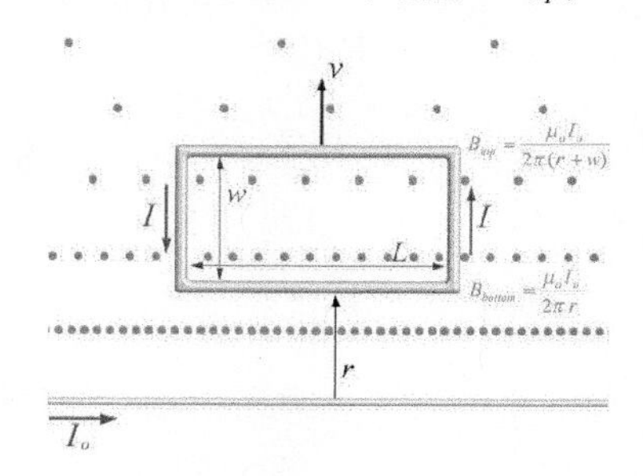

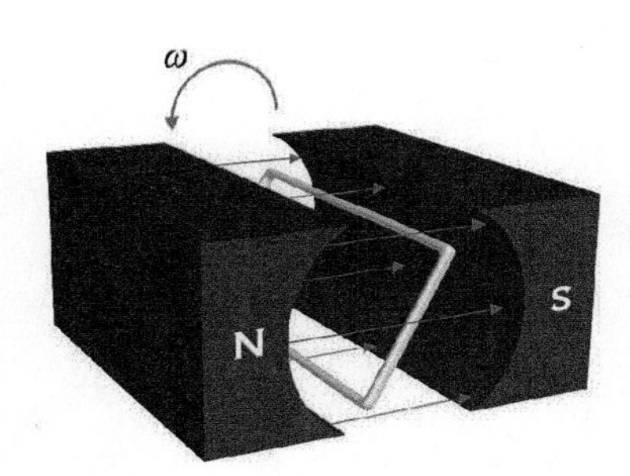

Generators

Motional EMF: $\mathcal{E} = \omega AB\cos(\omega t)$

PROBLEMS

1. Conducting Loop Moving through Constant Magnetic Field: A conducting loop is made in the form of two squares of sides $s_1 = 3.6$ cm and $s_2 = 6.5$ cm as shown. At time $t = 0$, the loop enters a region of length $L = 18.7$ cm that contains a uniform magnetic field $B = 1.7$ T, directed in the positive z-direction. The loop continues through the region with constant speed $v = 32$ cm/s. The resistance of the loop is $R = 2.9\ \Omega$. (a) At time $t = t_1 = 0.038$ s, what is the magnitude and direction of I_1, the induced current in the loop? (b) At time $t = t_2 = 0.765$ s, what is the magnitude and direction of I_2, the induced current in the loop? (c) What is $F_x(t_2)$, the x-component of the force that must be applied to the loop to maintain its constant velocity $v = 32$ cm/s at $t = t_2 = 0.765$ s? (d) At time $t = t_3 = 0.622$ s, what is the magnitude and direction of I_3, the induced current in the loop?

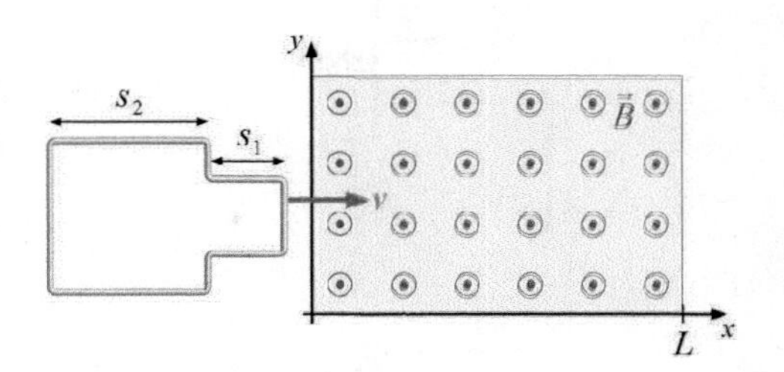

FIGURE 16.8 Problem 1

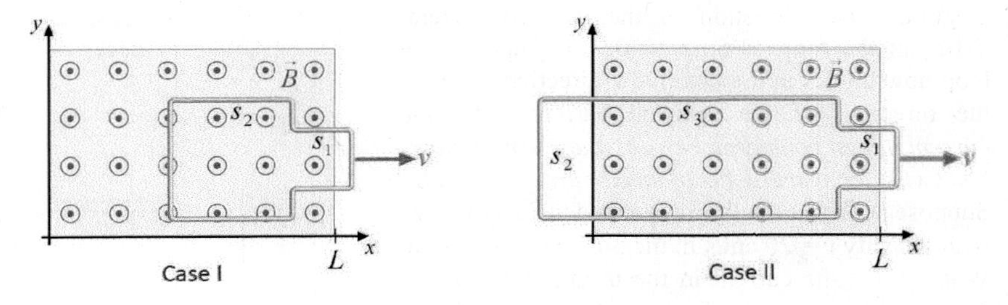

FIGURE 16.9 Problem 1 part (e)

(e) Consider the two cases shown above. How does I_{I}, the magnitude of the induced current in Case I, compare to I_{II}, the magnitude of the induced current in Case II?

(i) $I_{\text{I}} < I_{\text{II}}$
(ii) $I_{\text{I}} = I_{\text{II}}$
(iii) $I_{\text{I}} > I_{\text{II}}$

2. Conducting Loop and Current-Carrying Wire: An infinite straight wire carries current $I_1 = 3.9$ A in the positive y-direction as shown. At time $t = 0$, a conducting wire, aligned with the y-direction is located a distance $d = 52$ cm from the y-axis and moves with velocity $v = 20$ cm/s in the negative x-direction, as shown in Figure 16.10. The wire has length $w = 27$ cm. (a) What is $\mathcal{E}(0)$, the *emf* induced in the moving wire at $t = 0$? Define the *emf* to be positive if the potential at point a is higher than that at point b. (b) What is $\mathcal{E}(t_1)$, the *emf* induced in the moving wire at $t = t_1 = 1.7$ s? Define the *emf* to be positive if the potential at point a is higher than that at point b. (c) The wire is now

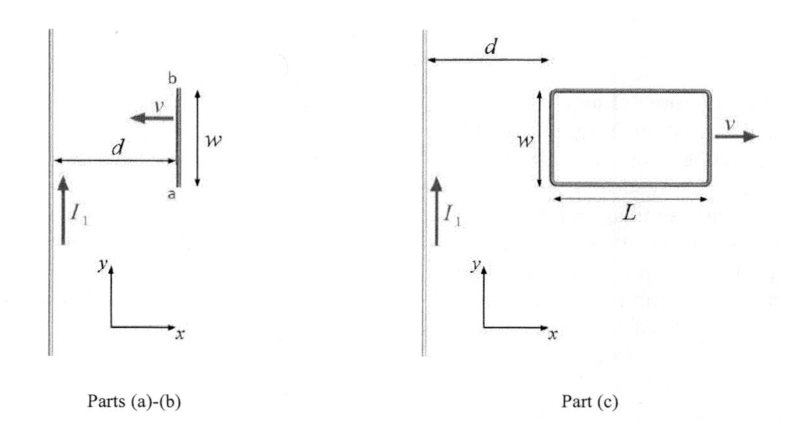

FIGURE 16.10 Problem 2

replaced by a conducting rectangular loop as shown. The loop has length $L = 60$ cm and width $w = 27$ cm. At time $t = 0$, the loop moves with velocity $v = 20$ cm/s with its left end located a distance $d = 52$ cm from the y-axis. The resistance of the loop is $R = 1.2\ \Omega$. What is the magnitude and direction of the induced current, $I(0)$, in the loop at time $t = 0$? (d) Suppose the loop now moves in the positive y-direction. What is the direction of the induced current now: *the current flows counterclockwise, the current flows clockwise*, or *there is no induced current now*? (e) Suppose now that the loop is rotated 90° and moves with velocity $v = 20$ cm/s in the positive x-direction. What is I_2, the current in the infinite wire, if the induced current in the loop at the instant shown ($d = 52$ cm) is the same as it was in the third part of this problem (i.e., when the left end of loop was at a distance $d = 52$ cm from the y-axis)?

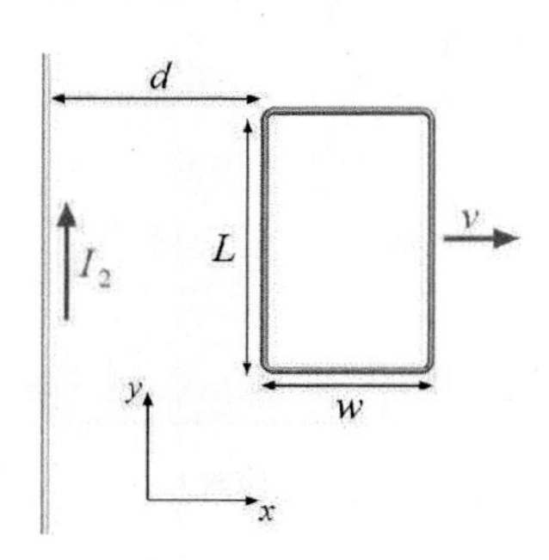

FIGURE 16.11 Problem 2 part (e)

3. Moving Coil (Interactive Example): At $t = 0$, a rectangular coil of resistance $R = 2\ \Omega$ and dimensions $w = 3$ cm and $L = 8$ cm enters a region of constant magnetic field $B = 1.6$ T directed into the screen as shown. The length of the region containing the magnetic field is $L_B = 15$ cm. The coil is observed to move at constant velocity $v = 5$ cm/s. What is the force required at time $t = 0.8$ s to maintain this velocity?

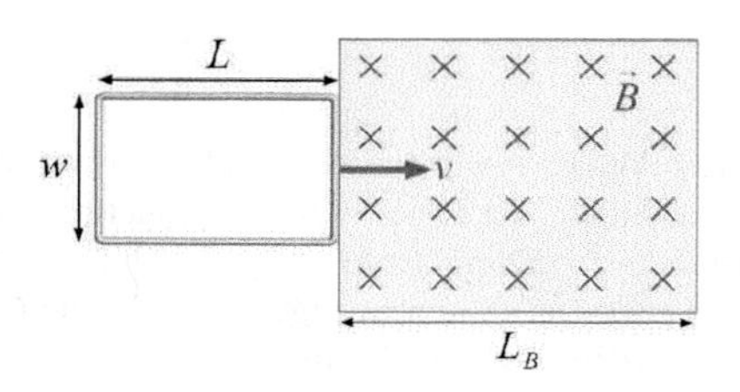

FIGURE 16.12 Problem 3

UNIT

17

FARADAY'S LAW

17.1 Overview

In this unit we will generalize the motional *emf* results from the last unit by introducing a new quantity, the magnetic flux, which will allow us to formulate Faraday's law, a new fundamental law of electricity and magnetism. Faraday's law attributes the production of these *emfs* to a changing magnetic flux. We call this a generalization of the motional *emfs* we saw last time because it is possible for the magnetic flux to change in time without needing a conductor to be in motion.

The sign of this *emf* is determined by energy conservation considerations. In particular, the *emf* itself creates a magnetic flux that opposes the change in flux that produced it. We will close by applying Faraday's law to situations in which no conductor is moving through a magnetic field.

17.2 Magnetic Flux

We can describe the motional *emf* produced in all three examples we discussed in the last unit in a simple, unified way by introducing a new quantity, the **magnetic flux**. The magnetic flux through the surface defined by the boundaries of a loop is defined exactly analogously to the electric flux. Namely, the magnetic flux Φ_B is defined as

$$\Phi_B \equiv \int_{surface} \vec{B} \cdot d\vec{A}$$

In the case of the moving conducting bar in Figure 16.2, $\vec{B}$ is parallel to $d\vec{A}$ so that the magnetic flux at any time t is simply given by the product of the magnetic field and the area. The time rate of change of the magnetic flux then is just equal to the product of the magnetic field and the time rate of change of the area swept out by the bar.

$$\Phi_B = BA \qquad \Rightarrow \qquad \frac{d\Phi_B}{dt} = B\frac{dA}{dt}$$

Now this time rate of change of the area is just equal to the product of the length of the bar and the speed of the bar. Consequently, we see that the time rate of change of the flux here is equal to the product of the magnetic field, the length of the bar and the speed of the bar.

$$\frac{d\Phi_B}{dt} = BLv$$

In the last unit, we demonstrated that the motional *emf* created in this circuit is equal to exactly this same combination of parameters. Therefore, we see for this case of the moving conducting bar in Figure 16.2, the *emf* induced in the circuit is just equal to the time rate of change of the magnetic flux.

We claim that the *emf* generated in the other two examples can also be written simply as the time rate of change of the magnetic flux and will demonstrate this fact in the next two sections.

17.3 Wire Loop Moving through a Non-Uniform Magnetic Field

In the case of the moving loop in the vicinity of an infinite straight current-carrying wire in Figure 16.4, we determined that the *emf* induced was proportional to ΔB, the difference in the magnetic field at the positions of the top and bottom wires.

If we define Δt to be the time it takes the bottom of the loop to move to the position of the top of the loop ($\Delta t = w / v$), we can rewrite the *emf* as

$$\mathcal{E} = A\frac{\Delta B}{\Delta t}$$

where A denotes the area of the loop. We will now demonstrate the quantity on the right-hand side of the equation is indeed the time derivative of the magnetic flux.

We start by applying the definition of the magnetic flux to this example.

$$\Phi_B \equiv \int_{surface} \vec{B} \cdot d\vec{A} = \int_{surface} B\, dA$$

Since the magnetic field is *not* a constant over the loop, we actually have to do the integral. Let's consider a small time *dt* during which the loop moves a distance equal to *v dt*. During this time, the magnetic flux through the loop has changed by an amount $d\Phi_B$, which can be calculated as the difference of the flux lost at the bottom ($B_{bottom} Lv\, dt$) and that gained at the top ($B_{top} Lv\, dt$).

$$d\Phi_B = Lv\, dt\, \Delta B \qquad \Rightarrow \qquad \frac{d\Phi_B}{dt} = Lv\Delta B$$

Replacing the constant velocity *v* as $w / \Delta t$, we obtain

$$\frac{d\Phi_B}{dt} = L\frac{w}{\Delta t}\Delta B = A\frac{\Delta B}{\Delta t}$$

which is once again the expression previously obtained for the *emf* induced in the loop.

17.4 Flux through a Rotating Loop (The Generator)

We will now look at the *emf* produced in a generator in terms of the magnetic flux that passes through the rotating loop. Let's assume the loop has area A and rotates at a constant angular frequency ω in a space containing a uniform magnetic field $\vec{B}$, as shown in Figure 16.6.

At time *t*, the plane of the coil makes an angle $\theta = \omega t$ with the magnetic field, as shown in Figure 16.7. The magnetic flux through the loop at time *t* is given by

$$\Phi_B \equiv \int \vec{B} \cdot d\vec{A} = \vec{B} \cdot \int d\vec{A} = \vec{B} \cdot \vec{A} = BA\sin(\omega t)$$

If we simply differentiate this expression with respect to time we obtain a form for the time rate of change of the flux:

$$\frac{d\Phi_B}{dy} = \omega BA\cos(\omega t)$$

This expression for the *emf* at time *t* is identical to the one we obtained in the last unit from considerations of the Lorentz force on the charge carriers in the loop.

What have we learned here? We have learned that in all three of these examples involving motional *emf*, we can write a form for the *emf* induced in the loop as just the time rate of change of the magnetic flux through the loop. In the next section we will discover that this expression is very general and indeed holds for induced *emfs* that are not generated by the motion of conductors through a magnetic field.

17.5 Faraday's Law

We've just seen that we can express the motional *emf* produced in all of these last three examples in terms of the time rate of change of the magnetic flux. The big question is: Is this fact just a coincidence or does it indicate that something more general is going on here. Namely, that anytime the magnetic flux changes, an *emf* will be produced?

We can address this question by re-examining the moving conducting loop example. Namely, in this case, the flux changed because the loop was moving through a non-uniform magnetic field. There is another way to change the flux through the loop, however. Suppose the loop is kept stationary, but the current in the infinite straight wire changes in time. This changing current gives rise to a changing magnetic field, which results in a changing magnetic flux through the loop. The exciting question is: Will this changing flux induce an *emf* in the stationary loop? The recasting of the motional *emf* into a form that makes no mention of motion, only of a changing flux, could indicate that a more general law is operating here, namely that a *changing flux* is all that is necessary to induce an *emf*!

To look at this situation in more detail, it's natural to ask the question of whether we can choose a time dependence for the current I so that the resulting magnetic flux through the stationary loop is identical to that when the loop was moving with speed v and the current in the straight wire was held constant at the value I_o.

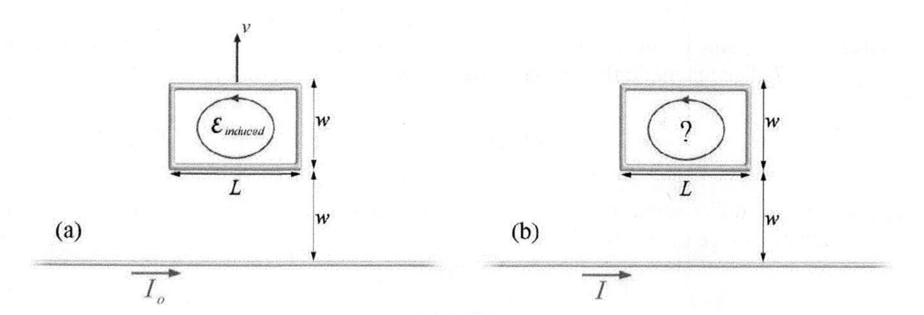

FIGURE 17.1 A comparison of two similar situations: (a) A rectangular conducting loop moves with velocity v away from a long straight wire carrying constant current I_o, inducing a motional *emf* in the loop. (b) The loop is stationary, but the current is not constant; it is chosen such that the time rate of change of the magnetic flux through the loop is the same as it was in (a).

The first step then is to calculate the magnetic flux through the moving loop. Let's assume that at $t = 0$, the bottom of the loop was at a distance w from the wire, as shown in Figure 17.1(a). The flux is determined by evaluating the integral of $\vec{B} \cdot d\vec{A}$ over the surface enclosed by the loop. B is just equal to $\mu_o I_o/(2\pi r)$ and dA is equal to $L\, dr$. We need to integrate r from the position of the bottom of loop at time t, which is given by $w + vt$, to the position of the top of the loop at time t, which is given by $2w + vt$. Doing the integral we obtain

$$\Phi_{moving}(t) = \int_{w+vt}^{2w+vt} \frac{\mu_o I_o}{2\pi r} L\, dr = \frac{\mu_o I_o L}{2\pi} \ln\left(\frac{2w+vt}{w+vt}\right)$$

For the stationary loop shown in Figure 17.1(b), the magnetic field at a distance r from the wire is given by $B = \mu_o I/(2\pi r)$, where I now will be a function of time t. dA is still $L\, dr$, but the limits of integration are now just w to $2w$, since the loop is stationary. Doing this integral, we obtain

$$\Phi_{stationary}(t) = \int_{w}^{2w} \frac{\mu_o I}{2\pi r} L\, dr = \frac{\mu_o I L}{2\pi} \ln 2$$

which indicates that that the time dependence of the flux is just that of the current I. If we equate the flux from the moving loop case with that of the stationary loop, we find

$$I = \frac{I_o}{\ln 2} \ln\left(\frac{2w+vt}{w+vt}\right)$$

While this time dependence looks unusual, if we plot it, as shown in Figure 17.2, we see that I is rather an unremarkable monotonically decreasing function of time.

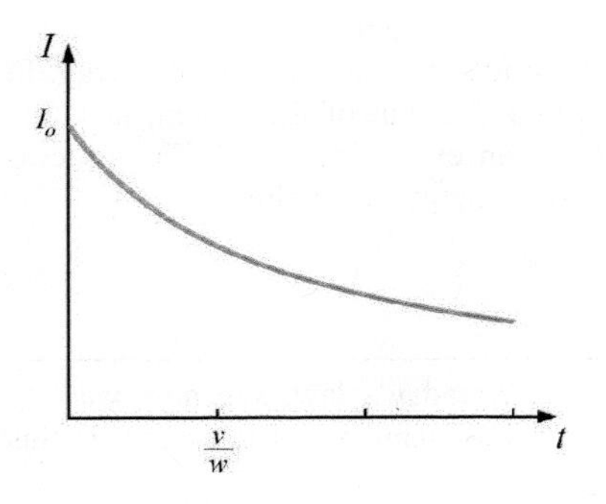

FIGURE 17.2 The time dependence of the current I in the long straight wire in Figure 17.1(b) that is needed to produce a flux in the stationary loop that is equivalent to that of the moving loop in Figure 17.1(a).

So what does all this mean? We've shown that by introducing a time-dependent current I in the straight wire, we can create exactly the same flux through the stationary loop as we did when the loop moved away from the straight wire when it carried a constant current. We now return to the big question: Is an *emf* induced in the stationary loop? The empirical answer is: Absolutely! This result indicates that the *emf* is really caused by the changing flux through the loop, no matter how we caused the flux to change. A reformulation of the motional *emf* result has led us to a new law of nature, that whenever magnetic flux changes, an *emf* is induced. The mathematical description

$$\mathcal{E} = -\frac{d\Phi_B}{dt}$$

is known as **Faraday's law** and is one of the four fundamental laws of electricity and magnetism. We will investigate this equation in more detail in the next section.

17.6 Lenz' Law

We've just introduced Faraday's law as

$$\mathcal{E} = -\frac{d\Phi_B}{dt}$$

We now want to investigate this mathematical expression in more detail. The first question is: What is the significance of the minus sign? This sign determines the sign of the *emf* which we identify as the direction of the current (clockwise or counterclockwise) that this *emf*, would produce if the loop were a complete circuit. This rule has its own name, **Lenz' law**, and can be stated simply that the *emf* is induced in a direction that opposes the change in flux that produced it.

Referring back to Figure 16.2, we see the conducting bar moving to the right on conducting rails in the region of a uniform magnetic field. In this case, the field is pointing out of the page, and as the bar moves, the area formed by the circuit is increasing. Therefore, the magnetic flux is increasing in the out-of-the-page direction. Then, by Lenz' law, we expect the flux produced by the induced *emf* to be into the screen to oppose the change in flux that brought it into being.

We should note that the basis of Lenz' law is just the idea of conservation of energy. In this case, for example, if the induced *emf* produced a flux out of the screen, this flux would add to the original flux, which would lead to an even bigger *emf*. This process could continue forever, giving rise to an ever-increasing energy in the circuit.

17.7 Faraday's Law: *E* and *B*

In the last section we focused on the minus sign in Faraday's law. We now want to address an even more important aspect of Faraday's law, namely, what exactly is this *emf*?

The answer is that this *emf* is a potential difference and we find it by integrating the electric field around the closed loop. In electrostatics, we know this integral is zero and that's what allows us to define an electric potential. Here, we have a non-zero result for the integral because we are no longer dealing with electrostatics: the changing magnetic field has induced an *emf*, which means it has actually created an electric field in the circuit, just as if we had inserted a battery into the circuit. If we replace the *emf* by the closed integral of $\vec{E} \cdot d\vec{l}$ around the loop, we obtain a new formulation of Faraday's law that only involves the fields $\vec{E}$ and $\vec{B}$. In particular, we say

$$\oint_{loop} \vec{E} \cdot d\vec{l} = -\frac{d\Phi_B}{dt}$$

where the integral is preformed around any closed loop. This formulation of Faraday's law is really amazing. It says that a changing magnetic flux will produce an electric field. It is a general statement relating electric and magnetic fields that applies even in empty space! This is the first step toward a more complete understanding of the interrelationship between electric and magnetic fields that will culminate in the production of electromagnetic waves.

17.8 Examples

We will close by looking at two more examples of the use of Faraday's law. We'll start with an example in which there is an induced *emf* that involves no motion of conductors through a magnetic field.

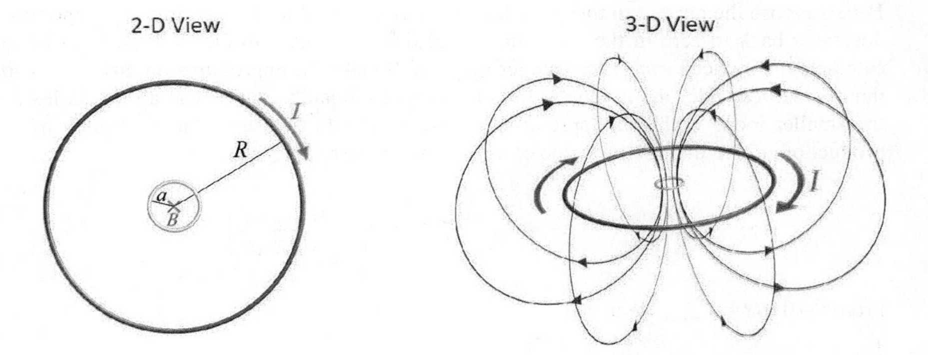

FIGURE 17.3 A circular conducting loop of radius R carries current I in the clockwise direction. A smaller circular conducting loop of radius a is centered on the center of the larger loop. An *emf* will be induced in the smaller loop if the current I changes in time.

Figure 17.3 shows a circular loop of radius R carrying current I in the clockwise direction. We know this current produces a magnetic field at the center of the loop, which points into the page and whose magnitude is proportional to the current and inversely proportional to the radius of the loop.

$$B_{center} = \frac{\mu_o I}{2R}$$

Also shown in Figure 17.3 is a smaller conducting circular loop of radius a at the center of the original loop. We want to know what *emf*, if any, will be induced in this smaller loop.

The magnetic field produced by the larger loop certainly gives rise to a non-zero magnetic flux through the smaller loop. From Faraday's law, we know an *emf* will be induced in the smaller loop if the flux changes in time. The flux will change in time if the current in the larger loop changes in time.

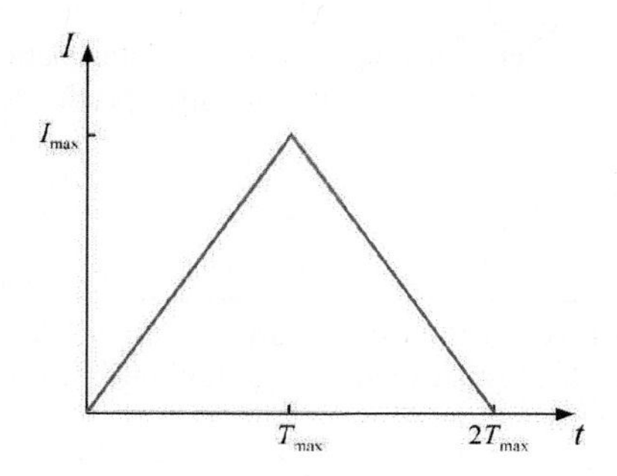

FIGURE 17.4 A plot of the time dependence of the current I in Figure 17.3. This changing current gives rise to a changing magnetic flux through the smaller ring, which results in an *emf* induced in the smaller ring.

Let's suppose the current in the outer loop increases from 0 to I_{max} in time T_{max} and then decreases back to zero in the same amount of time, as shown in Figure 17.4. In order to calculate the induced *emf* in the smaller loop, we'll make the approximation that $a << R$ so that we can consider the magnetic field to be approximately constant at all points inside the smaller loop. With this approximation, we obtain an *emf* that is proportional to the product of a^2 / R and the time rate of change of the current.

$$\mathcal{E} = -\frac{d\Phi_B}{dt} \approx -A\frac{dB_{center}}{dt} = -(\pi a^2)\left(\frac{\mu_o}{2R}\frac{dI}{dt}\right)$$

From $t = 0$ to $t = T_{max}$, we have

$$\mathcal{E} \approx -\frac{\mu_o \pi a^2}{2R}\frac{I_{max}}{T_{max}}$$

From $t = T_{max}$ to $t = 2T_{max}$, we have

$$\mathcal{E} \approx +\frac{\mu_o \pi a^2}{2R}\frac{I_{max}}{T_{max}}$$

What is the significance of the sign of the *emf* here? The sign of the *emf* indicates the direction of the current flow. As the magnitude of the clockwise current in the larger loop increases, a counterclockwise current is induced in the smaller loop. This induced current produces a field through the smaller loop that points out of the screen, opposing the change in flux that produced it. As the magnitude of the clockwise current in the larger loop decreases, a clockwise current is induced in the smaller loop to oppose the change in flux that produced it.

We'll now take one more look at an alternating-current generator to make sure we understand its use as an important device that transforms mechanical energy into electrical energy.

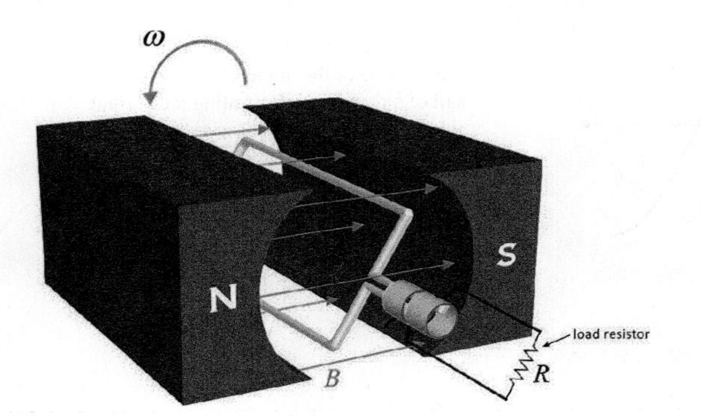

FIGURE 17.5 An alternating current generator. A conducting coil of many turns is made to rotate in a region of constant magnetic field. The coil is connected to a load using slip rings and brushes.

Figure 17.5 shows a generator consisting of a conducting coil of area A, with many turns (N) that is made to rotate in a region of space containing a uniform magnetic field $\vec{B}$ at a constant angular frequency ω. We know from Faraday's law that an *emf* will be induced in the rotating coil since the orientation of the coil relative to the magnetic field changes in time, creating a changing flux through the coil.

$$\Phi_B = NBA\sin(\omega t) \qquad \Rightarrow \qquad \mathcal{E} = -\omega NBA\cos(\omega t)$$

How exactly does this *emf* change in time? Figure 17.6 shows the side and front views of the coil at a moment when θ is increasing, which leads to an increase in flux (out of the screen in the front view). The induced *emf* will therefore oppose this increase and will be *clockwise*. Once the coil passes $\theta = 90°$, the flux begins to decrease, which will cause the induced *emf* to become counterclockwise.

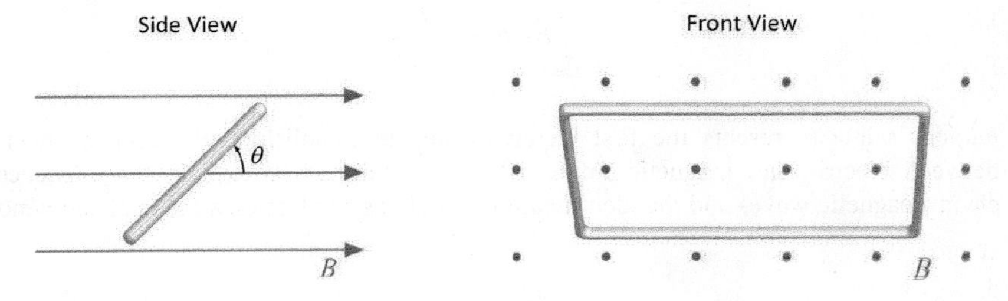

FIGURE 17.6 The side and front views of the rotating coil in Figure 17.5 at a time when θ is increasing, which leads to an increase in flux (out of the page in the front view).

A plot of the time dependence of this induced *emf* is shown in Figure 17.7, demonstrating this alternating feature between clockwise and counterclockwise directions for the *emf.* If this coil is connected to a load resistor through slip rings and brushes, for example, the current in this resistor will be an **alternating current** at angular frequency ω.

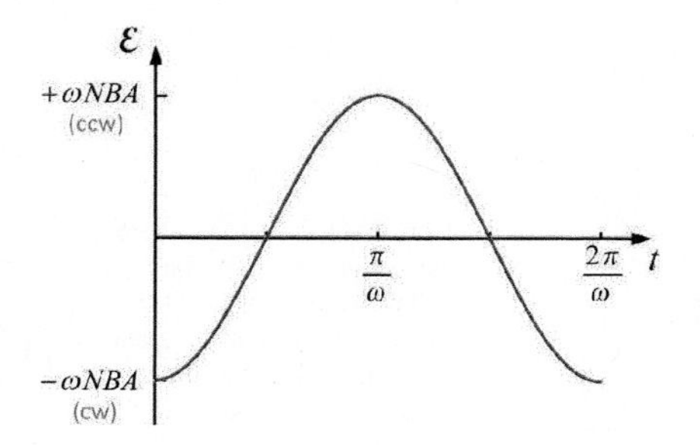

FIGURE 17.7 A plot of the time dependence of the induced current in the coil of a generator, illustrating the current alternating between clockwise and counterclockwise directions.

17.9 Summary

In this unit we first defined the concept of magnetic flux and found that the motional *emfs* produced in the three examples from the last unit could all be written simply as the time rate of change of the magnetic flux through the circuit.

We then introduced Faraday's law that states that whenever magnetic flux changes in time, *not* just in the case of a moving conductor, an *emf* will be produced. In particular, this induced *emf* will just be equal to minus the time rate of change of the magnetic flux.

$$\mathcal{E} = -\frac{d\Phi_B}{dt}$$

Finally, we observed that this induced *emf* is determined by integrating the electric field around the loop, so that Faraday's law can be written more generally only in terms of the electric and magnetic fields. A changing magnetic flux creates an electric field.

$$\oint_{loop} \vec{E} \cdot d\vec{l} = -\frac{d\Phi_B}{dt}$$

Faraday's law represents the first important step in establishing the deep connections between electric and magnetic fields, which ultimately will explain the existence of electromagnetic waves and the identification of light as an electromagnetic phenomenon.

Main Points

Magnetic Flux

The magnetic flux through a surface is defined as the integral of $\vec{B}\cdot d\vec{A}$ over that surface. The geometric interpretation of the magnetic flux is that it simply counts the number of field lines that pass through the surface.

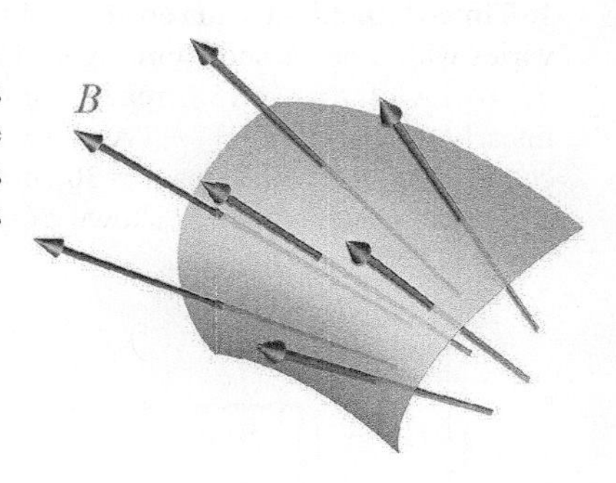

$$\Phi_B = \int_{surface} \vec{B}\cdot d\vec{A}$$

Faraday's Law

Whenever magnetic flux changes in time, not just in the case of a moving conductor, an *emf* will be produced.

$$\mathcal{E}_{induced} = -\frac{d\Phi_B}{dt}$$

Conducting Bar on Rails

$$|\mathcal{E}| = \frac{d\Phi_B}{dt} = vLB$$

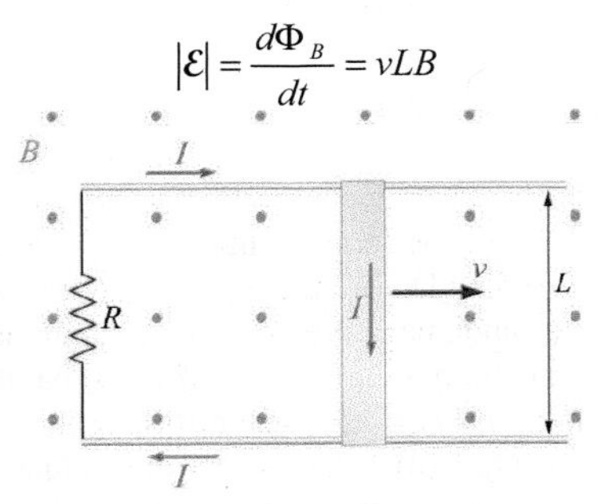

Wire Loop and a Straight Current

$$|\mathcal{E}| = \frac{d\Phi_B}{dt} = vL\left(B_{bottom} - B_{top}\right)$$

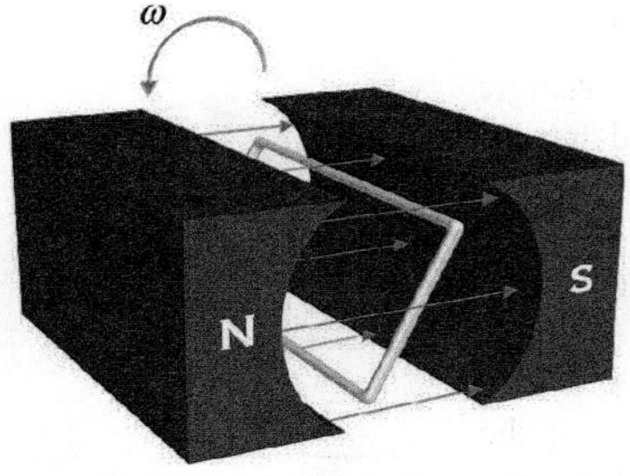

Generators

$$|\mathcal{E}| = \frac{d\Phi_B}{dt} = \omega AB\cos(\omega t)$$

Since the *emf* is the line integral of the electric field, Faraday's law fundamentally relates $\vec{E}$ and $\vec{B}$ fields.

$$\oint_{loop} \vec{E}\cdot d\vec{l} = -\frac{d}{dt}\int_{surface} \vec{B}\cdot d\vec{A}$$

PROBLEMS

1. Time-Dependent Current in a Wire: An infinite straight wire carries a current I that varies with time, as shown in Figure 17.8. It increases from 0 at $t = 0$ to a maximum value $I_1 = 4.1\,\text{A}$ at $t = t_1 = 15\,\text{s}$, remains constant at this value until $t = t_2$ when it decreases linearly to a value $I_4 = -4.1\,\text{A}$ at $t = t_4 = 31\,\text{s}$, passing through zero at $t = t_3 = 25.5$ s. A conducting loop with sides $w = 30$ cm and $L = 55$ cm is fixed in the x-y plane at a distance $d = 27$ cm from the wire as shown.

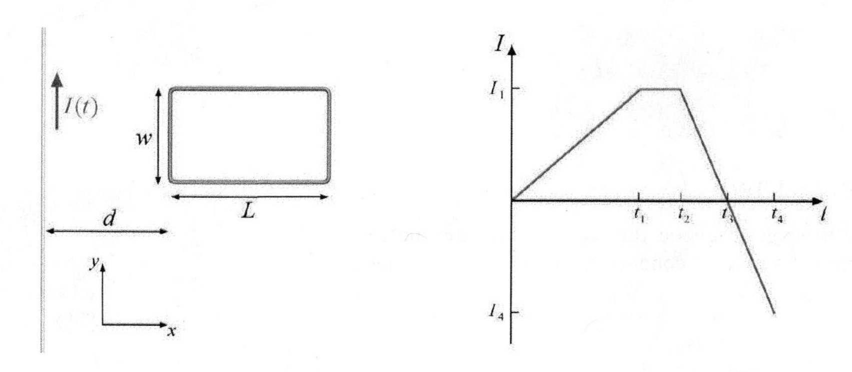

FIGURE 17.8 Problem 1

(a) What is the magnitude of the magnetic flux Φ through the loop at time $t = t_1 = 15\,\text{s}$? (b) What is $\mathcal{E}_1$, the induced *emf* in the loop at time $t = 7.5$ s? Define the *emf* to be positive if the induced current in the loop is clockwise and negative if the current is counterclockwise. (c) What is $\mathcal{E}_2$, the induced *emf* in the loop at time $t = 17$ s? (d) What is the direction of the induced current in the loop at time $t = t_3 = 25.5$ seconds: *clockwise*, *counterclockwise*, or *there is no induced current at* $t = t_3$? (e) What is $\mathcal{E}_4$, the induced *emf* in the loop at time $t = 28.25$ s?

2. Rotating Triangle in a Magnetic Field: A conducting wire formed in the shape of a right triangle with base $b = 26$ cm and height $h = 63$ cm and having resistance $R = 1.3\ \Omega$, rotates uniformly around the y-axis in the direction indicated by the arrow. The triangle makes one complete rotation in time $t = T = 1.9$ s. A constant magnetic field $B = 1.5\,\text{T}$ pointing in the positive z-direction (out of the screen) exists in the region where the wire is rotating. (a) What is ω, the angular frequency of rotation? (b) What is I_{max}, the magnitude of the maximum induced current in the loop? (c) At time $t = 0$, the wire is positioned as shown. What is the magnitude of the magnetic flux Φ_1 at time $t = t_1 = 0.7125$ s? (d) What is the magnitude and

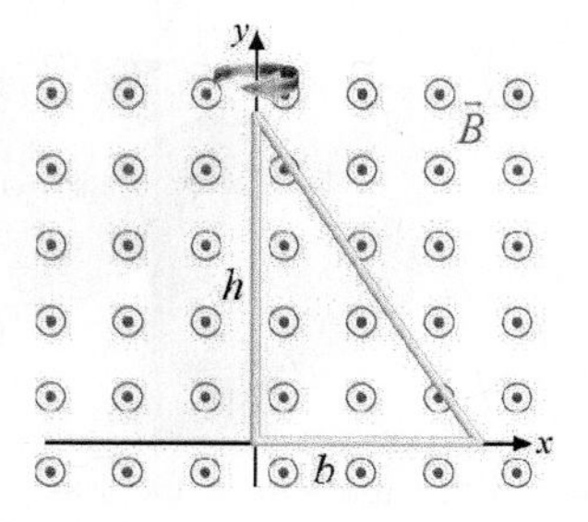

FIGURE 17.9 Problem 2

direction of I_1, the induced current in the loop at time $t = 0.7125$ s? (e) Which of the following statements about Φ_o, the magnitude of the flux through the loop at time $t = t_o = 0.475$ s, and I_o, the magnitude of the current through the loop at time $t = t_o = 0.475$ s, is true? $\Phi_{\max}$ and $I_{\max}$ are defined to be the maximum values these quantities achieve during the complete rotation.

(i) $\Phi_o = 0$ and $I_o = 0$
(ii) $\Phi_o = 0$ and $I_o = I_{\max}$
(iii) $\Phi_o = \Phi_{\max}$ and $I_o = 0$
(iv) $\Phi_o = \Phi_{\max}$ and $I_o = I_{\max}$

(f) Suppose the frequency of rotation is now doubled. How do $\Phi_{\max}$, the maximum value of the flux through the loop, and $I_{\max}$, the maximum value of the induced current in the loop change?

(i) $\Phi_{\max}$ and $I_{\max}$ both double.
(ii) $\Phi_{\max}$ doubles and $I_{\max}$ remains the same.
(iii) $\Phi_{\max}$ remains the same and $I_{\max}$ doubles.
(iv) Both $\Phi_{\max}$ and $I_{\max}$ remain the same.

PART IV: AC CIRCUITS

UNIT 18

INDUCTION AND *RL* CIRCUITS

18.1 Overview

In this unit, we will focus on how induced *emfs*, as described by Faraday's law, can be used in electrical circuits. We will notice a strong similarity here to our development of the idea of capacitance and the use of capacitors in electrical circuits.

We'll start with the concept of self-inductance, in which the act of changing the current through a loop induces an opposing current in that same loop. Our primary example of a physical inductor will be a long solenoid. We will calculate the inductance of the long solenoid in terms of its volume and the number of turns per unit length.

We will then examine the behavior of circuits that include resistors and inductors and discover exponential time dependences that are very similar to those in *RC* circuits.

Finally, we'll calculate the energy that is stored in the inductor when a current flows through it.

18.2 Self-Inductance

Consider a circular loop of wire carrying a constant current I in a clockwise direction, as shown in Figure 18.1. This current will produce a magnetic field (and hence magnetic flux) inside the loop that is directed into the page. Although in general the magnetic flux through the loop may be difficult to calculate, we know that the magnitude of the

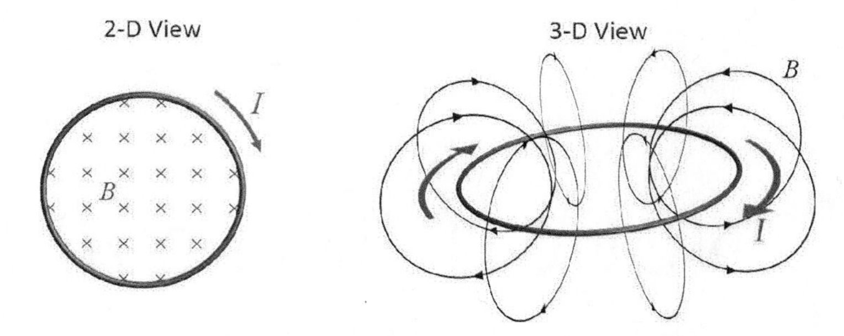

FIGURE 18.1 A circular loop carrying current I in the clockwise direction creates a magnetic field $\vec{B}$ directed into the page in the region enclosed by the loop.

magnetic field is directly proportional to the magnitude of the current. Therefore, the magnetic flux is also directly proportional to the current. Let's define the **self inductance** L to be the ratio of the magnetic flux through the loop, to the current going around the loop.

$$L \equiv \frac{\Phi_B}{I}$$

This self inductance depends only on the geometry of the loop. The SI unit for inductance is the henry (H), defined to be 1 T-m^2/A.

We'll do an explicit calculation of the inductance for a solenoid shortly, but first let's think about what happens if we steadily increase the current flowing around this loop. As the current increases, the magnetic field will also increase resulting in an increase in the magnetic flux through the loop. Faraday's law tells us this changing flux will induce an *emf* that opposes the change in flux. In particular, the changing flux will induce an *emf* of $-d\Phi_B / dt$. Replacing the flux Φ_B by the product of the self-inductance L with the current I, we obtain the expression for the induced *emf* in the inductor.

$$\mathcal{E} = -L\frac{dI}{dt}$$

Hence, we see that the inductance tells us how "hard" it will be to change the current through the device. A large inductance means it will be difficult to change the current quickly.

18.3 Solenoids

The archetypal **inductor** is a long solenoid, a uniform cylindrical coil of wire carrying current I and having n turns per unit length (see Figure 18.2). In order to determine its inductance, we must first determine the magnetic field produced by the solenoid. In general this would be a complicated calculation involving an integration of the Biot-Savart law for each element of current. Figure 18.3 shows the results of a numerical integration for a solenoid with 40 turns whose length is approximately equal 8 times its radius. The field outside the solenoid is small (e.g., the field at point P is only 2% of the field inside the solenoid, and the field inside the solenoid is large and is uniform to better than 0.1%.

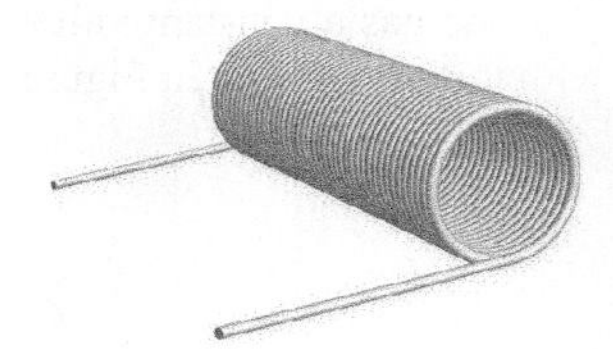

FIGURE 18.2 A solenoid consisting of a piece of wire wrapped in many turns.

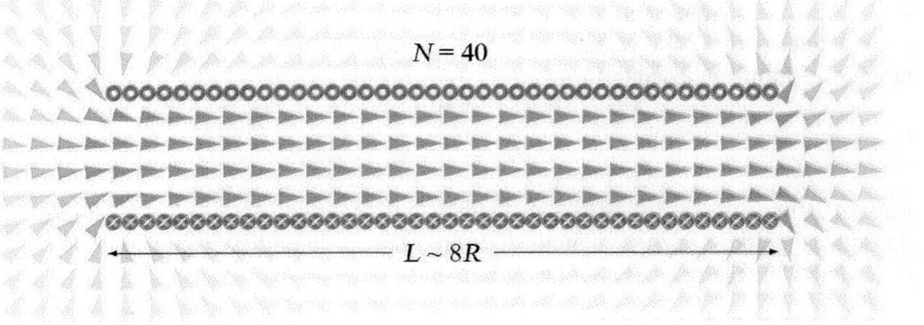

FIGURE 18.3 The magnetic field produced by a long solenoid, a uniform cylindrical coil of wire carrying current I and having 40 turns and a length that is 8 times its radius. The direction of the arrows indicate the direction of the field and the intensity of the arrows indicates the strength of the field. Note that the field is approximately uniform on the inside of the solenoid except at the entrance and exit and that the field outside the solenoid is small.

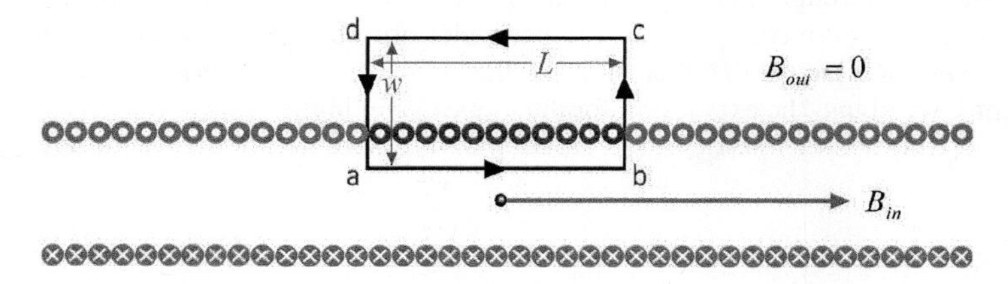

FIGURE 18.4 The magnitude of the magnetic field (B_{in}) inside a solenoid whose radius is small compared to its length can be calculated from Ampère's law using the rectangular path shown.

In the limit that the length of the solenoid is much larger than its radius, the field outside the solenoid is approximately zero and the field inside the solenoid is directed along the axis and has a constant value. We can determine this constant value using Ampère's law with the path shown in Figure 18.4.

$$\oint_{loop} \vec{B} \cdot d\vec{l} = \mu_o I_{enclosed}$$

Notice that the only non-zero segment of the integral is from *a* to *b*, where the magnetic field is constant and parallel to $d\vec{l}$. The other three segments yield zero, as the magnetic field is either perpendicular to the path, or zero. Therefore, the left-hand side of the Ampère's law equation is just equal to the product of the length of the solenoid and the magnetic field (*BL*).

$$\oint_{loop} \vec{B} \cdot d\vec{l} = \int_{a \to b} \vec{B} \cdot d\vec{l} = \int_{a \to b} B \, dl = B \int_{a \to b} dl = BL$$

To evaluate the right-hand side of the Ampère's law equation, we need to calculate the current enclosed by the rectangle, which is equal to the product of the current and the number of turns enclosed (*nL*).

$$I_{enclosed} = N_{enclosed} I = (nL)I$$

Equating these two sides of the Ampère's law equation, we obtain the important result that the magnitude of the magnetic field in a long solenoid is proportional to the number of turns per unit length (*n*) and the current (*I*).

$$B = \mu_o nI$$

We can use this value for the magnetic field to determine that the magnitude of the magnetic flux through a solenoid of length *z* and radius *r* to be

$$\Phi_B = NBA = nz(\mu_o nI)\pi r^2 = \mu_o n^2 z \pi r^2 I$$

We can now use this expression for the flux to determine the inductance L of a long solenoid is proportional to the square of the number of turns per unit length times the volume of the solenoid.

$$L \equiv \frac{\Phi_B}{I} = \mu_o n^2 z \pi r^2 = \mu_o n^2 \cdot Volume$$

You may recall that the archetypal capacitor is a parallel-plate capacitor in which the separation between the plates is much smaller than the length and width of the plate. In this case the capacitor contains a constant electric field and the value of the capacitance is also totally determined by the geometry ($C = \varepsilon_o A / d$). We will shortly see that there are many other similarities between these two devices.

18.4 *RL* Circuits: Qualitative

Inductors are often used in electronic circuits; the circuit symbol looks like a coil of wire, as shown in Figure 18.5. To use Kirchhoff's voltage rule, we write the voltage across the inductor as $V_L = L\,dI/dt$ where the higher potential side is specified by the labeled direction of positive current flow, as shown in Figure 18.5. We'll introduce the use of inductors in circuits with a qualitative description of the circuit shown in Figure 18.6.

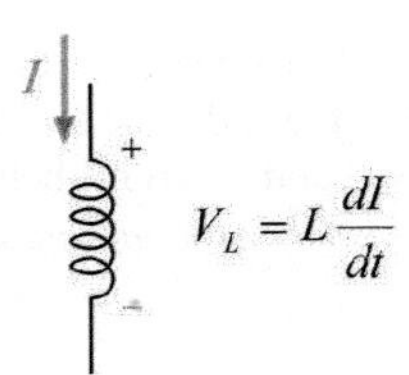

FIGURE 18.5 The use of inductors in circuits: Proceeding in the direction of the positive current flow, we encounter a voltage drop across the inductor given by $L\,dI/dt$.

At time $t = 0$, the switch is closed and current begins to flow in the circuit. We will focus first on the outer loop composed of the battery, a resistor, and the inductor. If the inductor were not in the circuit, the current in this loop would immediately attain the value V/R. The presence of the inductor, however, makes it impossible for the current to go from zero to V/R instantaneously. Why? We know the voltage across the inductor is proportional to the time rate of change of the current. An instantaneous change in current corresponds to dI/dt equal to infinity, which would lead to an infinite voltage across the inductor, which is impossible. Therefore, the current will increase from zero, but at a finite rate. Immediately after the switch is closed, the current then must be *zero* and the full voltage of the battery must appear across the inductor ($V_L(0) = V_b$). As time goes on, the current increases, which leads to an *increase* in the voltage across the resistor, which leads to a *decrease* in the voltage across the inductor. Consequently, the time rate of change of the current (dI/dt) must

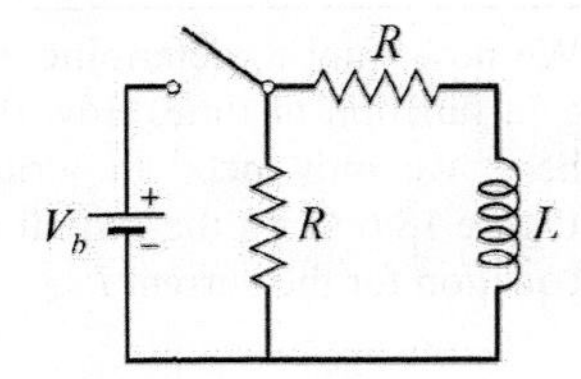

FIGURE 18.6 A circuit composed of two resistors and an inductor that can either be connected to or disconnected from a battery by a switch.

decrease with time. If we wait long enough, dI / dt approaches zero, giving asymptotic values for the current $I(\infty) = V_b / R$ and for the voltage across the inductor $V_L(\infty) = 0$.

After the switch has been closed for a long time, the switch is opened, removing the battery from the circuit (returning to the configuration shown in Figure 18.6). What will happen now? If the inductor were not in the circuit, the current would instantly go to zero. Once again, however, the presence of the inductor makes such a discontinuous change in the current impossible. Immediately, after the switch is opened, the current through the inductor remains at its value immediately before the switch was opened.

Applying Kirchhoff's voltage rule to the right loop (the only loop in the circuit now), we see that the voltage across the two resistors plus the voltage across the inductor must be zero.

$$IR + IR + V_L = 0 \qquad \rightarrow \qquad \left(\frac{V_b}{R}\right)R + \left(\frac{V_b}{R}\right)R + V_L = 0$$

Solving this equation we see that immediately after the switch is opened, the voltage across the inductor must be equal to *twice the battery voltage.*

$$V_L = -2V_b$$

As time goes on, the current decreases, which requires dI / dt to decrease as well. The asymptotic values for both the current and the voltage across the inductor are zero. Note the important differences we've just seen in the voltage and current behaviors of inductors; namely, that *current* through the inductor *must always be continuous*, but the *voltage* across an inductor *may be discontinuous*!

Does this description remind you of the charging and discharging of a capacitor that we studied in Unit 11? In fact, the mathematics for the *RL* and *RC* circuits are identical, as we will see when we solve the *RL* circuit in the next section.

18.5 *RL* Circuits: Quantitative

We now want to determine quantitatively the current and the voltage across the inductor as a function of time. How do we go about doing that? Well, there's nothing really new here; we only need to write down Kirchhoff's voltage rule for the outside loop in Figure 18.6 (with the switch closed). When we do that, we get the following differential equation for the current I:

$$IR + L\frac{dI}{dt} - V_b = 0$$

This equation may look familiar to you. In fact, it's identical in *form* to the equation we obtained for charging a capacitor:

$$\frac{q}{C} + R\frac{dq}{dt} - V_b = 0$$

Here are the differences: (1) our equation is an equation for *I*, not *q*, (2) the coefficient of the dI/dt term is *L*, while the coefficient of the dq/dt term is *R*, and (3) the coefficient of the *I* term is *R*, while the coefficient of the *q* term is $1/C$.

The time dependence for the charging capacitor was exponential with a time constant $\tau = RC$. Therefore, the time dependence for our current circuit will also be exponential, but with a time constant $\tau = L/R$, which we obtained by making the correspondences of *R* and $1/C$ in the capacitor equation with *L* and *R*, respectively, in the inductor equation.

We can write down the solution for *I* using this time-dependence, subject to the initial conditions we discussed in the last section.

$$I(t) = \frac{V_b}{R}\left(1 - e^{-\frac{R}{L}t}\right)$$

We can find the voltage across the inductor by simply differentiating this expression for *I* and multiplying by *L*.

$$V_L(t) = V_b e^{-\frac{R}{L}t}$$

The plots shown in Figure 18.7 display the expected behavior; the current starts from zero and increases exponentially with time constant $\tau = L/R$ toward an asymptotic value of V_b/R. The voltage across the inductor starts at the battery voltage V_b and decreases exponentially with the same time constant toward an asymptotic value of zero.

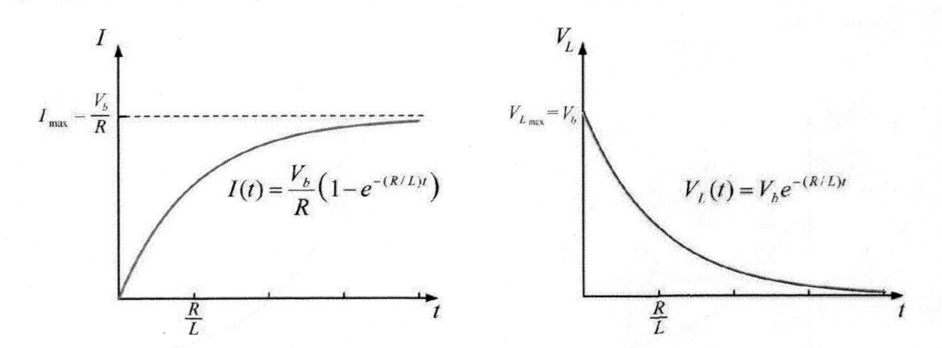

FIGURE 18.7 The time dependence of the current (I) and the voltage across the inductor (V_L) when the switch in Figure 18.5 is closed at $t = 0$.

We'll now determine the current and the voltage across the inductor when we open the switch in Figure 18.5 after it has been closed for a long time. Once again, we write down

Kirchhoff's voltage rule for the new loop and obtain a differential equation for the current *I*.

$$I(2R) + L\frac{dI}{dt} = 0$$

This equation is identical in form to what we found for discharging a capacitor (see Unit 11, section 4). In this case, though, the loop contains two resistors in series so that the equivalent resistance is 2*R* in this circuit. Consequently, when we make the substitutions, we determine that the time dependence of the current will be exponential with a time constant equal to the inductance divided by the equivalent resistance.

$$\tau = \frac{L}{R_{equivalent}} = \frac{L}{2R}$$

Applying the initial conditions we arrived at in our qualitative discussion of this circuit, we obtain expressions for the current through the inductor and voltage across the inductor as a function of time.

$$I(t) = \frac{V_b}{R}e^{-(R_{equivalent}/L)t}$$

$$V_L(t) = -2V_b e^{-(R_{equivalent}/L)t}$$

The plots shown in Figure 18.8 display the expected behavior; the current starts from its value just before the switch opened ($\sim V_b / R$) and decreases exponentially toward its asymptotic value of zero. Note that immediately after the switch is opened, the voltage across the inductor must jump to twice the battery voltage in order to maintain the same current. It then decreases exponentially with the same time constant toward its asymptotic value of zero.

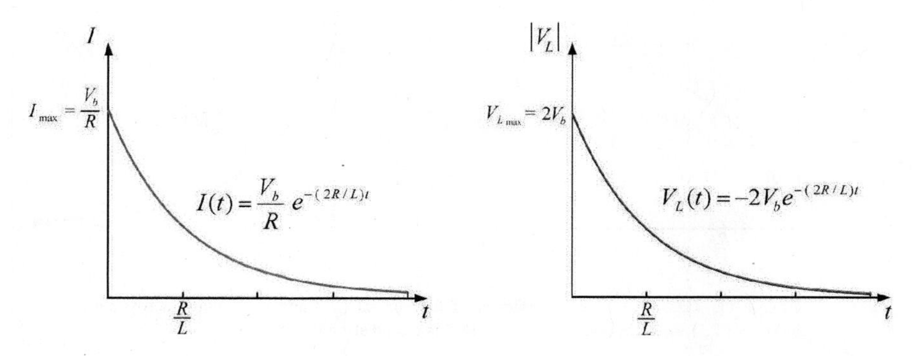

FIGURE 18.8 The time dependence of the current (I) and the absolute value of the voltage across the inductor (V_L) when the switch in Figure 18.5 is opened after a very long time (redefined to be $t = 0$).

18.6 Energy in an Inductor

We will now discuss the energy in an inductor. The claim is that inductors store energy when they carry current, much like capacitors store energy when they are charged.

We'll start by looking at the energy flow in the *RL* circuit, shown in Figure 18.6, when the switch is closed. Recall that the rate at which a device changes the energy in the circuit is the product of the voltage across that element with the current through that element. The battery supplies energy at the rate $P_b = IV_b$. The resistor dissipates energy at the rate $P_R = IV_R = I^2R$. By plotting the time dependence of these expressions in Figure 18.9, we see that not all of the energy that is being supplied by the battery is being dissipated by the resistor; we expect this difference is being stored in the inductor. Mathematically, the difference between P_b and P_R is certainly $P_L = IV_L$, as shown, but let's look at this energy in a bit more detail to see exactly how this works.

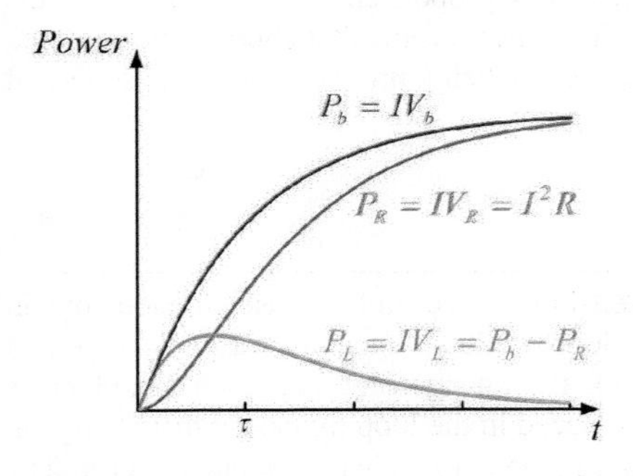

FIGURE 18.9 The power delivered by the battery (P_b), dissipated by the resistor (P_R), and stored in the inductor (P_L) as a function of time in the *RL* circuit shown in Figure 18.5 ($t = 0$ when switch is closed).

The rate at which energy is being stored in the inductor is given by

$$P_L = \frac{dU_L}{dt} = V_L I = LI\frac{dI}{dt}$$

If we integrate this equation, we obtain the form for the energy stored in the inductor at any time:

$$U_L = \frac{1}{2}LI^2$$

Note that this expression is independent of *R*; inductors store an amount of energy that depends only on the inductance and the square of the current flowing through them, just as capacitors store an amount of energy that depends only on the capacitance and the square of the charge on them ($U_C = Q^2 / 2C$).

Where is this energy stored? In the capacitor, we claimed it was stored in the electric field contained within the capacitor and calculated the energy density for the particular case of the parallel-plate capacitor. For the inductor, we will make the claim that the energy is stored in the magnetic field contained within the inductor. We can calculate this density for the specific case of the long solenoid.

Earlier in this unit, we calculated the inductance L of the long solenoid to be $L = \mu_o n^2 \pi r^2 z$. The energy density inside the solenoid is equal to the total energy of the inductor ($\frac{1}{2} LI^2$) divided by the volume of the solenoid.

$$u = \frac{\frac{1}{2} LI^2}{\pi r^2 z} = \frac{1}{2}\mu_o n^2 I^2$$

We can rewrite this expression in terms of the magnitude of the magnetic field inside the solenoid ($B = \mu_o nI$):

$$u_B = \frac{1}{2}\frac{B^2}{\mu_o}$$

Recall that the energy density of the capacitor was proportional to the square of the electric field inside the capacitor ($u_E = \frac{1}{2}\varepsilon_o E^2$). Thus, it seems that energy can be stored in electric and magnetic fields, with the energy density being proportional to the square of the strength of the field.

18.7 Summary

In this unit, we introduced the concept of self-inductance and the use of inductors in circuits. We began by noting that a current in a loop produces a magnetic flux through that loop. If the current changes in time, then the flux also changes in time. If the flux changes in time, then by Faraday's law, an *emf* will be induced in the loop to oppose that change in flux. In other words, the loop itself provides a hindrance to a change of current in it. The self-inductance L is a measure of this hindrance. The self-inductance of a conducting loop is defined as the ratio of the magnetic flux through the loop to the current that produced it.

$$L \equiv \frac{\Phi_B}{I}$$

This quantity is determined totally by the geometry of the loop. We used Faraday's law to determine that the *emf* induced in the loop:

$$\mathcal{E} = -L\frac{dI}{dt}$$

We then calculated the self-inductance of a long solenoid and found it to be proportional to the product of the volume of the solenoid and the square of the number of turns per unit length ($L = \mu_o n^2 \pi r^2 z$).

We then went on to explore the behavior of *RL* circuits. We found that the behavior of an *RL* circuit was identical to that of the corresponding *RC* circuit. In particular, we found that if we replaced (q, $1/C$, and R) in the KVR equation for an *RC* circuit by (I, R, and L), we obtained the KVR equation for the corresponding *RL* circuit. The time constant describing the behavior of *RL* circuits, therefore, was given by $\tau = L/R$. We determined complete expressions for the time dependence of the current and voltage across the inductor after a switch is closed/opened that inserted/removed a battery from the circuit, shown in Figure 18.8, by examining the asymptotic values for these parameters.

Finally, we determined that energy is stored in an inductor when current flows through it, much in the same way that energy is stored in a capacitor when it is charged. In the capacitor, the energy is stored in the electric field between the plates, and in the inductor the energy is stored in the magnetic field in its volume. The total energy stored in an inductor is proportional to the product of its inductance L and the square of the current flowing through it.

$$U_L = \frac{1}{2} L I^2$$

We used a long solenoid to calculate the energy density in this magnetic field and found it was proportional to the square of the magnetic field, much like we found that the energy density in the electric field was proportional to the square of the electric field.

MAIN POINTS

Inductance

Self-inductance of a loop is a measure of the hindrance provided by that loop to a change in the current in that loop.

Self-Inductance

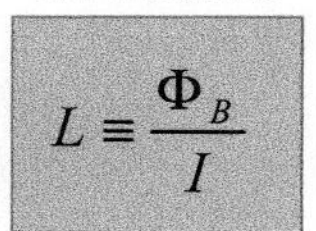

Inductor Voltage

$$\mathcal{E} = -L\frac{dI}{dt}$$

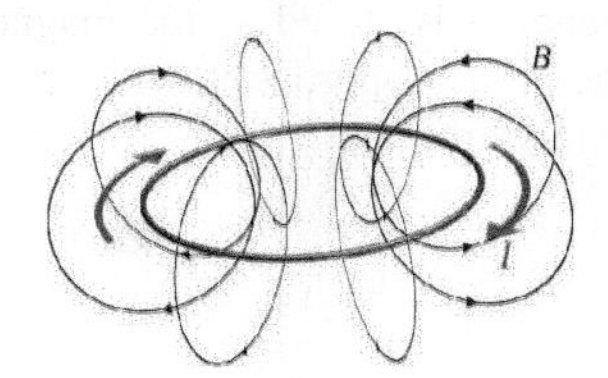

RL Circuits

The time dependence of the current in an *RL* circuit is exponential with a time constant given by the ratio of the inductance to the resistance in the circuit.

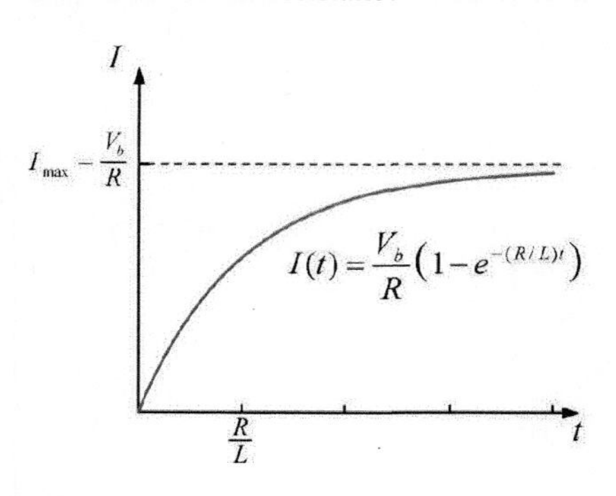

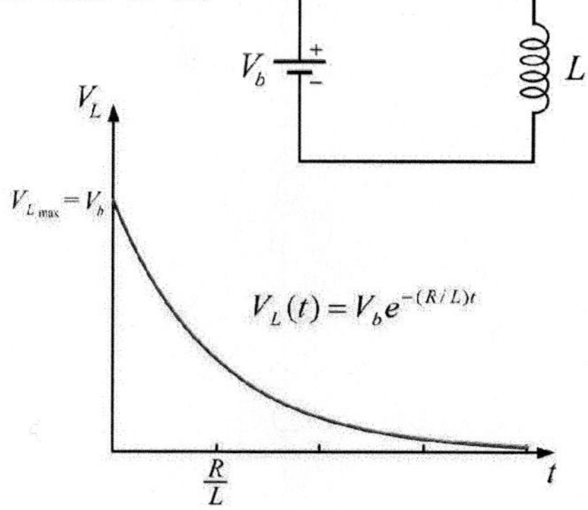

Energy

Inductor Energy: The energy stored in an inductor is proportional to the square of the current.

$$U_L = \frac{1}{2}LI^2$$

Magnetic Energy Density: Any magnetic field carries energy with an energy density that is proportional to the square of the magnetic field.

$$u_B = \frac{B^2}{2\mu_o}$$

PROBLEMS

1. Two Loop *RL* Circuit 1: A circuit is constructed with four resistors, one inductor, one battery, and a switch as shown. The values for the resistors are: $R_1 = R_2 = 80\ \Omega$, $R_3 = 73\ \Omega$, and $R_4 = 123\ \Omega$. The inductance is $L = 683$ mH and the battery voltage is $V_b = 24$ V. (a) The switch has been open for a long time when at time $t = 0$, the switch is closed. What is $I_1(0)$, the magnitude of the current through the resistor R_1 just after the switch is closed? (b) What is $I_1(\infty)$, the magnitude of the current that flows through the resistor R_1 after the switch has been closed for a long time? (c) What is $V_L(0)$, the magnitude of the voltage across the inductor just after the switch is closed? (d) What is $I_L(\infty)$, the magnitude of the current through the inductor after the switch has been closed for a very long time? (e) What is $I_2(0)$, the magnitude of the current through the resistor R_2 just after the switch is closed?

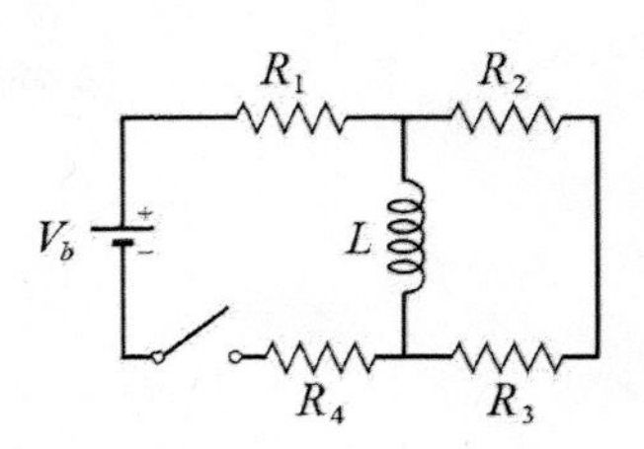

FIGURE 18.10 Problem 1

2. Two Loop *RL* Circuit 2: A circuit is constructed with four resistors, one inductor, one battery, and a switch as shown. The values for the resistors are $R_1 = R_2 = 30\ \Omega$, $R_3 = 95\ \Omega$, and $R_4 = 105\ \Omega$. The inductance is $L = 201$ mH and the battery voltage is $V_b = 12$ V. The positive terminal of the battery is indicated with a + sign. (a) The switch has been open for a long time when at time $t = 0$, the switch is closed. What is $I_4(0)$, the magnitude of the current through the resistor R_4 just after the switch is closed? (b) What is $I_4(\infty)$, the magnitude of the current through the resistor R_4 after the switch has been closed for a very long time? (c) What is $I_L(\infty)$, the magnitude of the current through the inductor after the switch has been closed for a very long time? (d) After the switch has been closed for a very long time, it is then opened. What is $I_3(t_{open})$, the current through the resistor R_3 at a time $t_{open} = 4.5$ ms after the switch was opened? The positive direction for the current is indicated in the figure. (e) What is $V_{L_{\max}}(closed)$, the magnitude of the maximum voltage across the inductor during the time when the switch is closed? (f) What is $V_{L_{\max}}(open)$, the magnitude of the maximum voltage across the inductor during the time when the switch is open?

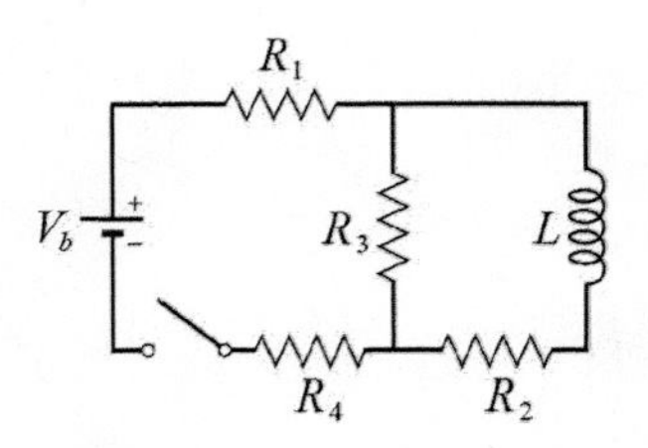

FIGURE 18.11 Problem 2

3. *LR* Time (INTERACTIVE EXAMPLE): Three resistors ($R_1 = 120\ \Omega$, $R_2 = 330\ \Omega$, and $R_3 = 240\ \Omega$) and an ideal inductor ($L = 1.6$ mH) are connected to a battery ($V_b = 9$ V) through a switch, as shown. The switch has been open for a long time before it is closed at $t = 0$. At what time does the current through the inductor (I_3) reach a value that is 63% of its maximum value?

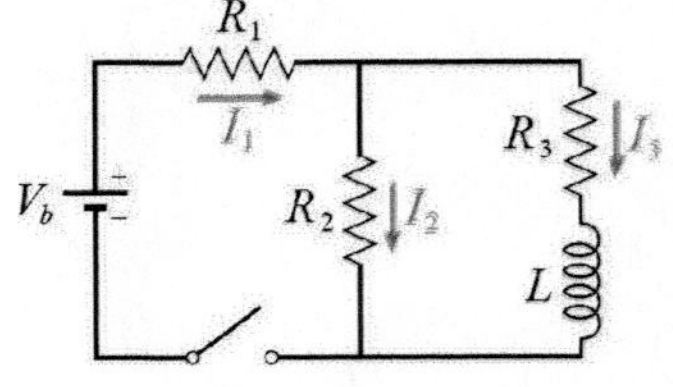

FIGURE 18.12 Problem 3

UNIT

19

LC AND *RLC* CIRCUITS

19.1 Overview

In this unit, we will discuss electrical circuits that contain inductors and capacitors. We'll start with a qualitative description of a circuit containing an inductor and a capacitor that has been given an initial charge. We will see that this circuit is an oscillator, with the energy being continuously transferred back and forth between the capacitor and the inductor.

We will then use Kirchhoff's rules to determine the quantitative behavior of the circuit. We will calculate the characteristic frequency of the oscillation and note the close analogy to the case of a mass hanging on a spring. We will also explicitly discuss the energy transfer between the inductor and the capacitor.

Finally we'll add a resistor to the circuit and examine the resulting behavior. We will find that energy will be dissipated in this resistor, which will cause the oscillations to become damped.

19.2 *LC* Circuits: Qualitative

We'll start with a qualitative description of the circuit shown in Figure 19.1. We'll first move the switch to position *a*. In this position, we have a battery of voltage V_b charging a capacitor *C* through a resistor *R*. We know that the charge on the capacitor will approach its asymptotic value of CV_b with a time constant $\tau = RC$. If we wait several time constants, the capacitor will be essentially fully charged.

At that point, we move the switch to position *b*. We now have removed the battery and resistor from the circuit and are left with only a capacitor and inductor. Initially the capacitor is fully charged and we expect it to discharge through the inductor in the completed circuit. However, unlike the discharge of a capacitor through a resistor, the presence of the inductor will prevent the current from changing abruptly and the initial current will be *zero*. The current increases in the clockwise direction until the capacitor is fully discharged, as shown in the top-right part of Figure 19.2 (note we have labeled this current as *negative*, since $I \equiv dQ/dt$ is positive when Q increases).

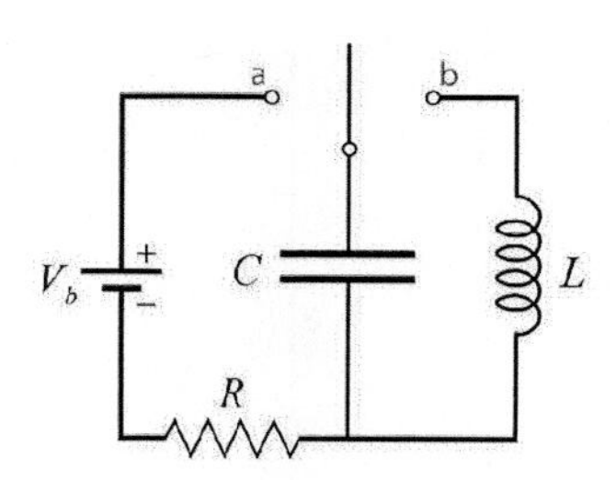

FIGURE 19.1 The switch is initially moved to position *a* to charge the capacitor. After a long time, the switch is moved to position *b* to remove the battery and resistor to create an *LC* circuit with a stored energy of $\frac{1}{2} CV_b^2$.

At this point we might expect the current to stop. The presence of the inductor, however, prevents that from happening. Since the inductor opposes a change in current, it produces an *emf* that keeps the current flowing in the clockwise direction, causing the capacitor to become charged with the opposite sign. This charging continues as the inductor slowly allows the current to go to zero, as shown in the bottom-right part of Figure 19.2. At this point the capacitor is fully charged with the opposite polarity from which it started. Hence, the capacitor will once again begin to discharge, but this time the current will be in the counterclockwise direction, reaching a maximum value when the capacitor is fully discharged, as shown in the bottom-left part of Figure 19.2. Again, the inductor will prevent the current from immediately stopping, and provide a potential to keep the current flowing in the counterclockwise direction and charge the capacitor. The current will decrease coming to zero when the capacitor has returned to its fully charged state, as shown in the top-left part of Figure 19.2, exactly where we started from.

This circuit acts as an oscillator. Both the current and charge on the capacitor oscillate back and forth between positive and negative values. We can think of this process also in terms of energy transfer. Initially, all of the energy in the circuit is stored in the capacitor. As current begins to flow, the energy is transferred from the capacitor to the inductor until the capacitor is fully discharged. At that point, all of the energy is stored in the inductor. As the current begins to decrease, though, energy is transferred from the inductor to the capacitor as it begins to charge again. This conversion of energy between electrical (in the capacitor) and magnetic (in the inductor) may remind you of the oscillating mass on a spring from mechanics where we had the conversion of energy between potential (in the spring) and kinetic (of the mass). In fact, the mathematics is identical, as we shall see in the next section.

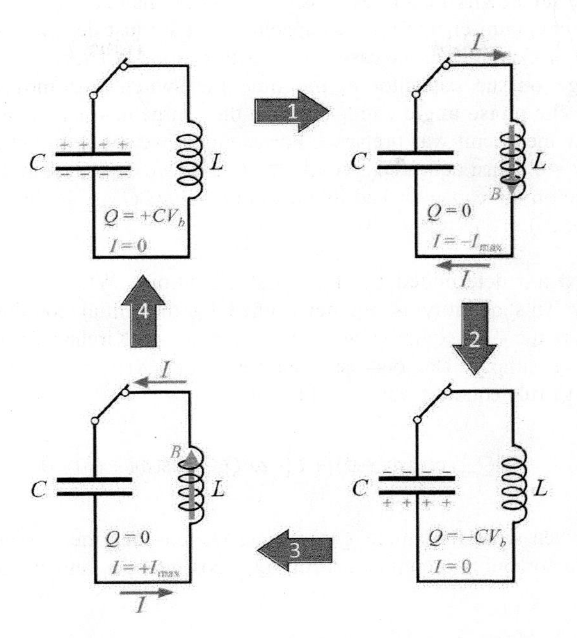

FIGURE 19.2 The oscillation cycle in an *LC* circuit.

19.3 *LC* Circuits: Quantitative

We now want to determine quantitatively the current and the charge on the capacitor as a function of time. How do we go about doing that? Once again, there's nothing really new here; since there is just one loop in the circuit, we only need to write down Kirchhoff's voltage rule. The result is

$$\frac{Q}{C} + L\frac{dI}{dt} = 0$$

This becomes a single differential equation for Q, once we substitute dQ/dt for I.

$$\frac{Q}{C} + L\frac{d^2Q}{dt^2} = 0$$

To solve this equation, we need a function $Q(t)$, which when differentiated twice, has the same time dependence as the function itself. We know that sines and cosines have this property. Therefore, we will assume a solution of the form

$$Q(t) = Q_{\max} \cos(\omega t + \phi)$$

There are three unknowns in this expression for $Q(t)$, namely, $Q_{\max}$, ω, and ϕ. $Q_{\max}$ represents the maximum charge on the capacitor, which must depend on how the circuit was prepared. For example, in the case we described in the last section, $Q_{\max}$ would have been the charge on the capacitor at the time the switch was moved to position b ($Q_{\max} \sim CV_b$). The **phase angle** ϕ tells us when the charge is at its maximum, which also depends on how the circuit was prepared. For example, we see from the equation that the charge at time $t = 0$ is just equal $Q_{\max} \cos\phi$. In the case we described in the last section, at $t = 0$, the charge on the capacitor had its maximum value, $Q_{\max}$, so that $\cos\phi$ had to be 1 (i.e., ϕ would be 0°).

So $Q_{\max}$ and ϕ are determined by the initial conditions. What about the **oscillation frequency** ω? This quantity is not determined by the initial conditions; it must be determined from the one equation we have, courtesy of Kirchhoff's voltage rule. To determine ω, we simply take our assumed form for $Q(t)$ and substitute it into the Kirchhoff voltage rule equation above and obtain

$$\frac{1}{C}[Q_{\max} \cos(\omega t + \phi)] + L[-\omega^2 Q_{\max} \cos(\omega t + \phi)] = 0$$

As promised, when we differentiate $Q(t)$ twice, we get back the $\cos(\omega t + \phi)$ term. We can therefore factor out the common term, $Q_{\max} \cos(\omega t + \phi)$, and we are left with the equation

$$\frac{1}{C} - \omega^2 L = 0$$

Therefore, we arrive at the result

$$\omega = \frac{1}{\sqrt{LC}}$$

This whole exercise may be somewhat familiar to you. Recall that in mechanics, we considered a block of mass m, resting on a frictionless floor, and connected to a wall by a spring with spring constant k, as shown in Figure 19.3. When we stretch the spring and release it, the block undergoes an oscillatory motion with a frequency equal to the square root of the spring constant divided by the mass of the block.

$$\omega = \sqrt{\frac{k}{m}}$$

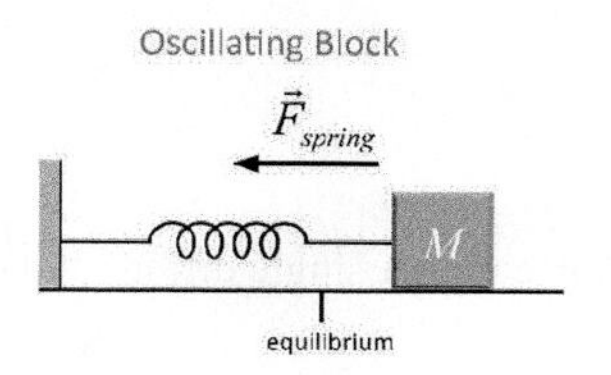

FIGURE 19.3 An oscillator from mechanics: a block of mass m on a frictionless surface is attached to a wall by a spring with spring constant k. The resulting angular frequency is given by $\omega = \sqrt{k/m}$.

We determined this frequency by applying Newton's second law for the system. Namely, the force ($-kx$) is set equal to the product of the mass and the acceleration.

$$-kx = m\frac{d^2x}{dt^2}$$

This equation of motion for the block-spring system is absolutely equivalent to the one we just obtained from Kirchhoff's voltage rule with the substitutions of k for $1/C$, m for L, and x for Q. The behavior of *LC* circuits and the mass-spring system are described by exactly the same equation.

We now plot the charge Q and current I as a function of time in Figure 19.4 and get the expected results. Q begins with its maximum value and decreases to zero as the current decreases from zero to attain its maximum negative value just as Q passes through zero. The cycle then continues with the charge reaching a maxima or minima whenever the current passes through zero and vice versa.

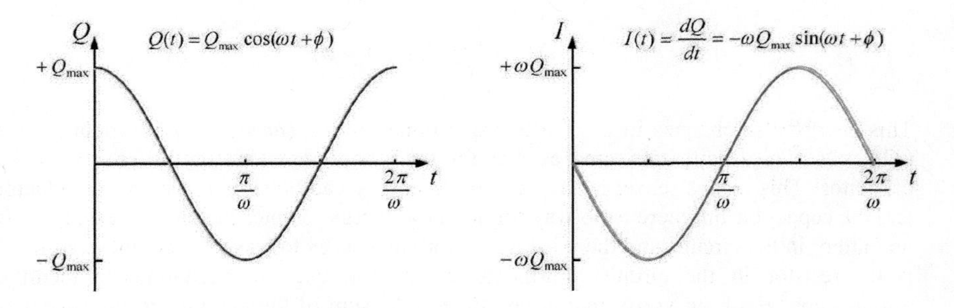

FIGURE 19.4 The time dependence for Q, the charge on the capacitor, and I, the current in an *LC* circuit. In these graphs, the phase angle ϕ has been set to 0°.

19.4 *LC* Circuits and Energy

In our qualitative discussion of the *LC* circuit, we claimed that the energy was constantly being transferred between the capacitor and the inductor. Since we have now obtained quantitative expressions for the charge on the capacitor and the current in the circuit, we can verify this claim.

Energy can be stored by both capacitors (in the electric field between the plates) and in inductors (in the magnetic field inside the coil). The energy stored in the capacitor is proportional to the square of its charge ($U_C = Q^2/2C$). Substituting our expression for $Q(t)$ into this equation, we obtain a form for the energy stored in the capacitor.

$$U_C(t) = \frac{Q_{max}^2}{2C}\cos^2(\omega t + \phi)$$

This energy changes in time, being proportional to $\cos(\omega t+\phi)$. Its maximum value is $Q_{\max}^2/2C$, the energy from the initial charge. This makes sense since, once the battery has been removed from the circuit, there is no way to add more energy to the *LC* circuit.

The energy stored in the inductor is proportional to the square of the current ($U_L = ½\,LI^2$). We can find the current *I* by differentiating the form for the charge *Q*.

$$I(t)=\frac{dQ}{dt}=-\omega Q_{\max}\sin(\omega t+\phi)$$

Substituting this expression for $I(t)$ into the equation for U_L, we obtain

$$U_L(t)=\frac{1}{2}LI^2=\frac{1}{2}L\omega^2Q_{\max}^2\sin^2(\omega t+\phi)$$

We can simplify this expression by replacing ω^2 by $1/LC$. When we make this substitution, we find the result

$$U_L(t)=\frac{1}{2}\frac{Q_{\max}^2}{C}\sin^2(\omega t+\phi)$$

This energy also changes in time; it is proportional to $\sin^2(\omega t+\phi)$. Its maximum value ($Q_{\max}^2/2C$) is exactly the same as that for the maximum value of the energy in the capacitor! This makes sense; in the *LC* circuit, energy can be stored in both the inductor and the capacitor, but there is no way for any new energy to enter the circuit (e.g., there is no battery in the circuit) and there is no way for any energy to leave the circuit (e.g., there is no resistor in the circuit). Furthermore, we can use the trigonometric identity, $\sin^2\theta+\cos^2\theta=1$, to verify that at any time *t*, the sum of the energies in the capacitor and the inductor is equal to $Q_{\max}^2/2C$. The total energy in the circuit does not change; the only thing that changes is the fraction of the energy that is in the inductor and the capacitor at any time, as illustrated in the plot of U_L and U_C as a function of time displayed in Figure 19.5.

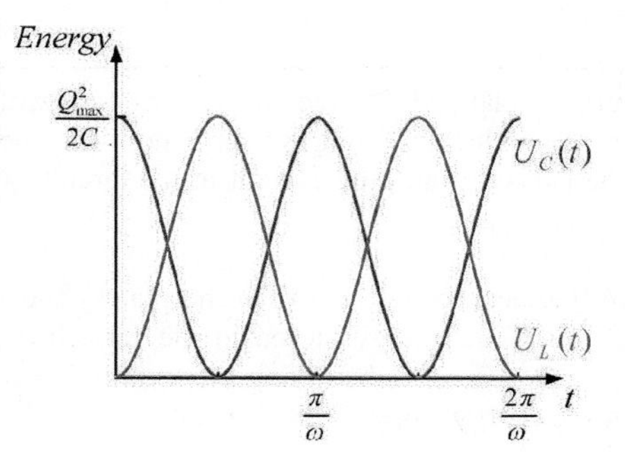

FIGURE 19.5 The time dependence of the energy stored in the capacitor (U_C) and the energy stored in the inductor (U_L) in an *LC* circuit.

This result can be understood in terms of the analogous block and spring result. Namely, the energy in the inductor (½ LI^2) plays the role of the kinetic energy (½ mv^2) of the block, while the energy in the capacitor ($Q^2/2C$) plays the role of the potential energy in the spring (½ kx^2).

19.5 *RLC* Circuits

We now want to look at what happens if we add a resistor R to our LC circuit, as shown in Figure 19.6. Clearly, by adding this resistance, we have supplied a means to remove energy from the circuit. Namely, as current I flows in this circuit, energy will be dissipated in the resistor at the rate $P = I^2 R$.

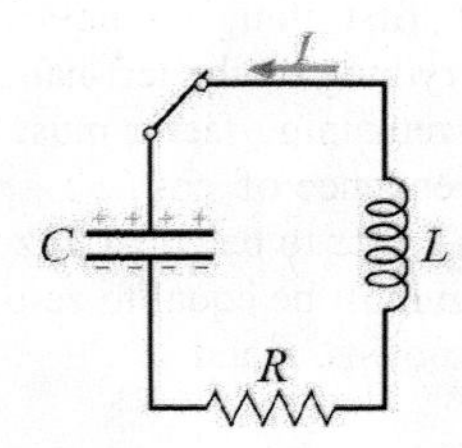

FIGURE 19.6 An *RLC* circuit.

If R is sufficiently small, we expect its effect to be small. Namely, we expect the oscillations in charge and current to persist, but with a continuously decreasing amount of total energy in the circuit. Therefore, we expect the amplitude of the oscillations to become smaller as time goes on. We call this behavior **damped oscillations**. The amount of damping depends on the value of the resistance. If R is sufficiently large, then the damping will become dominate, with the energy being removed so fast that we don't see the oscillations.

How can we calculate this circuit behavior? Well, once again, the only thing we can do is to write down Kirchhoff's voltage rule for the circuit. In this case, we have yet another differential equation.

$$\frac{Q}{C} + R\frac{dQ}{dt} + L\frac{d^2Q}{dt^2} = 0$$

So far we have been able to guess the solutions to the KVR equations because we knew functions that when differentiated once (the exponential) or twice (sines and cosines) give back the same time dependence. Here, we have some sort of combination of both effects going on. It may be reasonable, but certainly not obvious, that a good guess here would be the product of an exponential and a cosine function. In particular, let's assume

$$Q(t) = Ae^{-\beta t} \cos(\omega' t + \phi)$$

Here, A and ϕ must be determined from the initial conditions. The **damping factor** β and the oscillating frequency ω', however, must be determined from the KVR equation. This may seem a bit daunting since we have two unknowns and just one equation, but it will all work out as we shall soon see.

Our starting point here is that we have the KVR equation and an assumed form for $Q(t)$, the charge on the capacitor as a function of time. Our procedure is as before: We substitute this form for $Q(t)$ into the KVR equation and see what happens.

In this case, we need to recall the rule for differentiating the product of two functions (here, an exponential times the cosine) and we obtain the rather lengthy equation

$$Ae^{-\beta t}\left\{\cos(\omega' t+\phi)\left[\frac{1}{C}-R\beta-L\left(\omega'^2-\beta^2\right)\right]+\sin(\omega' t+\phi)\left[-R\omega'+2L\beta\omega'\right]\right\}=0$$

The first thing to notice is that there is an overall factor of $Ae^{-\beta t}$ that multiplies everything on the left-hand side of the equation. Since this factor cannot be zero, we know the remaining factor must be zero. This remaining factor has two terms: one has the time dependence of $\cos(\omega' t+\phi)$ and the other has the time dependence of $\sin(\omega' t+\phi)$. For this factor to be equal to zero at all times *t*, the coefficients of the cosine term and the sine term must be equal to zero, separately. Therefore, we now have two equations for our two unknowns, β and ω'.

$$\frac{1}{C}-R\beta-L(\omega'^2-\beta^2)=0$$

$$-R\omega'+2L\beta\omega'=0$$

Setting the coefficient of the sine term equal to zero, we obtain an expression for the damping factor β

$$\beta=\frac{R}{2L}$$

This result is reasonable in the sense that if R is small, the damping will be small, and if R is large, the damping will be large. Setting the coefficient of the cosine term equal to zero, we can obtain an expression for the oscillation frequency ω', which we can rewrite more compactly in terms of the **natural frequency**

$$\omega_o\equiv\frac{1}{\sqrt{LC}}$$

and the damping factor β as

$$\omega'^2=\omega_o^2-\beta^2$$

This is an interesting result; in the limit of small damping (R small and therefore β small), we get the expected result that ω' approaches ω_o. As R increases, β increases, which results in a decrease of ω'.

The remaining unknowns, A and ϕ, are found using our initial conditions. Before the switch is closed in our circuit, the capacitor is fully charged with a charge Q_o and there is no current flowing in the circuit. Once the switch is closed, the inductor will keep the current from changing instantaneously so that the initial current is zero. Setting $Q(0) = Q_o$ in our expression for $Q(t)$, we obtain

$$Q_o = A\cos\phi$$

We can then find the current $I(t)$ by differentiating $Q(t)$:

$$I(t) = -Ae^{-\beta t}\left[\beta\cos(\omega' t + \phi) + \omega'\sin(\omega' t + \phi)\right]$$

Using the initial condition of $I(0) = 0$, we can solve for ϕ:

$$\phi = \tan^{-1}\left(-\frac{\beta}{\omega'}\right)$$

Finally, plugging this expression for ϕ into our expression for Q_o, we can solve for our last unknown *A*:

$$A = \frac{Q_o}{\cos\left[\tan^{-1}\left(-\frac{\beta}{\omega'}\right)\right]}$$

In the limit of small damping ($\beta << \omega_o$), the constant *A* reduces to simply Q_o and $Q(t)$ takes the same form as it had in the previous section for *LC* circuits.

We now want to get a more concrete understanding of this behavior by examining the time dependence of *Q*, *I*, U_L, and U_C for various values of β. We'll start with a minimal amount of damping by choosing $\beta = .01\omega_o$. In Figure 19.7, we see the plots for *Q* and *I* are very similar to those of the *LC* circuit shown previously in Figure 19.4. The oscillation frequency is essentially the same and the damping is small; after two complete cycles, the charge has decreased from its initial value by about 12%. The energy plots for $\beta = .01\omega_o$ in Figure 19.7 also look similar to those of the *LC* circuit shown in Figure 18.5. Since the

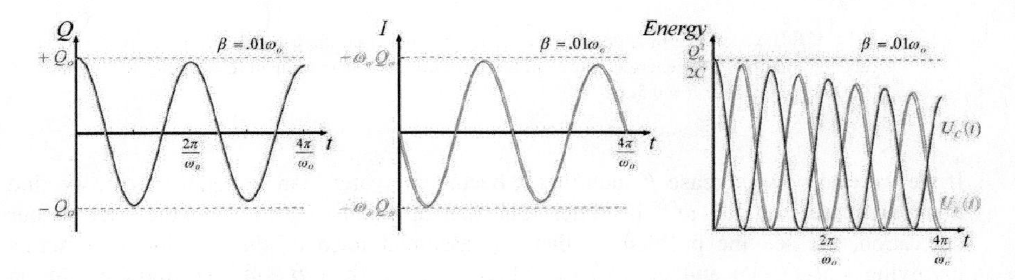

FIGURE 19.7 The time dependence for *Q*, the charge on the capacitor, *I*, the current and U_C and U_L, the energies stored in the capacitor and inductor in an *RLC* circuit with the damping factor $\beta = .01\omega_o$.

energy in the capacitor is proportional to the square of the charge, we see that U_C has decreased from its initial value by about 22%.

As we increase β, we expect the damping to increase. The plots for $\beta = 0.1\omega_o$ are shown in Figure 19.8. Here we see that after two complete cycles, the charge has decreased from its initial value by about 72% while U_C has decreased by about 92%. The oscillation frequency has only decreased by about a half a percent.

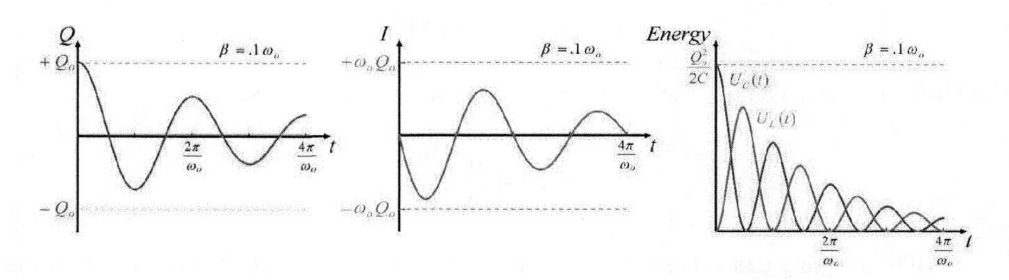

FIGURE 19.8 The time dependence for Q, the charge on the capacitor, I, the current and U_C and U_L, the energies stored in the capacitor and inductor in an *RLC* circuit with the damping factor $\beta = 0.1\omega_o$.

If we increase β to be equal to $0.5\omega_o$, we see severe damping, as shown in Figure 19.9. Indeed it's difficult to see the oscillatory piece beyond about one cycle. The effect on energy is, as expected, larger still. The oscillatory frequency has decreased by about 13%.

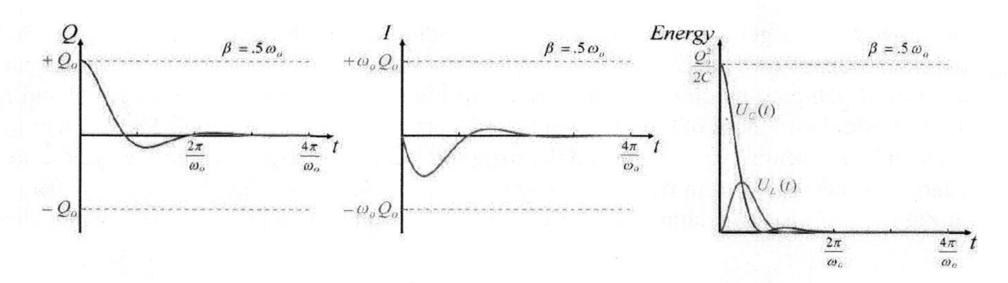

FIGURE 19.9 The time dependence for Q, the charge on the capacitor, I, the current and U_C and U_L, the energies stored in the capacitor and inductor in an *RLC* circuit with the damping factor $\beta = 0.5\omega_o$:

If we were to now increase β such that it becomes greater than or equal to ω_o, we find that our expression for ω'^2 becomes less than or equal to zero! Looking back at our derivation, we see the problem is that our assumed form of the solution (two terms involving $\sin(\omega' t + \phi)$ and $\cos(\omega' t + \phi)$) breaks down when $\beta \geq \omega_o$. To find the solution for $\beta \geq \omega_o$, we need more advanced techniques that are beyond the scope of this course, but the end result is not particularly remarkable. All oscillatory behavior vanishes, leaving a purely exponential behavior.

The onset of this exponential behavior occurs at $\beta = \omega_o$, and is called **critical damping**. The time dependence for Q, I, U_C, and U_L are shown in Figure 19.10. The value of R for critical damping is given by

$$R_{critical} = 2\sqrt{\frac{L}{C}}$$

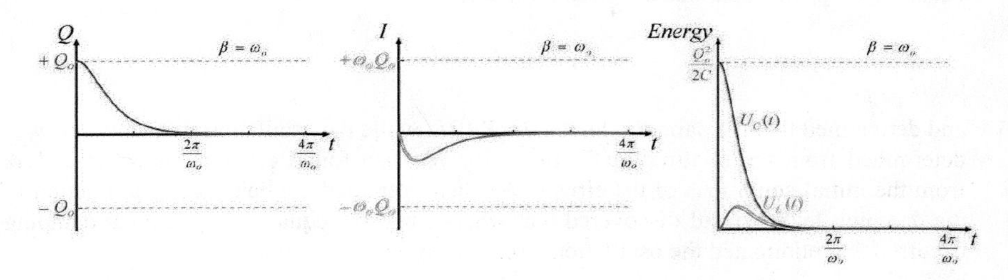

FIGURE 19.10 The time dependence for Q, the charge on the capacitor, I, the current and U_C and U_L, the energies stored in the capacitor and inductor in an *RLC* circuit with the damping factor $\beta = \omega_o$.

19.6 Summary

In this unit, we studied circuits that contained inductors and capacitors. We began by making a qualitative analysis of a circuit containing one inductor and one capacitor. The initial charge on the capacitor would begin to decrease as current would start to flow in the circuit. The inductor would maintain the current once the capacitor was discharged, resulting in the capacitor becoming charged with the opposite sign. This process would continue, resulting in both the charge on the capacitor and the current in the circuit oscillating in time. From an energy viewpoint, we could see that the energy in the system would also oscillate back and forth between the capacitor and the inductor.

We then went on to the quantitative analysis of the circuit by writing down Kirchhoff's voltage rule for the single loop and found a second order differential equation for Q, the charge on the capacitor. We assumed a solution of the form

$$Q(t) = Q_{\max}\cos(\omega t + \phi)$$

and determined that the oscillation frequency ω was given by

$$\omega = \frac{1}{\sqrt{LC}}$$

We verified that the sum of the energies stored in the capacitor and the inductor at any time t was a constant equal to a $Q_{\max}^2/2C$, the initial energy that was given to the capacitor. We noted that the mathematics of this situation is identical to that of the harmonic oscillator, e.g., a mass m connected by a spring of constant k to a fixed position in x, with the following substitutions: $x \to Q$, $m \to L$, and $k \to 1/C$.

We then investigated what would happen if a resistor were added to the circuit. We noted the main difference would be that energy would now be dissipated in the resistor when current flowed through it, leading to a damping of the oscillations. Once again we wrote down Kirchhoff's voltage rule for the single loop and found a second order differential equation for Q. We assumed a solution of the form

$$Q(t) = Ae^{-\beta t}\cos(\omega' t + \phi)$$

and determined that the damping factor $\beta = R/2L$, while the oscillation frequency ω' was determined from the relation: $\omega'^2 = \omega_o^2 - \beta^2$. We also found expressions for A and ϕ from the initial conditions of the circuit. We then examined the behavior as a function of the damping factor β and discovered that when β became equal to ω_o, critical damping occurred that eliminated the oscillations from the system.

MAIN POINTS

LC Circuit Is an Oscillator

An *LC* circuit oscillates at the natural frequency, ω_o.

$$\omega_o = \frac{1}{\sqrt{LC}}$$

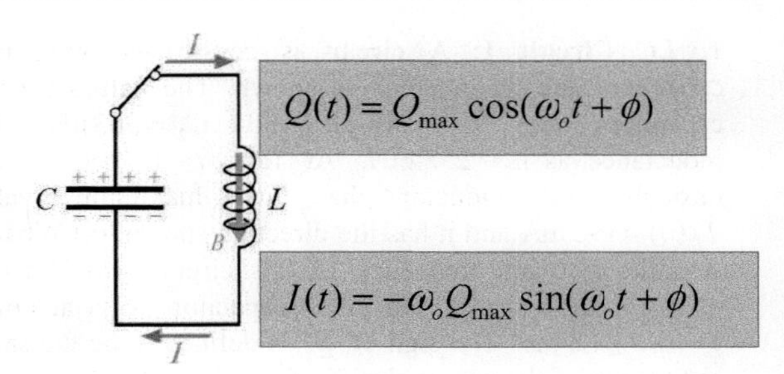

I
C
L
B
I
$Q(t) = Q_{\max}\cos(\omega_o t + \phi)$
$I(t) = -\omega_o Q_{\max}\sin(\omega_o t + \phi)$

Energy Oscillates in *LC* Circuit

Energy oscillates back and forth between the capacitor and inductor exactly as the kinetic energy and potential energy oscillate in a block and spring.

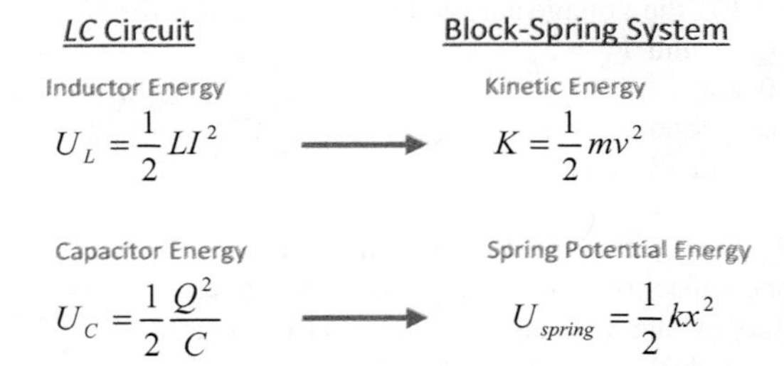

LC Circuit
Block-Spring System
Inductor Energy
$U_L = \frac{1}{2}LI^2$
Kinetic Energy
$K = \frac{1}{2}mv^2$
Capacitor Energy
$U_C = \frac{1}{2}\frac{Q^2}{C}$
Spring Potential Energy
$U_{spring} = \frac{1}{2}kx^2$

RLC Circuit Is a Damped Oscillator

The resistor removes energy from the system, thereby damping the oscillator.

A larger resistor damps more.

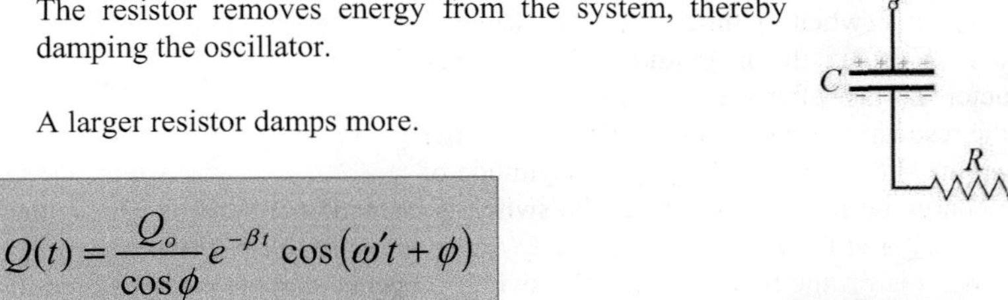

$$Q(t) = \frac{Q_o}{\cos\phi}e^{-\beta t}\cos(\omega' t + \phi)$$

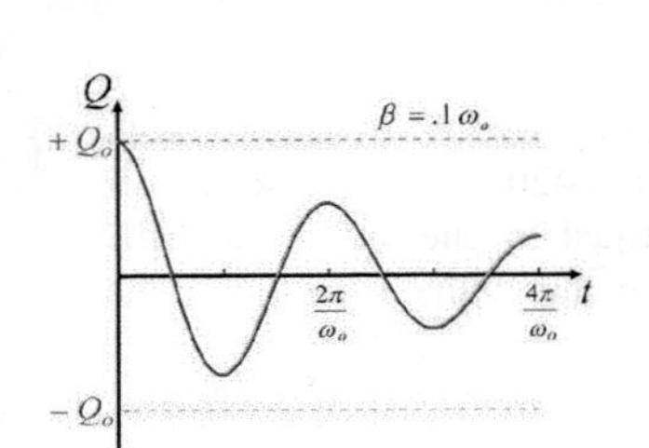

Damping Factor

$$\beta = \frac{R}{2L}$$

Oscillation Frequency

$$\omega'^2 = \omega_o^2 - \beta^2$$

PROBLEMS

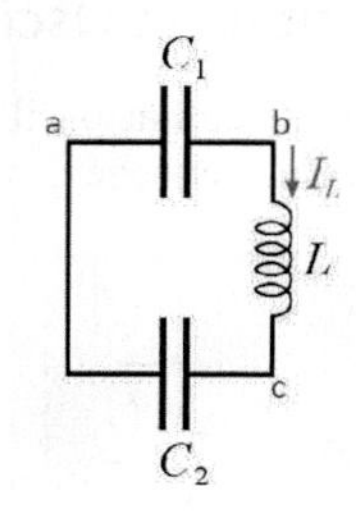

FIGURE 19.11 Problem 1

1. *LC* Circuit 1: A circuit is constructed with two capacitors and an inductor as shown. The values for the capacitors are: $C_1 = 347\ \mu\text{F}$ and $C_2 = 363\ \mu\text{F}$. The inductance is $L = 258$ mH. At time $t = 0$, the current through the inductor has its maximum value $I_L(0) = 198$ mA and it has the direction shown. (a) What is ω_o, the resonant frequency of this circuit? (b) What is $Q_1(t_1)$, the charge on the capacitor C_1 at time $t = t_1 = 24.8$ ms? The sign of Q_1 is defined to be the same as the sign of the potential difference $V_{ab} = V_a - V_b$ at time $t = t_1$. (c) What is $V_{bc}(t_1) = V_b - V_c$, the voltage across the inductor at time $t_1 = 24.8$ ms? Note that this voltage is a signed number. (d) What is $Q_{1,\max}$, the magnitude of the maximum charge on capacitor C_1? At time $t = t_2$, the magnitude of the current through the inductor has its maximum value. What are the magnitudes of Q_1, the charge on capacitor C_1, and V_L, the voltage across the inductor at this time?

(i) $Q_1 = Q_{\max}$ and $V_L = V_{\max}$
(ii) $Q_1 = 0$ and $V_L = V_{\max}$
(iii) $Q_1 = Q_{\max}$ and $V_L = 0$
(iv) $Q_1 = 0$ and $V_L = 0$

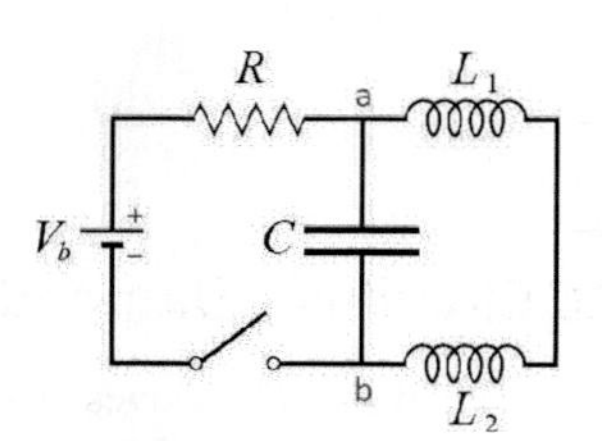

FIGURE 19.12 Problem 2

2. *LC* Circuit 2: A circuit is constructed with a resistor, two inductors, one capacitor, one battery, and a switch as shown. The value of the resistance is $R = 244\ \Omega$. The values for the inductances are $L_1 = 379$ mH and $L_2 = 134$ mH. The capacitance is $C = 191\ \mu\text{F}$ and the battery voltage is $V_b = 12$ V. The positive terminal of the battery is indicated with a + sign. (a) The switch has been closed for a long time when at time $t = 0$, the switch is opened. What is $U_{L,1}(0)$, the magnitude of the energy stored in inductor L_1 just after the switch is opened? (b) What is ω_o, the resonant frequency of the circuit just after the switch is opened? (c) What is $Q_{\max}$, the magnitude of the maximum charge on the capacitor after the switch is opened? (d) What is $Q(t_1)$, the charge on the capacitor at time $t = t_1 = 5.94$ ms. $Q(t_1)$ is defined to be positive if $V_a - V_b$ is positive. (e) What is t_2, the first time after the switch is opened that the energy stored in the capacitor is a maximum? (f) What is the total energy stored in the inductors plus the capacitor at time $t = t_2$?

3. *LC* Circuit (INTERACTIVE EXAMPLE): A simple *LC* circuit consists of a capacitor with capacitance $C = 0.05\ \mu\text{F}$ and an inductor with inductance $L = 420$ mH. Suppose that at time $t = 0$, the stored electric and magnetic energies are equal to one another and the instantaneous current is 75 mA. What is $Q_{\max}$, the maximum charge that is stored on the capacitor in this situation?

4. *LCR* Energy (INTERACTIVE EXAMPLE): Four resistors ($R_1 = 60\ \Omega$, $R_2 = 220\ \Omega$, $R_3 = 330\ \Omega$, and $R_4 = 480\ \Omega$), an ideal inductor ($L = 8$ mH), and a capacitor ($C = 250\ \mu$F) are connected to a battery ($V_b = 9$ V) through a switch, as shown in the figure below. The switch has been open for a long time before it is closed at $t = 0$. What is U_{stored}, the total stored energy in the circuit elements (not including the battery) a long time after the switch is closed?

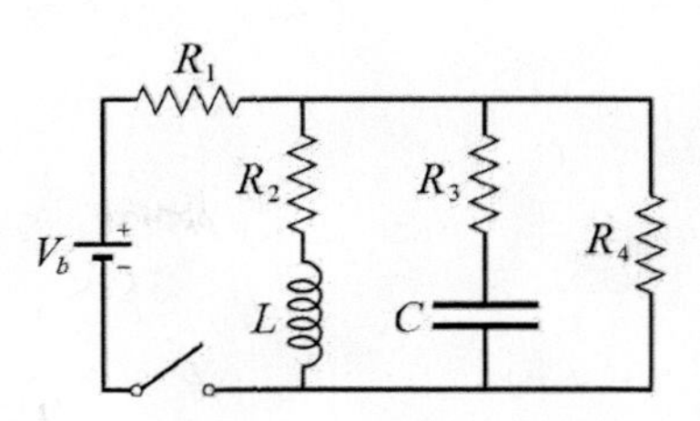

FIGURE 19.13 Problem 2

UNIT

20

AC CIRCUITS

20.1 Overview

In this unit we will begin our study of **alternating-current circuits** (AC circuits). These circuits are of immense practical importance. Almost all electrical power is produced and distributed using AC circuits. Generators naturally produce AC voltages that can be easily and efficiently stepped up or down as needed with the use of transformers. Our focus today will be the solution of the driven series *LCR* circuit. Next time we will focus on properties and applications of this solution, namely, resonant conditions, power considerations, and the use of transformers.

We'll start by introducing the use of an AC generator to sustain oscillations in a series *LCR* circuit. In order to develop a new technique for the solution of this circuit, we will investigate the behavior of simple circuits consisting of an AC generator and one other component (*R*, *C*, or *L*).

We will then introduce phasors, a graphical representation that explicitly represents the phase information in the circuit. We will then represent the voltages across each of the elements in the circuit as phasors so that we can develop a completely general graphical solution of the driven series *LCR* circuit.

20.2 The Driven Series *LCR* Circuit

In the last unit, we discovered that an *LC* circuit was a natural oscillator. Unfortunately, a pure *LC* circuit is impossible to construct since any real inductor has resistance. This resistance removes energy from the circuit so that the natural oscillations will eventually damp out.

It is possible, however, to construct a circuit in which oscillations can be sustained. The idea is to supply energy at the same rate it is being dissipated by the resistance. This energy can be supplied by the addition of an AC generator to the circuit, as shown in Figure 20.1. You may recall that we introduced generators as an example of Faraday's law in Unit 17. The idea is that if a conducting loop is rotated through a constant magnetic field, an *emf* is induced in the loop. This *emf* oscillates in time with the frequency of rotation of the loop.

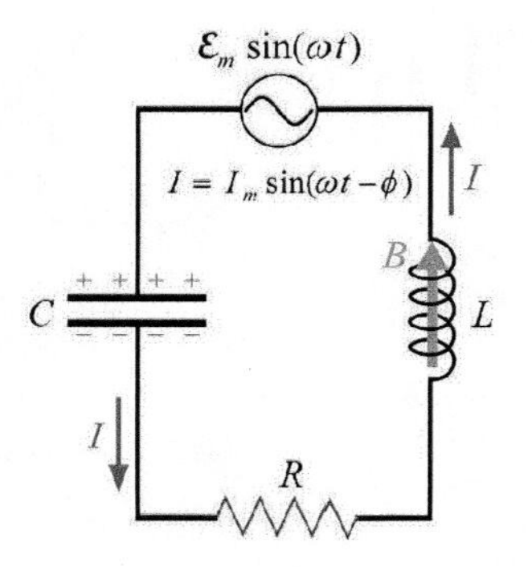

FIGURE 20.1 A series *LCR* circuit driven by an AC generator.

We can solve this driven series *LCR* circuit by writing down Kirchhoff's voltage rule for the loop and we will obtain a second order differential equation. With a wise assumption for the current in this circuit, we could slog through a lot of algebra to obtain the solution. Instead, we will look at a series of simple circuits containing only one element plus the AC generator to develop an understanding of the relation between the resulting current and the voltage across each element. We can then use this understanding to build a graphical solution to the general driven series *LCR* circuit.

20.3 Circuits with an AC Generator and One Other Element (*R*, *C*, or *L*)

We'll start with a circuit composed only of a resistor R and an AC generator with $\mathcal{E}(t) = \mathcal{E}_m \sin(\omega t)$, as shown in Figure 20.2. We start by writing down Kirchhoff's voltage rule for this circuit. Following the current shown, we first encounter a voltage drop across the resistor and then a voltage gain across the generator: $IR - \mathcal{E}_m \sin(\omega t) = 0$.

The solution is simple; at all times, the current is just equal to the voltage across the generator divided by *R*. In Figure 20.3, we plot V_R, the voltage across the resistor, and I_R, the current through the resistor, as a function of time. We see that V_R is in phase with I_R (i.e., at any time *t*, I_R is proportional to V_R).

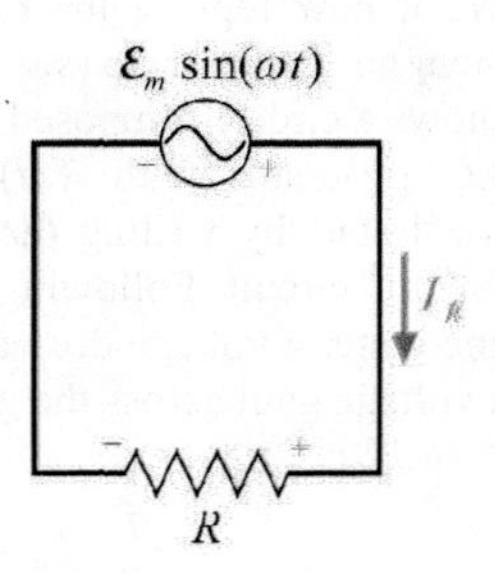

FIGURE 20.2 A circuit composed of a resistor *R* and an AC generator.

You may recall that if a vector rotates uniformly about the origin, its projection along any axis passing through the origin oscillates sinusoidally with time. Consequently, we can graphically represent the value of the current as a function of time as the vertical component of a vector with length I_{max} that rotates counterclockwise with frequency ω, as shown in Figure 20.4. The vector is horizontal at $t = 0$, giving zero current, increases to a positive maximum when $\omega t = \pi / 2$, and then decreases to zero again at $\omega t = \pi$ and becomes negative before completing the cycle at $\omega t = 2\pi$. This rotating vector is called a **phasor**.

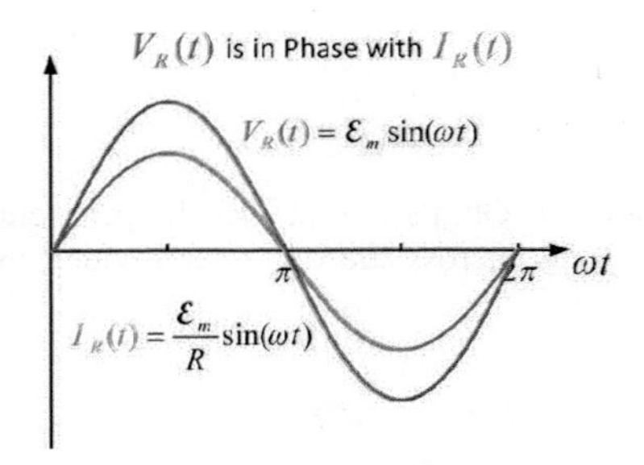

FIGURE 20.3 The time dependence of V_R, the voltage across the resistor, and I_R, the current for the circuit shown in Figure 20.2.

Since the voltage across the resistor is also sinusoidal, it too may be represented as a phasor. The length of the phasor corresponds to the maximum voltage across the resistor, and its initial orientation is horizontal just like the current. The two phasors are always on top of each other since the voltage across the resistor is in-phase with the current.

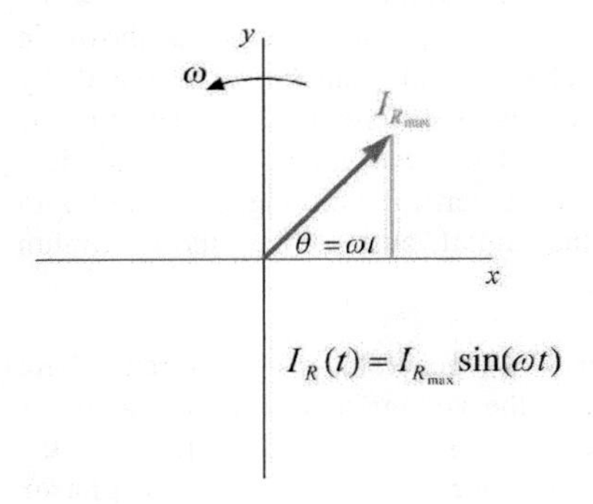

FIGURE 20.4 The current I_R represented as a phasor, a vector that rotates counterclockwise with frequency ω. The value of the current at any time is given by the projection of this phasor along the vertical axis: $I_R(t) = I_{R_{max}} \sin(\omega t)$.

We'll now replace the resistor with a capacitor and then an inductor to see what changes. Figure 20.5 shows a circuit composed only of a capacitor C and an AC generator with $\mathcal{E}(t) = \mathcal{E}_m \sin(\omega t)$. Once again, we'll start by writing down Kirchhoff's voltage rule for this circuit. Following the current shown, we first encounter a voltage drop across the capacitor and then a voltage gain across the generator.

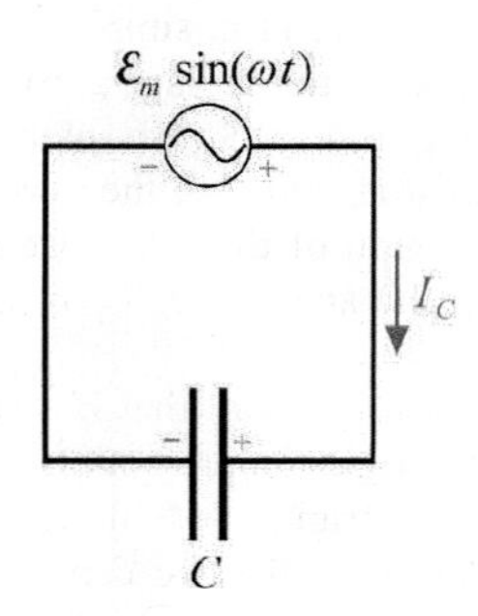

FIGURE 20.5 A circuit composed of a capacitor C and an AC generator.

$$\frac{q}{C} - \mathcal{E}_m \sin(\omega t) = 0$$

Solving this equation for q, we find that $q(t) = C\mathcal{E}_m \sin(\omega t)$. To find the current I_C, we just differentiate this expression, obtaining $I_C(t) = \omega C\mathcal{E}_m \cos(\omega t)$. Note that the peak current through the circuit is proportional to the peak voltage across the capacitor, just as it was for the resistor. If we define the **reactance** of the capacitor, X_C, as

$$X_C \equiv \frac{1}{\omega C}$$

we are left with an expression that is analogous to Ohm's law in that the peak current through the capacitor is equal to the peak voltage across the capacitor divided by its reactance.

$$(I_C)_{\max} = \omega C\mathcal{E}_m = \frac{\mathcal{E}_m}{X_C}$$

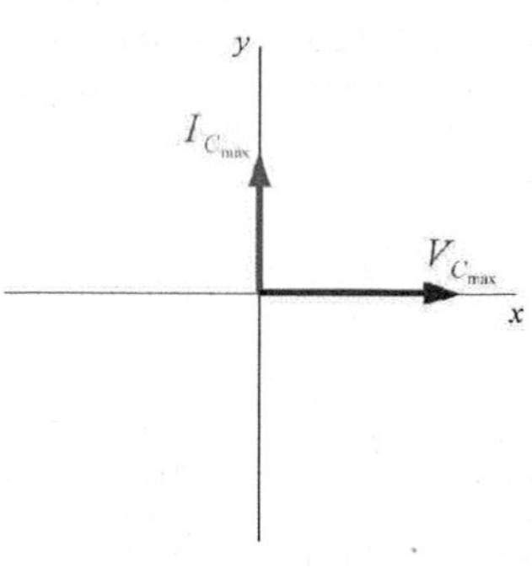

We can represent the voltage across the capacitor at any time as the projection along the vertical axis of a phasor whose length is equal to the maximum voltage across the capacitor. The initial orientation of the phasor is horizontal in order to insure that the initial voltage of the capacitor is zero, as shown in Figure 20.6. The current can be represented by another phasor whose length is equal to the maximum value of the current, but whose original orientation will be vertical pointing up in order to insure that the initial current has its maximum positive value.

FIGURE 20.6 The phasor diagram at $t = 0$ for the circuit shown in Figure 20.5. The current phasor leads the voltage phasor by 90°.

Unlike the previous generator-resistor circuit, here we see that V_C, the voltage across the capacitor, is *not* in phase with I_C, the current through the capacitor. Indeed, we can see from this **phasor diagram** or the plot of V_C and I_C as a function of time, as shown in Figure 20.7, that V_C lags I_C by a quarter cycle, or 90°.

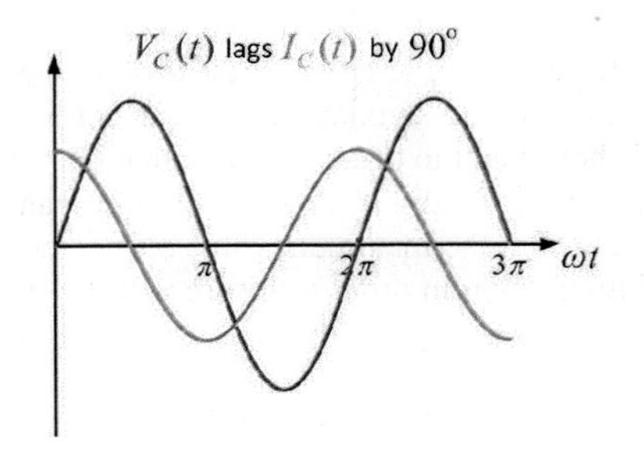

FIGURE 20.7 The time dependence of V_C, the voltage across the capacitor, and I_C, the current for the circuit, shown in Figure 20.5.

Qualitatively, we can understand this phase relationship between V_C and I_C in terms of the fact that it takes time for the charge from the current to accumulate on the plates of the capacitor to produce a voltage. This result is totally general for any sinusoidally-driven circuit; the voltage across a capacitor will always lag the current through the capacitor by 90°.

We'll now replace the capacitor with an inductor to see what changes. Figure 20.8 shows a circuit composed only of an inductor L and an AC generator with $\mathcal{E}(t) = \mathcal{E}_m \sin(\omega t)$. Once again, we'll start by writing down Kirchhoff's voltage rule for this circuit. Following the current shown, we first encounter a voltage drop across the inductor and then a voltage gain across the generator.

$$L\frac{dI_L}{dt} - \mathcal{E}_m \sin(\omega t) = 0$$

Solving this equation for dI_L / dt, we find that

$$\frac{dI_L}{dt} = \frac{\mathcal{E}_m}{L}\sin(\omega t)$$

To find the current I_L, we need to integrate this expression to obtain

$$I_L = \int dI_L = \int_0^t \frac{\mathcal{E}_m}{L}\sin(\omega t)\, dt = -\frac{\mathcal{E}_m}{\omega L}\cos(\omega t)$$

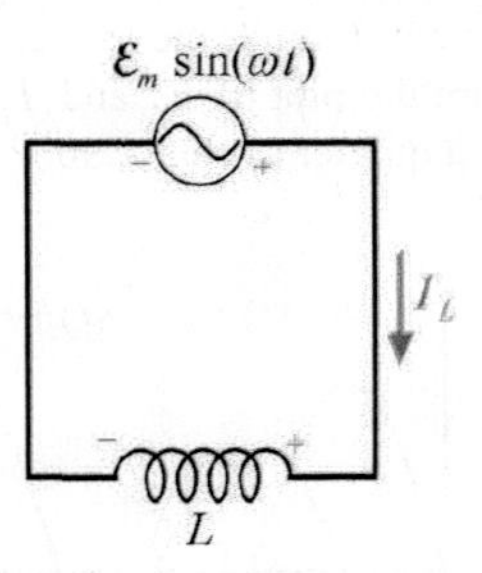

FIGURE 20.8 A circuit composed of an inductor L and an AC generator.

Note that the peak current through the inductor is proportional to the peak voltage across the inductor, just as it was for the resistor and capacitor. If we define the reactance of the inductor, X_L, as

$$X_L \equiv \omega L$$

We are left with an expression analogous to Ohm's law in that the peak current through the inductor is equal to the peak voltage across the inductor divided by its reactance.

$$(I_L)_{\max} = \frac{\mathcal{E}_m}{\omega L} = \frac{\mathcal{E}_m}{X_L}$$

We can represent the voltage across the inductor at any time as the projection along the vertical axis of a phasor whose length is equal to the maximum voltage across the inductor. The initial orientation of the phasor is horizontal in order to insure that the initial voltage across the inductor is zero, as shown in Figure 20.9. The current can be represented by another phasor whose length is equal to the maximum value of the current, but its original orientation will be vertical pointing down in order to insure that the initial current has its maximum negative value.

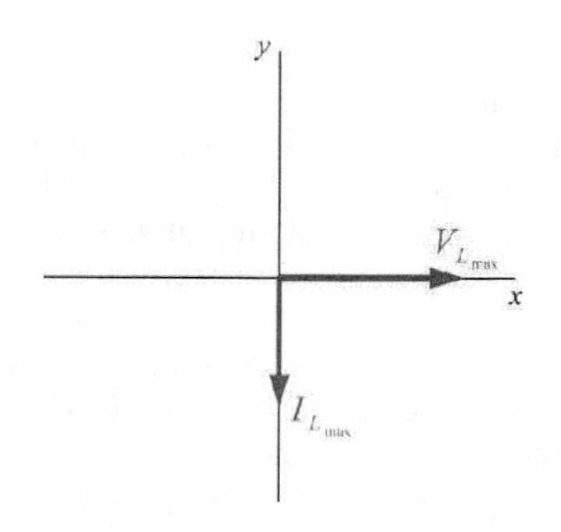

FIGURE 20.9 The phasor diagram at $t = 0$ for the circuit shown in Figure 20.8. The current phasor lags the voltage phasor by 90°.

Once again we see that V_L, the voltage across the inductor is *not* in phase with I_L, the current through the inductor. In this case, though, we can see from this phasor diagram or from the plot of V_L and I_L as a function of time shown in Figure 20.10 that V_L leads I_L by a quarter cycle, or 90°.

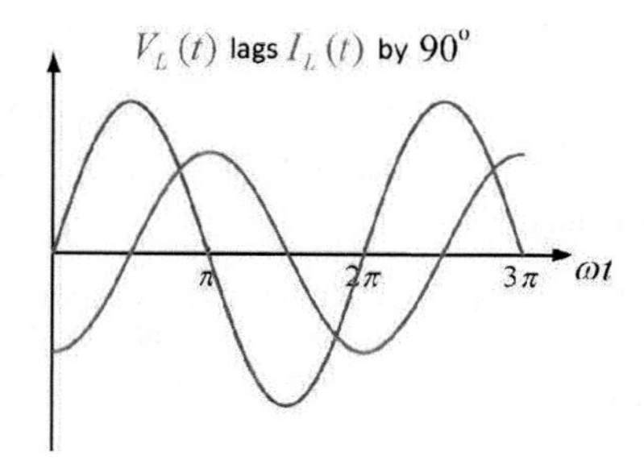

FIGURE 20.10 The time dependence of V_L, the voltage across the inductor, and I_L, the current for the circuit, shown in Figure 20.8.

Qualitatively, we can understand this phase relationship between V_L and I_L in terms of the fact that the voltage across the inductor can respond immediately to a small change in the current. This result is totally general for any sinusoidally-driven circuit; the voltage across an inductor will always lead the current through it by 90°.

We'll now look again at the driven series *LCR* circuit, armed with this knowledge about the phase relationships between current and voltage for each of the circuit elements.

20.4 The Driven *LCR* Circuit Solution from Phasors

We now return to our original *LCR* circuit driven by an AC generator, shown in Figure 20.1. How do we go about solving this circuit quantitatively? I guess the first question is: When we say we want to solve this circuit, what do we want to find? This circuit has just one loop; therefore, there is only one current. Our job is to find this current as a function of time.

From our recent experience with the circuits containing an AC generator and a single element, we saw that the resulting current always oscillated at the driving frequency ω, but that it was not always in phase with the generator voltage. Therefore, we can assume the current in this case to be of the form

$$I = I_m \sin(\omega t - \phi)$$

Solving the circuit, therefore, means determining I_m, the maximum amplitude of the current oscillations, and ϕ, the phase difference between the AC generator and the current.

To solve for I_m and ϕ, we do as we always do; we write down Kirchhoff's voltage rule for this loop and obtain the differential equation

$$L\frac{dI}{dt} + IR + \frac{q}{C} - \mathcal{E}_m \sin(\omega t) = 0$$

This equation contains an explicit time-dependent term that refers to the driving voltage $\mathcal{E}_m \sin(\omega t)$. If we substitute our form for *I* (and *q*) into this equation, we would eventually arrive at an equation of the form $A\sin(\omega t) + B\cos(\omega t) = 0$. For this equation to hold for all times *t*, we would require both *A* and *B* to be zero, giving us two equations in the two unknowns, I_m and ϕ. Rather than follow this tedious path, we will use the phasors that we introduced earlier to build a simple visual representation of the circuit behavior that will result in a direct solution for I_m and ϕ.

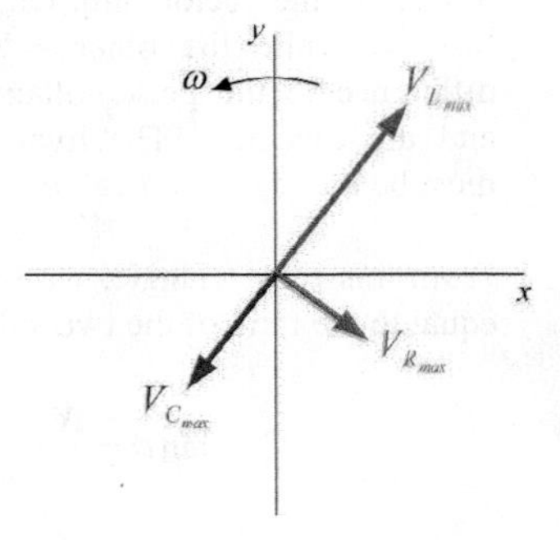

FIGURE 20.11 The voltage phasors for each element in the AC circuit shown in Figure 20.1 rotate together at frequency *w* with the fixed-phase relationship shown.

We will now use the phasors that we introduced earlier to actually solve for I_m and ϕ. The real power in the phasor representation comes from the fact that we know the phase relations between the current and the voltages across the resistor, inductor, and capacitor. In particular, we know the voltage across the resistor is always in phase with the current, while the voltage across the inductor leads the current by 90° and the voltage across the capacitor lags the current by 90°.

Consequently, these three voltage phasors can be considered as a single entity, as shown in Figure 20.11. The inductor phasor leads the resistor phasor by 90°, which in turn leads the capacitor phasor by 90°. The actual voltages at any time t can be determined by taking the projections along the vertical axis.

We know the phase relationships between the three phasors and their relative lengths (R, X_L, and X_C). We do not, however, know their absolute lengths since we do not know *the peak current.* We also do not know the phase angle ϕ between the driving voltage and the current, as shown in Figure 20.12. We will now determine these two unknowns from Kirchhoff's voltage rule.

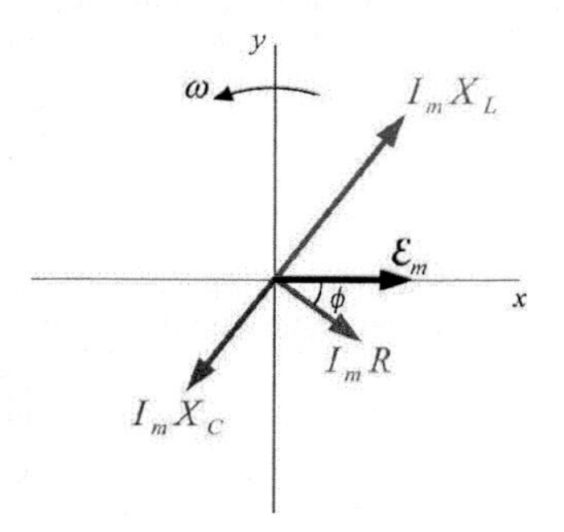

FIGURE 20.12 The voltage phasors at $t = 0$ for the AC circuit shown in Figure 20.1. The phase angle ϕ is defined as the angle between the driving *emf* ($\mathcal{E}_m$) and the voltage across the resistor ($I_m R$).

In terms of the phasors, Kirchhoff's voltage rule becomes a vector equation: namely, the vector sum of the three phasors must be equal to the phasor that represents the generator voltage.

$$\vec{V}_R + \vec{V}_L + \vec{V}_C = \vec{\mathcal{E}}_m$$

To find the solution, we simply draw the right triangle, shown in Figure 20.13, that represents the vector sum. One side of this triangle is simply the peak voltage across the resistor, while the other side corresponds to the difference of the peak voltages across the inductor and the capacitor. The hypotenuse of the triangle must be the maximum value of the generator voltage.

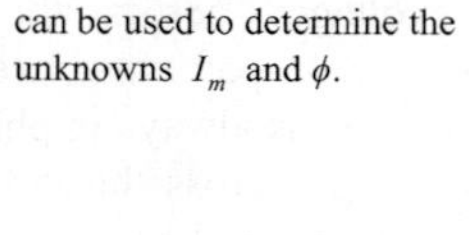

FIGURE 20.13 The right triangle obtained from the vector sum of the voltage phasors in Figure 20.12 that can be used to determine the unknowns I_m and ϕ.

From this right triangle, we see that the $\tan\phi$ is just equal to the ratio of the two sides of triangle.

$$\tan\phi = \frac{X_L - X_C}{R}$$

The maximum value of the current, I_m, can be found by applying the Pythagorean theorem to this triangle.

$$I_m = \frac{\mathcal{E}_m}{\sqrt{R^2 + (X_L - X_C)^2}} \equiv \frac{\mathcal{E}_m}{Z}$$

It is usual to define the denominator in this expression for I_m to be the **impedance**, Z, of the circuit, since it is equal to the ratio of the peak generator voltage to the peak current through the circuit. Since this last development has been pretty abstract, we'll close this unit with a specific example that demonstrates this phasor solution.

20.5 Example: Phasor Solution for Specific Driven *LCR* Circuit

Consider the driven *LCR* circuit shown in Figure 20.14. The generator has a maximum voltage of 10 V and oscillates at a frequency of 60 Hz, which converts to an angular frequency of 377 radians / second. The values for the circuit components are $20\,\Omega$, $20\,\text{mH}$, and $150\,\mu\text{F}$, respectively.

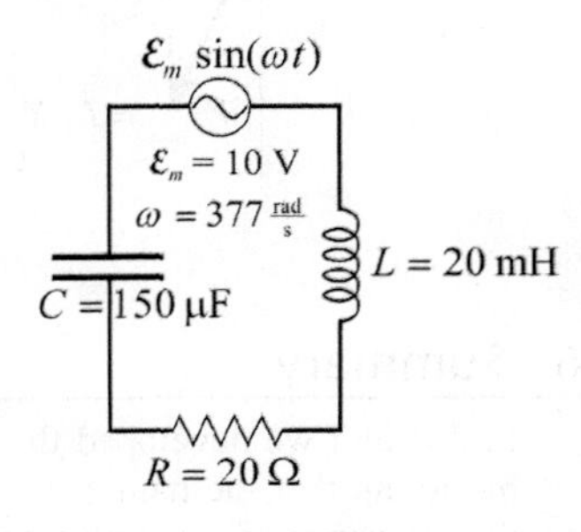

FIGURE 20.14

Our goal is to determine I_m, the maximum current, and ϕ, the phase angle between the current and the generator voltage. Once we know these two quantities, we can determine the values of any circuit parameter at any time. We can determine both of these quantities by simply constructing the **impedance triangle**, shown in Figure 20.15.

One side of the triangle corresponds to the resistance (20 Ω), while the other side is given by the difference in the reactances of the inductor ($\omega L = 7.5\,\Omega$) and the capacitor ($1/\omega C = 17.7\,\Omega$). The negative value ($-10.2\,\Omega$) indicates that the net impedance of the inductor and capacitor carries the phase of a capacitor. Consequently, the generator voltage will lag the current by a phase angle $\phi = -27°$, which is determined from the expression

$$\tan\phi = -10.2\,\Omega / 20\,\Omega = -0.51$$

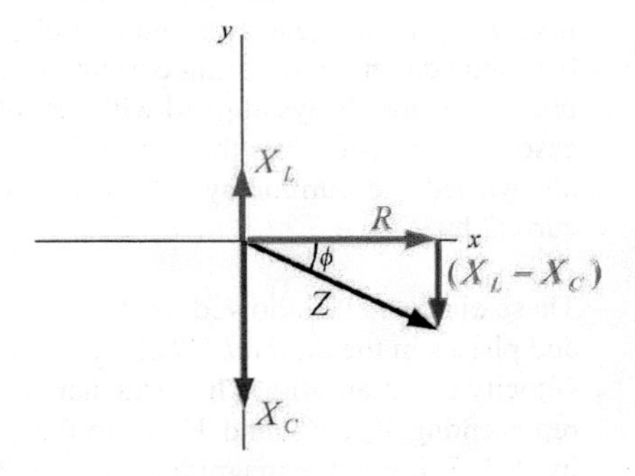

FIGURE 20.15 The impedance triangle for the circuit shown in Figure 20.14. The net reactance ($X_L - X_C$) is negative, indicating that the current leads the driving voltage in this circuit.

The impedance of the circuit is just equal to the hypotenuse of this triangle, which corresponds to 22.5 Ω. We can now determine the maximum current, by dividing the generator voltage by this impedance to obtain $I_m = 0.44$ A.

We can now construct the voltage phasor diagram at $t = 0$ by multiplying each component of our impedance triangle by our value for I_m, as shown in Figure 20.16. All components rotate together at frequency $\omega = 377$ radians / second. The voltage across any device can then be obtained by taking the projection of the appropriate component along the vertical axis.

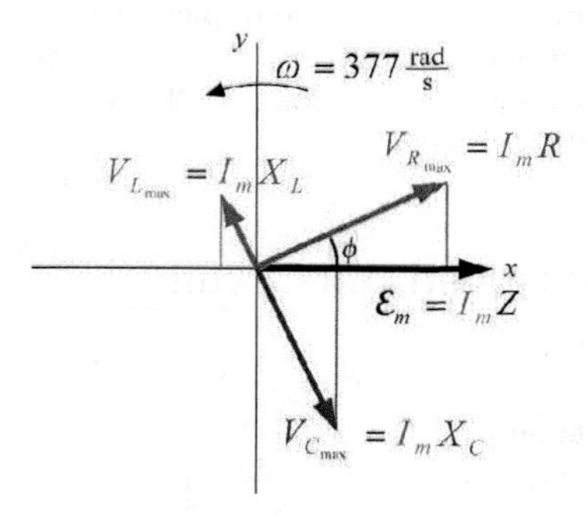

FIGURE 20.16 The voltage phasor diagram at $t = 0$ for the circuit shown in Figure 20.14. The voltages across each element are given by the vertical projections of the appropriate phasor as shown.

20.6 Summary

In this unit we developed the general solution to the driven series *LCR* circuit. We began by noting that the natural oscillations of a real *LC* circuit will always be damped due to the resistance of any real inductor. This damping can be eliminated, however, if we add an AC generator to the circuit. The generator will provide energy to the circuit at the same rate that it is dissipated by the resistance in the circuit.

Our initial steps in solving this driven series *LCR* circuit involved studying circuits involving an AC generator and one circuit element (*L*, *C*, or *R*). From these studies, we determined that the resulting current always oscillated at the driving frequency, but that its phase was not always aligned with that of the voltage of the generator. Indeed, only in the case of the resistor, was the current in phase with the voltage. For the inductor, the voltage always led the current by 90°, while for the capacitor, the voltage always lagged the current by 90°.

These observations allowed us to introduce a new graphical representation of the voltages and phases in the driven *LCR* circuit as phasors, vectors that rotate with a constant angular velocity about an origin. In particular, we found that the relative directions of the phasors representing V_R, V_L, and V_C were fixed, with V_L leading V_R by 90° and V_R leading V_C by 90°. The relative magnitudes of the V_R, V_L V_L, and V_C phasors were determined by the driving frequency ω and the values for *R*, *L*, and *C*, respectively. In particular, we characterized the frequency-dependent impedances of the inductor and capacitor by introducing the reactances, $X_L \equiv \omega L$ and $X_C \equiv 1/\omega C$.

For a generator voltage described by $\mathcal{E}(t) = \mathcal{E}_m \sin(\omega t)$, the current *I* can be written as $I(t) = I_m \sin(\omega t - \phi)$. The two unknowns, I_m, the maximum value of the current, and ϕ, the phase difference between the current and the generator voltage, can be determined from a graphical construction using the phasors. In particular, from the impedance triangle, we obtain

$$\tan\phi = \frac{X_L - X_C}{R}$$

$$I_m = \frac{\mathcal{E}_m}{\sqrt{R^2 + (X_L - X_C)^2}} \equiv \frac{\mathcal{E}_m}{Z}$$

The voltage across any element in the circuit can be determined at time t by taking the vertical projection of the corresponding voltage phasor.

MAIN POINTS

Series *LCR* Circuit Driven by an AC Generator Sustains Oscillations

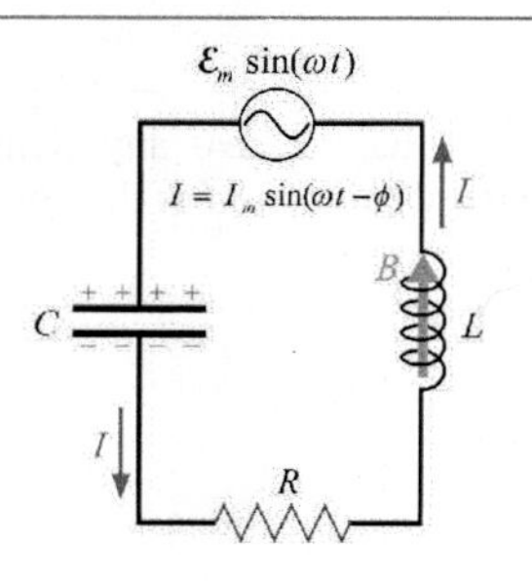

Energy supplied by a generator is dissipated by the resistor and stored in the capacitor and inductor.

Phasors

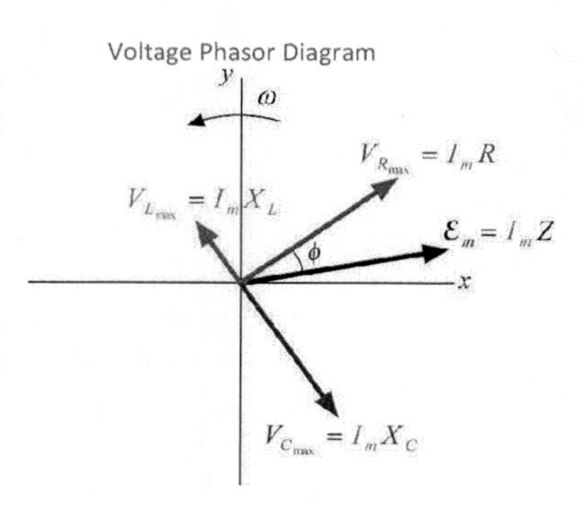

Phasors are vectors that rotate at the generator frequency and can be used to determine the current and the voltages in the circuit.

Reactances are defined to relate the maximum voltage across each element to the maximum current in the circuit.

$$X_L \equiv \omega L$$

$$X_C \equiv \frac{1}{\omega C}$$

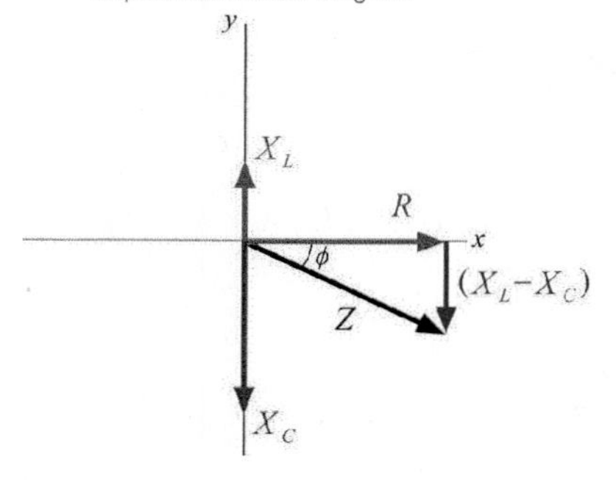

The maximum current (I_m) and the phase (ϕ) between the current and the driving voltage can be determined from the impedance triangle.

$$\tan\phi = \frac{X_L - X_C}{R}$$

$$I_m = \frac{\mathcal{E}_m}{\sqrt{R^2 + (X_L - X_C)^2}} \equiv \frac{\mathcal{E}_m}{Z}$$

PROBLEMS

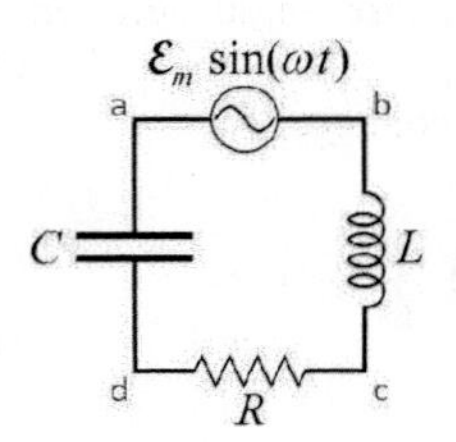

FIGURE 20.17 Problems 1-3

1. AC Circuit 1: A circuit is constructed with an AC generator, a resistor, a capacitor, and an inductor, as shown in Figure 20.17. The generator voltage varies in time as $\mathcal{E} = V_a - V_b = \mathcal{E}_m \sin(\omega t)$, where $\mathcal{E}_m = 120$ V and $\omega = 671$ rad/s. The values for the remaining circuit components are $R = 77\ \Omega$, $L = 169.6$ mH, and $C = 10.2\ \mu$F. (a) What is Z, the impedance of the circuit? (b) What is $I_{\max}$, the magnitude of the maximum value of the current in the circuit? (c) What is ϕ, the phase angle between the generator voltage and the current in this circuit? The phase ϕ is defined to be positive if the current leads the generator voltage, and negative otherwise. (d) What is t_1, the first time after $t = 0$ when the voltage across the inductor is zero? (e) Which of the following statements is true?

(i) The voltage across the generator is zero when the magnitudes of the voltages across the inductor and the capacitor are maximum.
(ii) The current in the circuit is zero when the magnitudes of the voltages across the inductor and the capacitor are maximum.
(iii) The magnitude of the voltage across the generator is maximum when the magnitudes of the voltages across the inductor and the capacitor are maximum.
(iv) The magnitude of the current in the circuit is maximum when the magnitudes of the voltages across the inductor and the capacitor are maximum.
(v) There is no time when the magnitudes of the voltages across the inductor and capacitor are maximum.

(f) What is $V_C = V_d - V_a$, the voltage across the capacitor, at time $t = 0$? Note that V_C is a signed number.

2. AC Circuit 2: A circuit is constructed with an AC generator, a resistor, a capacitor, and an inductor, as shown in Figure 20.17. The generator voltage varies in time as $\mathcal{E} = V_a - V_b = \mathcal{E}_m \sin(\omega t)$, where $\mathcal{E}_m = 120$ V and $\omega = 548$ rad/s. The values for the remaining circuit components are $R = 88\ \Omega$, $C = 372\ \mu$F. The value for L is unknown. What is known is that the voltage across the generator leads the current in the circuit by $\phi = 62°$. (a) What is t_1, the first time after $t = 0$ when the magnitude of the voltage across the inductor is maximum? (b) What is Z, the impedance of the circuit? (c) What is L, the value of the inductor? (d) What is $V_{L_{\max}}$, the magnitude of the maximum voltage across the inductor? (e) What is $V_L = V_b - V_c$, the voltage across the inductor, at time $t = 0$? Note that V_L is a signed number. (f) In order to make the voltage across the generator become in phase with the current in the circuit (i.e., make the phase ϕ equal to zero), how would the value of C have to change, keeping all other circuit parameters the same?

(i) The value of C would need to be decreased.
(ii) The value of C would need to be increased.
(iii) It is impossible to bring the voltage across the generator in phase with the current in the circuit by only changing the value of C, keeping all other circuit parameters fixed.

3. AC Circuit I (INTERACTIVE EXAMPLE): A series *RLC* circuit ($L = 24$ mH, $C = 40$ μF, and R is unknown) has an AC generator with frequency $f = 310$ Hz and amplitude $\mathcal{E}_{max} = 120$ V. The peak instantaneous current in the circuit is $I_{max} = 1.4$ A. What is ϕ, the phase angle between the driving *emf* and the current in the circuit? Define ϕ to be positive if the voltage leads the current and ϕ to be negative if the current leads the voltage.

UNIT

21

AC CIRCUITS: RESONANCE AND POWER

21.1 Overview

In the last unit we developed a general solution for the driven series *LCR* circuit. We now want to examine in more detail some properties of this solution. We'll start by looking at how the amplitude of the current oscillations varies with the driving frequency. We will discover that the maximum current amplitude occurs when the current is in phase with the generator voltage. This condition is called *resonance* and occurs at the natural frequency of the corresponding *LC* circuit, namely,

$$\omega_o = \frac{1}{\sqrt{LC}}$$

We will then examine the power that is delivered to the circuit. We will introduce the quality factor Q of the circuit that describes the sharpness of the resonance. We will close with a discussion of the use of transformers, an important application of these ideas. Transformers are used to step up or step down AC voltages so that power can be efficiently delivered to a circuit at a voltage different from that of the generator.

21.2 The Driven *LCR* Circuit

In the last unit we obtained the general solution for the driven series *LCR* circuit, shown in Figure 21.1. We used the method of phasors in which we represented impedances and voltages as vectors that rotated in a two-dimensional plane with the frequency of the generator. What follows is a brief review of that solution.

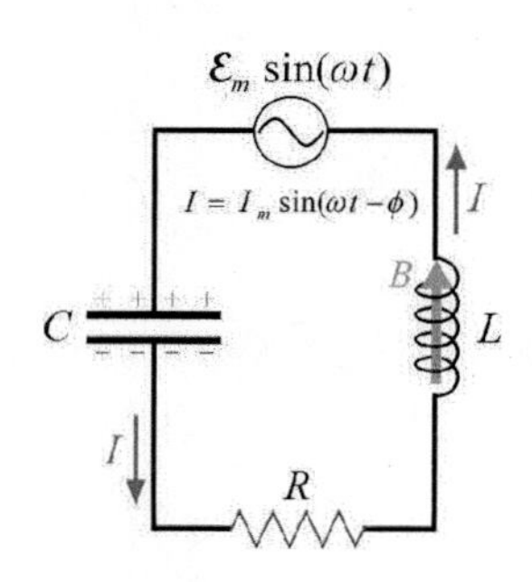

FIGURE 21.1 A series *LCR* circuit driven by an AC generator.

The voltage triangle we developed that allows us to read off the solution is shown in Figure 21.2. Namely, the hypotenuse of the right triangle is just $\mathcal{E}_m$, the amplitude of the generator voltage oscillations; one side is the amplitude of the voltage across the resistor ($I_m R$) and the other side is given by the difference in the amplitudes of the voltages across the inductor and capacitor ($I_m(X_L - X_C)$). The tangent of the phase angle ϕ between the current and driving voltage is just given by the ratio of the difference of the inductive and capacitive reactances, $X_L - X_C$, to the resistance *R*.

$$\tan\phi = \frac{X_L - X_C}{R}$$

If X_L is larger than X_C, ϕ is positive, indicating that the current is lagging the generator voltage. If X_L is smaller than X_C, ϕ is negative, indicating that the current is leading the generator voltage. We define *Z*, the impedance of the circuit, to be the magnitude of the vector sum of the phasors that represent the reactance difference ($X_L - X_C$) and the resistance *R*. We can then determine that I_m, the amplitude of the current oscillations, is just given by the ratio of $\mathcal{E}_m$ to the impedance *Z*.

FIGURE 21.2 The right triangle obtained from the vector sum of the voltage phasors in Figure 20.12 that can be used to determine the unknowns I_m and ϕ.

$$I_m = \frac{\mathcal{E}_m}{\sqrt{R^2 + (X_L - X_C)^2}} \equiv \frac{\mathcal{E}_m}{Z}$$

If this brief review of the driven series *LCR* circuit is not clear, you may want to look back at the end of the last unit in which we stepped through a concrete example of the solution. We will now use these features of the general solution to the driven series *LCR* circuit to explore the important phenomenon of resonance.

21.3 Resonance

It is clear from our phasor solution to the driven series *LCR* circuit that the amplitude of the current oscillations (I_m) depends on the driving frequency (ω), since the inductive reactance ($X_L = \omega L$) increases with frequency and the capacitive reactance ($X_C = 1/\omega C$) decreases with frequency. I_m has its maximum value at a frequency ω_o for which the impedance of the circuit

$$Z = \sqrt{R^2 + (X_L - X_C)^2}$$

is a minimum. We call this condition **resonance** and we see that it will occur when $X_L - X_C = 0$. Expanding X_L and X_C to make the frequency dependences explicit, we obtain the **resonant frequency** ω_o:

$$\omega_o = \frac{1}{\sqrt{LC}}$$

Note that this resonant frequency is exactly the same as the natural frequency of the oscillations of a circuit composed only of the inductor *L* and the capacitor *C*!

Therefore, we see that that the maximum current is delivered to the *LCR* circuit when the driving frequency is set equal to the natural oscillation frequency of the corresponding *LC* circuit. At this frequency the circuit behaves as though the resistor is the only element present. In other words, at the resonant frequency, the maximum current in the circuit is simply given by the maximum voltage of the generator divided by the resistance, and the current through the resistor is in phase with the generator.

$$I_m(\omega_o) = \frac{\mathcal{E}_m}{R}$$

We'll now examine the general frequency dependence of I_m, the amplitude of the current oscillations in a driven *LCR* circuit. We just discovered that I_m has a maximum at $\omega = \omega_o$, the natural oscillation frequency of the *LC* circuit. What is I_m at other frequencies?

We know in general that the peak current is just equal to the maximum voltage of the generator divided by the impedance of the circuit.

$$I_m(\omega) = \frac{\mathcal{E}_m}{Z}$$

This impedance can be rewritten in terms of the resistance in the circuit and the phase between the driving voltage and the current by simply looking at the impedance triangle shown in Figure 21.2. Clearly, the impedance is just equal to the resistance divided by $\cos\phi$.

$$Z = \frac{R}{\cos\phi}$$

Consequently, we see that the peak current at any frequency ω is simply equal to $\cos\phi$ times its value at the resonant frequency ω_o.

$$I_m(\omega) = I_m(\omega_o)\cos\phi$$

The frequency dependence of the peak current then is absolutely identical to the frequency dependence of $\cos\phi$. We can obtain this dependence by simply expanding the reactance terms in terms of the frequency.

$$\cos\phi = \frac{R}{\sqrt{R^2 + (X_L - X_C)^2}} = \frac{R}{\sqrt{R^2 + \left(\omega L - \dfrac{1}{\omega C}\right)^2}}$$

Further, since we know I_m has a maximum at the resonant frequency ω_o, it makes sense to introduce the dimensionless quantity $x \equiv \omega / \omega_o$. When we make this substitution, we can obtain, with some work, the expression

$$I_m = \frac{\mathcal{E}_m}{R}\cos\phi = \frac{\mathcal{E}_m}{R}\frac{1}{\sqrt{1 + Q^2\dfrac{(x^2-1)^2}{x^2}}}$$

where we have introduced a new quantity $Q^2 \equiv L / R^2 C$. We will see this factor again shortly when we discuss the power delivered to the circuit. Figure 21.3 shows a plot of the *peak current* (I_m) as a function of frequency (ω) for $Q = 2$. Had we chosen a larger value for Q, the peak at $\omega = \omega_o$ would have been narrower. We will begin our discussion of the power delivered to the circuit in the next section.

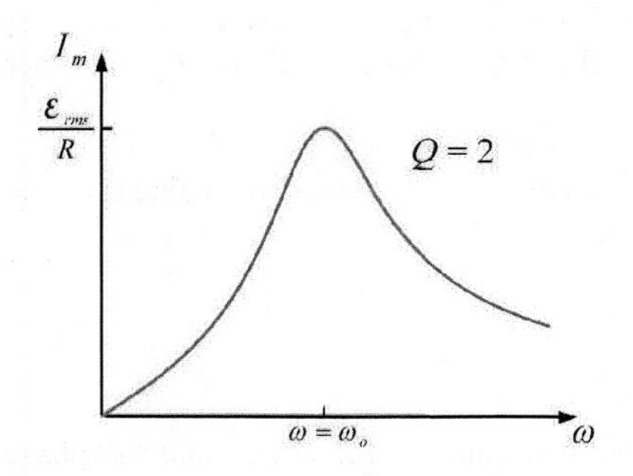

FIGURE 21.3 The maximum current (I_m) as a function of the driving frequency (ω) for a circuit with $Q = 2$.

21.4 Power

In the driven series *LCR* circuit, energy is delivered to the circuit by the generator. Some of this energy is dissipated in the resistor and some is stored and released in the capacitor and in the inductor.

The relevant quantity here is the average power delivered to the circuit by the generator in one complete cycle. This quantity is also equal to the average power dissipated in the resistor during one cycle. We can see that the inductor and capacitor contribute nothing to the average power in the circuit by simply noting that the voltage across each element is 90° out of phase with the current through the element; when we average the power (the product of the voltage and the current), we will be averaging the product of a sine and a cosine, which is equal to zero as can easily be seen from the plot in Figure 21.4.

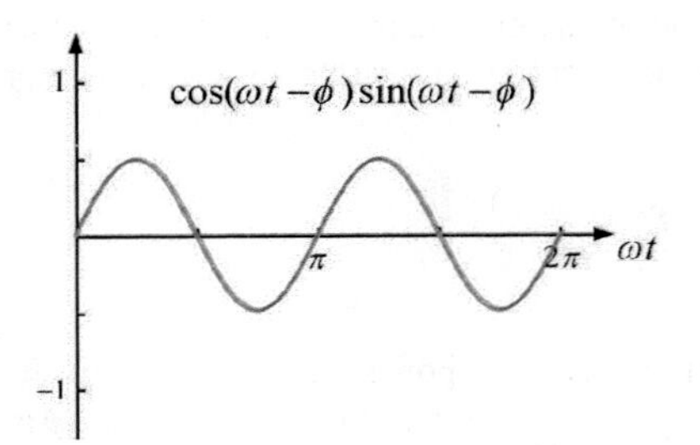

FIGURE 21.4 The time dependence of the power in an inductor in a driven series *LCR* circuit. The average value over one cycle ($\omega t = 2\pi$) is zero.

$$\langle P_L \rangle = I_m^2 X_L \langle \cos(\omega t - \phi)\sin(\omega t - \phi) \rangle = 0\,.$$

$$\langle P_C \rangle = I_m^2 X_C \langle -\cos(\omega t - \phi)\sin(\omega t - \phi) \rangle = 0$$

The average power dissipated by the resistor is the average of the product of the square of the current and the resistance. The average of the square of the current is just equal to the square of the peak current times the average of $\sin^2(\omega t - \phi)$, which is equal to one-half, as can be seen from the plot in Figure 21.5.

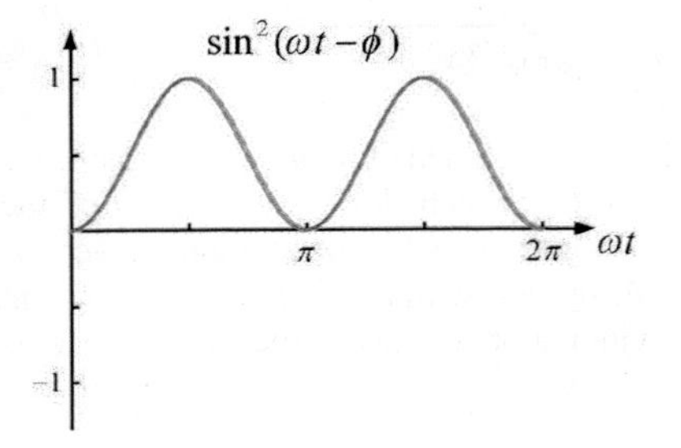

FIGURE 21.5 The time dependence of the power in a resistor in a driven series *LCR* circuit. The average value over one cycle ($\omega t = 2\pi$) is ½.

We can express this result in terms of the peak voltage supplied by the generator by noting that the resistance is just equal to the impedance of the circuit times the $\cos\phi$, and that the product of the impedance and the peak current is equal to the peak voltage supplied by the generator.

$$\langle P_R \rangle = I_m^2 R \langle \sin^2(\omega t - \phi) \rangle = I_m^2 R \left(\frac{1}{2} \right) = \frac{1}{2} I_m^2 Z \cos\phi = \frac{1}{2} I_m \mathcal{E}_m \cos\phi$$

We arrive, then, at the very simple expression shown for the average power delivered to the circuit in one cycle. The factor of $\cos\phi$ is usually called the **power factor**. Often this expression is rewritten in terms of the **root mean square**, or *rms*, values of the driving voltage and the current. For sinusoidal oscillations, the *rms* value is equal to the maximum value (the amplitude) divided by $\sqrt{2}$. Using the *rms* values, we see that the average power is just equal to the product of generator voltage times the *rms* current times the power factor.

$$\langle P_R \rangle = I_{rms} \mathcal{E}_{rms} \cos\phi$$

Both the peak current and the power factor depend on the driving frequency. To make this dependence more transparent, we rewrite the peak current in terms of the generator voltage, the resistance and the power factor.

$$\langle P_R \rangle = \langle P_{generator} \rangle = \frac{\mathcal{E}_{rms}^2}{R} \cos^2\phi$$

We now see that the frequency dependence of the average power is just given by the frequency dependence of $\cos^2\phi$. In the last section, we derived an expression for $\cos\phi$ in terms of the dimensionless variable $x \equiv \omega / \omega_o$, where ω_o is the resonant frequency of the circuit

$$\cos\phi = \frac{1}{\sqrt{1 + Q^2 \dfrac{(x^2 - 1)^2}{x^2}}}$$

When we replace $\cos\phi$ by this expression, we obtain an expression for the average power delivered to the circuit as a function of x

$$\langle P_{generator} \rangle = \frac{\mathcal{E}_{rms}^2}{R} \cos^2\phi = \frac{\mathcal{E}_{rms}^2}{R} \frac{x^2}{x^2 + Q^2 (x^2 - 1)^2}$$

This expression has its maximum value at $x = 1$ (namely, when $\omega = \omega_o$). Note that once again we have introduced the quantity $Q^2 = L / R^2 C$, which defines sharpness of the peak. For example, the peak in the plot shown in Figure 21.6 for the average power for $Q = 2$ looks about twice as broad as that for $Q = 4$. In fact, with a bit of algebra, you can show that in the limit that $Q >> 1$, the full width at half maximum of the peak is approximately equal to $1 / Q$!

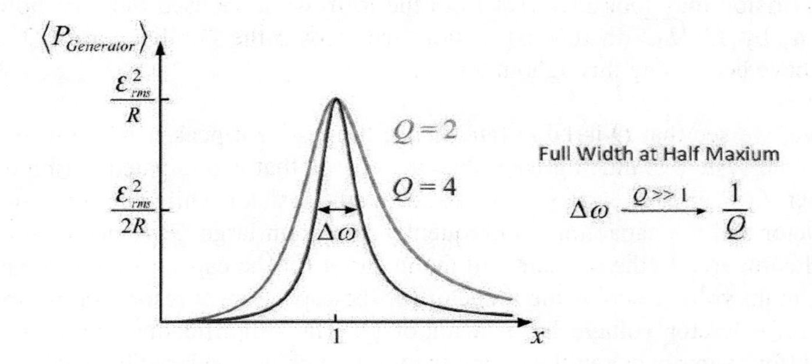

FIGURE 21.6 The average power supplied by the generator to the circuit in Figure 21.1 as a function of x, the ratio of the driving frequency to the resonant frequency. Plots are made for two values of $Q=\sqrt{L/R^2C}$.

21.5 Q Factor

Up to this point, we have introduced the quantity $Q^2 \equiv L/R^2C$ as a means to simplify expressions for both the maximum current and average power delivered to a driven series *LCR* circuit. We have just seen that Q is a measure of the sharpness of the resonant peak in the average power delivered to the circuit.

In fact, this quantity that we have been calling Q is defined more generally for any oscillating system in terms of the energy stored and dissipated in the system. In particular, the **quality factor** Q for any oscillating system is defined in terms of the ratio of U_{max}, the maximum energy stored in the system, to ΔU, the energy that is dissipated in the system during one cycle, i.e.,

$$Q \equiv 2\pi \frac{U_{max}}{\Delta U}$$

In this expression, both U_{max} and ΔU are meant to be evaluated at the resonance frequency.

In our case of the driven series *LCR* circuit, we know both the maximum energy stored in the circuit and also the energy dissipated in the resistor in each cycle. In particular, the maximum energy stored in the circuit is equal to the maximum energy stored in either the inductor or the capacitor ($U_{max} = ½ LI_{max}^2$) and the energy dissipated in the system is equal to the product of the product of the average power in the resistor ($½ I_{max}^2 R$) and the period of the oscillation ($2\pi / \omega_o$). Combining these equations, we obtain

$$Q = \frac{\omega_o L}{R}$$

This expression may look different from the form we have used thus far, however, if we replace ω_0 by $1/\sqrt{LC}$ in this expression, we recover the familiar form ($Q^2 = L/R^2C$) that we have been using throughout this unit.

Therefore, we see that Q is large (and hence the resonant peak narrow) when the energy stored in the system is much greater than the energy that is dissipated during one cycle. In our driven *LCR* circuit, energy is dissipated in the resistor while energy is stored in both the inductor and the capacitor. Consequently, to obtain large Q we need the resistance to be small compared to the reactance of the inductor (or the capacitor) at resonance. In fact, the maximum voltage across the inductor (or the capacitor) at resonance is larger than the maximum generator voltage by a factor of Q! This amplification factor of Q becomes essential, for example, when the circuit is used to pick up weak radio signals.

21.6 Transformers

We'll close this unit with a discussion of a very important application, namely, **transformers**. Transformers are used to step up or step down AC voltages so that power can be efficiently delivered to a circuit at a voltage different from that of the generator. For example, transformers are used at the power generation stage to increase the voltage so that the power can be transmitted efficiently, and at the user stage to decrease the voltage to allow it to be used more safely.

The ideal transformer consists of an iron core around which are wrapped two sets of coils. One set of coils, the primary, consists of N_P turns and is connected to an AC generator, as shown in Figure 21.7. The alternating current in the primary coil produces a changing magnetic flux in the iron. This flux is trapped by the iron and extends through the secondary coil, consisting of N_S turns. The "ideal" piece here refers to the assumptions that the coils have negligible resistance and that the entire flux is contained in the iron.

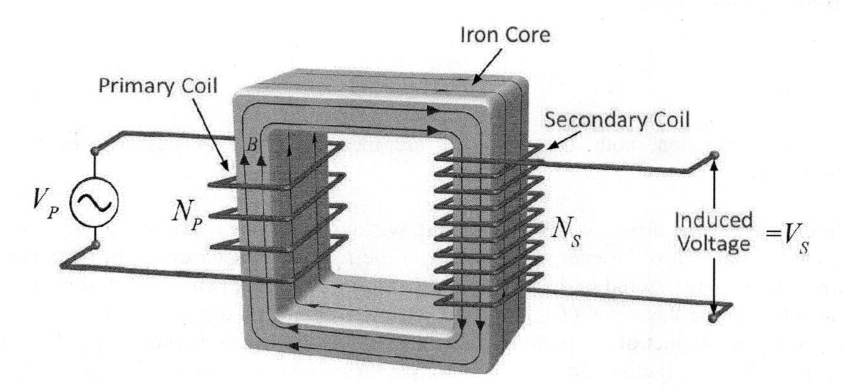

FIGURE 21.7 An ideal transformer with N_P turns on the primary and N_S turns on the secondary connected to an AC generator with voltage V_P produces an induced voltage on the secondary, V_S.

This changing magnetic flux through the secondary coil will induce an *emf* in that coil in accordance with Faraday's law. In particular, the *emf* induced per turn in the secondary is just equal to the *emf* induced per turn in the primary. Consequently, we obtain the relation that V_S, the voltage across the secondary coil, is equal to the product of V_P, the generator voltage and the ratio of N_S to N_P:

$$V_S = V_P \frac{N_S}{N_P}$$

If $N_S > N_P$, the transformer can be used to step up the primary voltage, while if $N_S < N_P$, the transformer can be used to step down the primary voltage.

If we now connect a resistive load to the secondary coils, current will be drawn and energy will be dissipated in the load. Once current begins to flow in the secondary coils, a new alternating magnetic field is created in the iron. This changing magnetic flux then extends to the primary coil. This flux opposes the initial flux produced by the generator. The voltage across the primary, however, cannot change since it is determined by the generator. Consequently, a new current, I_P, is generated in the primary with a phase such that the *emf* it induces in the primary exactly cancels the *emf* induced in the primary by I_S. The current I_P is in phase with the voltage across the transformer.

This process sounds (and is) quite complicated. However, to determine I_P, we can avoid looking at the details by simply applying conservation of energy. Namely, we know energy is being dissipated in the load at a rate equal to the voltage across the secondary times the current through it. In the ideal case that we have assumed, there are no other energy losses so that the power dissipated in the load must be equal to the power transferred from the generator to the primary, namely $I_P V_P$. Consequently, we see that I_P is given by

$$I_P = I_S \frac{V_S}{V_P} = I_S \frac{N_S}{N_P}$$

Therefore, in this ideal case, if we step the voltage up by a factor of N_S / N_P, the current that is drawn from the generator is larger than that drawn by the load by exactly this same ratio (N_S / N_P).

21.7 Summary

In this unit we examined resonance and power considerations in the driven series *LCR* circuit. We began by noting that I_m, the amplitude of the current oscillations, obtains its maximum value when the inductive reactance is equal to the capacitive reactance so that the total impedance of the circuit is just resistive. This condition occurs at the natural oscillation frequency of the corresponding *LC* circuit:

$$\omega_o = \frac{1}{\sqrt{LC}}$$

When we plotted I_m as a function of ω, we saw that the sharpness of the peak was determined by a factor defined to be $Q \equiv \sqrt{L/R^2C}$. The larger the Q factor, the sharper the peak.

We then discussed the average power delivered to the circuit by the generator during one cycle. We found that this average power could be expressed simply as the product of a power factor ($\cos\phi$) and the *rms* values of the current and voltage.

$$\langle P_{generator} \rangle = I_{rms} \mathcal{E}_{rms} \cos\phi$$

Once again we see that the peak in the plot of the average power as a function of ω occurs at ω_o, with the full-width at half maximum approximately equal to $1/Q$. We then found that this Q factor is a general property of all oscillating systems, being defined from

$$Q \equiv 2\pi \frac{U_{\max}}{\Delta U}$$

where $U_{\max}$ is the maximum energy stored in the system, and ΔU is the energy dissipated during one cycle when the system is at resonance.

Finally, we discussed the operation of a transformer. The key to understanding this device is Faraday's law, in that the flux produced by a primary coil being driven by a generator induces an *emf* in the secondary coil. We found that, for an ideal transformer (ideal in the sense that it has negligible resistance and that all magnetic flux remains in the iron core) the ratio of the induced *emf* in the secondary coil to that of the generator is just equal to the ratio of the number of turns in secondary to the number of turns in the primary, i.e.,

$$V_S = V_P \frac{N_S}{N_P}$$

We then used conservation of energy to determine that when a resistive load is connected to the secondary coil, the ratio of the induced current in the primary to that in the secondary is also equal to the ratio of the number of turns in secondary to the number of turns in the primary, i.e.,

$$I_P = I_S \frac{N_S}{N_P}$$

MAIN POINTS

Resonance in an AC Circuit

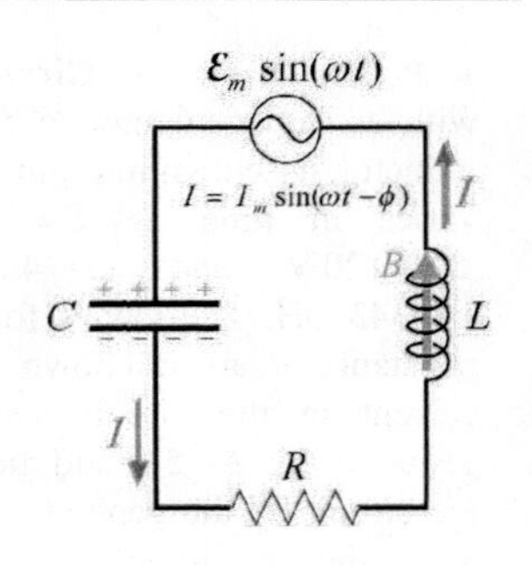

The resonant frequency of a driven series *LCR* circuit is defined to be the frequency at which the peak current in the circuit is a maximum.

$$X_L = X_C \longrightarrow Z = R \longrightarrow I_m = \frac{\mathcal{E}_m}{R}$$

Resonant Frequency

$$\omega_o = \frac{1}{\sqrt{LC}}$$

Power in an AC Circuit

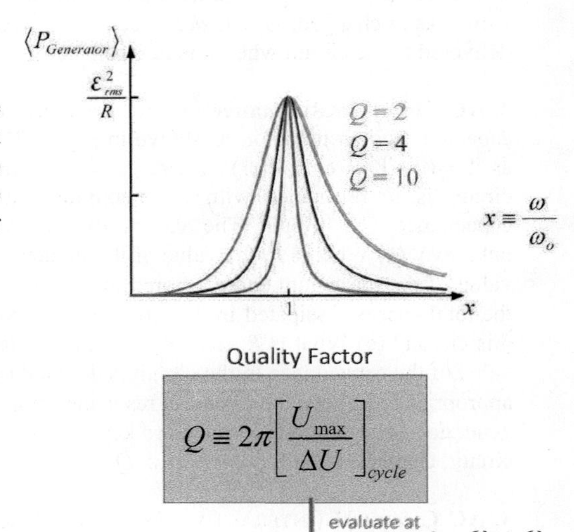

The frequency dependence of the average power delivered to a driven series *LCR* circuit is proportional to $\cos^2\phi$, where ϕ is the phase angle between the current and the generator voltage.

$$\langle P_{generator} \rangle = I_{rms} \mathcal{E}_{rms} \cos\phi$$

Quality Factor

$$Q \equiv 2\pi \left[\frac{U_{max}}{\Delta U} \right]_{cycle}$$

evaluate at $\omega = \omega_o$

Transformers

Transformers are used to induce a voltage in a secondary coil via Faraday's law that steps up or steps down an AC voltage presented to the primary coil.

$$\frac{V_S}{V_P} = \frac{N_S}{N_P} \qquad \frac{I_P}{I_S} = \frac{N_S}{N_P}$$

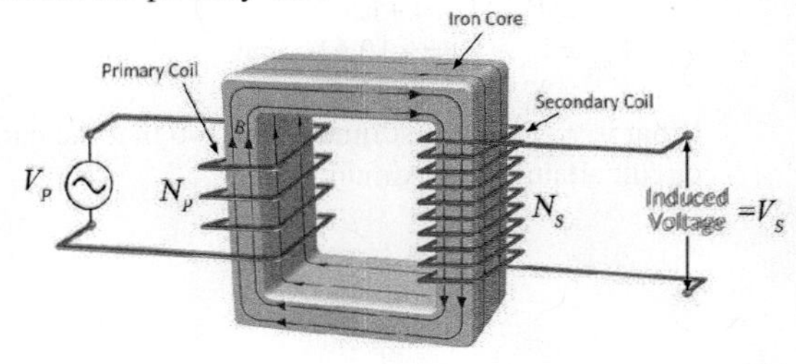

PROBLEMS

1. Power in an AC Circuit: A circuit is constructed with an AC generator, a resistor, a capacitor, and an inductor, as shown in Figure 21.8. The generator voltage varies in time as $\mathcal{E} = V_a - V_b = \mathcal{E}_m \sin(\omega t)$, where $\mathcal{E}_m = 120$ V and $\omega = 404$ rad/s. The inductance $L = 343$ mH. The values for the capacitance C and the resistance R are unknown. What is known is that the current in the circuit leads the voltage across the generator by $\phi = 54°$ and the average power delivered to the circuit by the generator is $P_{avg} = 89$ W. (a) What is $I_{\max}$, the amplitude of the current oscillations in the circuit? (b) What is R, the value of the resistance of the circuit? (c) What is C, the value of the capacitance of the circuit? (d) The value of ω is now changed, keeping all other circuit parameters constant, until resonance is reached. How was ω changed: *ω was decreased*, or *ω was increased*? What is the average power delivered to the circuit when it is in resonance?

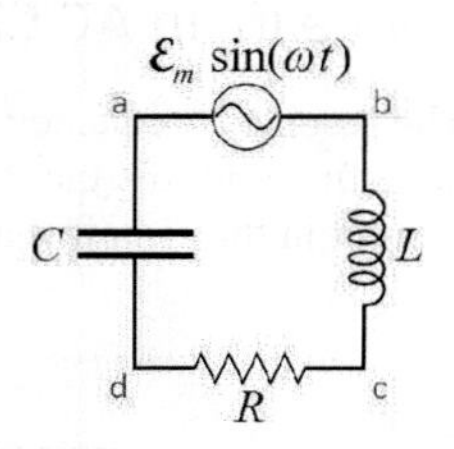

FIGURE 21.8 Problems 1-2

2. AC Circuit in Resonance: A circuit is constructed with an AC generator, a resistor, a capacitor, and an inductor, as shown in Figure 21.8. The generator voltage varies in time as $\mathcal{E} = V_a - V_b = \mathcal{E}_m \sin(\omega t)$, where $\mathcal{E}_m = 24$ V and $\omega = 193$ rad/s. At this frequency, the circuit is in resonance with the maximum value of the current $I_{\max} = 0.61$ A. The capacitance $C = 140$ μF. The values for the resistance R and the inductance L are unknown. (a) What is L, the value of the inductance of the circuit? (b) What is $U_{C_{\max}}$, the value of the maximum energy stored in the capacitor during one cycle? (c) What is ΔU, the total energy dissipated in the circuit in one cycle? (d) What is Q, the quality factor of this circuit? (e) What is R, the value of the resistance of the circuit? (f) Suppose now the value of the capacitance in the circuit is doubled ($C' = 2C$) and the inductance is changed appropriately to keep the circuit in resonance at angular frequency $\omega = 193$ rad/s while the generator voltage and resistance are kept constant. How does Q, the quality factor of the circuit, change, if at all: *Q increases*, *Q decreases*, or *Q stays the same*?

3. AC Circuit II (INTERACTIVE EXAMPLE): A series *RLC* circuit ($L = 340\times10^{-3}$ H, $C = 25\times10^{-6}$ F, and $R = 280\ \Omega$) has an AC generator with amplitude $\mathcal{E}_{\max} = 120$ V and unknown frequency f. At time $t = 0$, the following voltages are measured:

$V_{ad} \equiv V_a - V_d = +120$ V,
$V_{bc} \equiv V_b - V_c = -10.1$ V, and
$V_{cd} \equiv V_c - V_d = +60.6$ V.

What is $t_{I_{\max}}$, the first time after $t = 0$ that the current in the circuit attains its maximum value?

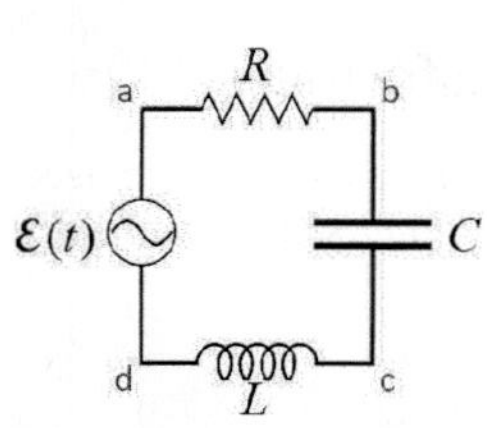

FIGURE 21.9 Problem 3

PART V: LIGHT AND OPTICS

UNIT 22

MAXWELL'S DISPLACEMENT CURRENT AND ELECTROMAGNETIC WAVES

22.1 Overview

In this unit we will introduce a new quantity, the displacement current, in order to make a necessary modification to Ampère's law, one of the fundamental laws of electricity and magnetism. With this modified version of Ampère's law, we will begin to develop an understanding of one of the most pervasive and important features of our modern world, electromagnetic waves.

We'll begin by identifying a lack of symmetry between Faraday's law and Ampère's law that will manifest itself in a real problem when applying Ampère's law to a circuit that is used to charge a capacitor. We will solve this problem by introducing Maxwell's displacement current and modifying Ampère's law to include its ability to create magnetic fields.

After a very brief review of traveling waves, we will demonstrate that Faraday's law and the new Ampère's law can be used to extract the wave equation for both electric and magnetic fields. We'll close by examining some properties of these electromagnetic waves. In particular we will evaluate the velocity of these waves and the relationship between the phase and amplitudes of these electric and magnetic waves.

22.2 The Fundamental Laws of Electricity and Magnetism

We'll start with a review of the four fundamental equations of electricity and magnetism that we have developed so far.

First, we have Gauss' law, which states that the integral of the electric flux through a closed surface is proportional to the charge enclosed by that surface.

$$\oint_{surface} \vec{E} \cdot d\vec{A} = \frac{Q_{enclosed}}{\varepsilon_o}$$

The corresponding equation for magnetic fields states that the integral of the magnetic flux through a closed surface is equal to zero.

$$\oint_{surface} \vec{B} \cdot d\vec{A} = 0$$

Since the integral of the magnetic flux through a closed surface should be equal to the magnetic charge, we say that magnetic charge does not exist. Next, we have Ampère's law that states that the integral of $\vec{B} \cdot d\vec{l}$ around a closed loop is proportional to the current that passes through that loop.

$$\oint_{loop} \vec{B} \cdot d\vec{l} = \mu_o I_{enclosed}$$

Finally, we have Faraday's law that states that the line integral of $\vec{E} \cdot d\vec{l}$ around a closed loop is proportional to the time rate of change of the magnetic flux through that loop.

$$\oint_{loop} \vec{E} \cdot d\vec{l} = -\frac{d\Phi_B}{dt}$$

These four equations are called **Maxwell's equations** and do form the basis of all of electricity and magnetism. You may wonder why they are called Maxwell's equations when we have not yet mentioned his name in connection with any of these laws. The reason we assign Maxwell's name to these equations is that he was the first to point out that the equations as we have written them are inconsistent with the conservation of charge, and offered a modification to one of these equations to rectify that situation.

It should not be obvious to you that these equations are inconsistent with the conservation of charge, but you may notice a lack of symmetry in the way $\vec{E}$ and $\vec{B}$ fields are represented in these equations. In particular, Faraday's law states that an electric field can be induced from a magnetic flux that changes in time, but Ampère's law as written does not allow for the analogous situation (i.e., that magnetic fields might be induced from an electric flux that changes in time). In fact, Maxwell did propose an addition to Ampère's law that explicitly predicted exactly how a changing electric flux could induce a magnetic field.

22.3 The Problem

To understand the difficulty with the current form of Ampère's law, we need to be more careful with our interpretation of the equation. Namely, we have said that

$$\oint_{loop} \vec{B} \cdot d\vec{l} = \mu_o I$$

where I is the current enclosed by that loop. We now want to discuss more carefully what is meant by the "current enclosed by the loop." Up until now, to evaluate Ampère's law, we always used a circle for the closed loop and called the "current enclosed by the loop" the current that passed through a finite plane defined as the interior of the circle. Now, in fact, the current can be defined by any bounding surface of the circle. Why? Well, just imagine the bounding surface to first be represented by a circular film stretched tightly from all points on the closed loop. If we let this film expand outward, as shown in Figure 22.1, we can see that whatever current passed through the initial film will eventually also pass through the expanded surface.

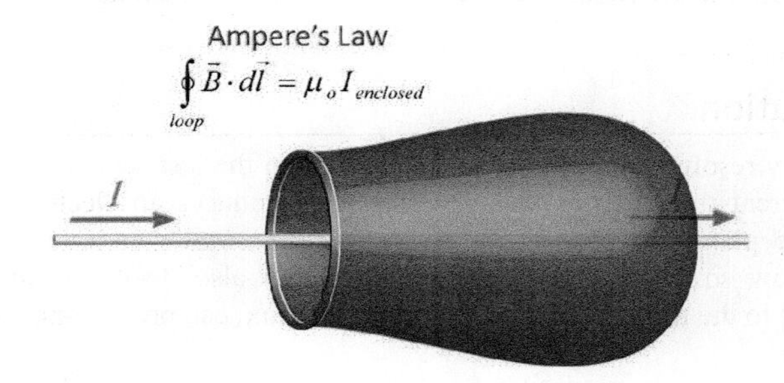

FIGURE 22.1 To determine the "enclosed current" in Ampère's law, any bounding surface (here, the circle or the expanded surface) can be used.

Fair enough, but how can this new understanding of "enclosed current" help us understand the problem with our original form of Ampère's law? Well, to see how this works, consider the circuit shown in Figure 22.2 in which we are charging a parallel-plate capacitor. If we define our closed loop to be the circle labeled *1* that surrounds the wire leading to the positive plate of the capacitor, we can now say that the "current enclosed" by this loop can be determined by using any bounding surface. If we use bounding surface 2 or 4, for example we will get that the current enclosed is *I*. If we choose bounding surface 3, however, we will get the enclosed current to be zero since no current moves between the plates of the capacitor.

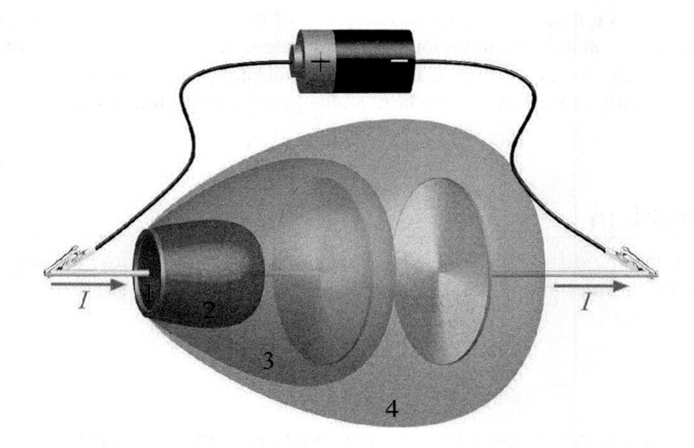

FIGURE 22.2 A charging circuit for a capacitor used to illustrate ambiguous applications of the original form of Ampère's law.

The integral of $\vec{B}\cdot d\vec{l}$ around closed path *1* is what it is and should not depend on what bounding surface we choose to determine the enclosed current. In this case, we expect the current enclosed to be equal to *I*, but if we choose a bounding surface that passes through the gap between the plate of the capacitor, we get zero. Maxwell fixed this problem by changing Ampère's law such that the "enclosed" current in this case, for example, will always be numerically equal to the real current *I* for any bounding surface.

22.4 The Solution

We will now resolve the problem we introduced in the last section. The key idea here is that the current that charges the capacitor also produces an electric flux in the region between the plates that *does change in time*. Maxwell's contribution was to modify Ampère's law so that not only real currents, but also displacement currents that are proportional to the time rate of change of electric flux, can produce magnetic fields.

To see how a current in a wire and a changing flux between the plates of a capacitor are related, we simply take a step back and remember that a current is defined to be the time rate of change of charge:

$$I \equiv \frac{dQ}{dt}$$

This quantity is well defined for both the wire and the capacitor. In the wire it refers to the charge that flows past any point per unit time, and in the capacitor it refers to the charge flowing onto the positive plate (and hence off the negative plate) per unit time. Since charge is conserved, these quantities have to be the same.

In the wire, dQ/dt is simply the current I, but what does it mean in the capacitor? The easiest way to illustrate the answer is to imagine a capacitor whose parallel plates have an area A, as shown in Figure 22.3. The electric flux between the plates is simply equal to the electric field E between the plates times the area A.

$$\Phi_E = EA = \left(\frac{Q}{\varepsilon_o A}\right)A = \frac{Q}{\varepsilon_o}$$

We have already shown that the electric field between parallel plates is proportional to the charge density on each plate. Since the charge density is just the total charge Q divided by the area A, we see that the flux between the plates is just the charge on the plates divided by ε_o. In other words, the charge on the capacitor plates is related to the flux between the capacitor plates in a very simple way. This means that time rate of change of the charge is related in the same simple way to the time rate of change of the flux. This is the result that we need.

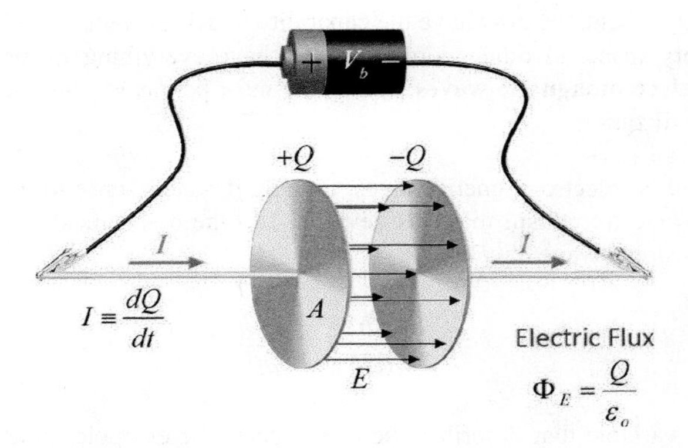

FIGURE 22.3 A charging circuit for a capacitor used to demonstrate that the electric flux in the region between the plates of the capacitor is proportional to the charge on the capacitor.

Just as we associate the time rate of change of charge in the wire with the current I, we associate the time rate of change of charge in the capacitor with a new kind of current I_D, called the **displacement current**.

$$I_D \equiv \varepsilon_o \frac{d\Phi_E}{dt}$$

The final step is to modify Ampère's law to include both of these currents.

$$\oint_{loop} \vec{B} \cdot d\vec{l} = \mu_o (I + I_D)$$

This elegantly resolves the problem we found in the previous section. Since the displacement current only exists between the plates of the capacitor, if we choose our bounding surface to be 1, 2, or 4, we find that $I_D = 0$ and that the "enclosed current" is just equal to *I*, as before. If we now choose our bounding surface to be 3, *I* is zero but I_D is not. In fact, since both represent the same dQ/dt, the current I in the wire is numerically equal to I_D in the capacitor gap, so that no matter what bounding surface we choose, the "enclosed current" is always the same, as it should be.

We have now resolved the problem of the charging capacitor, but at what price? We have changed Ampère's law, a fundamental law that is meant to always be true. This change has dramatic consequences that we will explore in the next section.

22.5 Electromagnetic Waves

We have now changed Ampère's law so that a changing electric field can now create a magnetic field. When coupled with Faraday's law in which a changing magnetic field can create an electric field, we now have the capability of self-sustaining electric and magnetic fields in empty space. In other words, we now have everything we need to explain the existence of **electromagnetic waves**, one of the most pervasive and important features of our current civilization.

Before we discuss electromagnetic waves, though, it makes sense to do a brief review of the relevant wave formalism that was developed in the mechanics course. We start with the one-dimensional wave equation:

$$\frac{d^2h}{dx^2} = \frac{1}{v^2}\frac{d^2h}{dt^2}$$

Here *h* is the variable that describes the disturbance, for example, in a wave on a string, *h*(*x*,*t*) represents the height of the rope at a given position *x* at a time *t*. The general solution to this equation has the form

$$h(x,t) = h_1(x - vt) + h_2(x + vt)$$

where h_1 describes the waveform traveling in the positive x direction and h_2 describes the shape of a wave traveling in the negative x direction.

The functions h_1 and h_2 can take any form, but the most common example we will use in this course is the harmonic plane wave. For example, the solution for a **harmonic plane wave** moving in the positive x direction is given by

$$h(x,t) = A\cos(kx - \omega t)$$

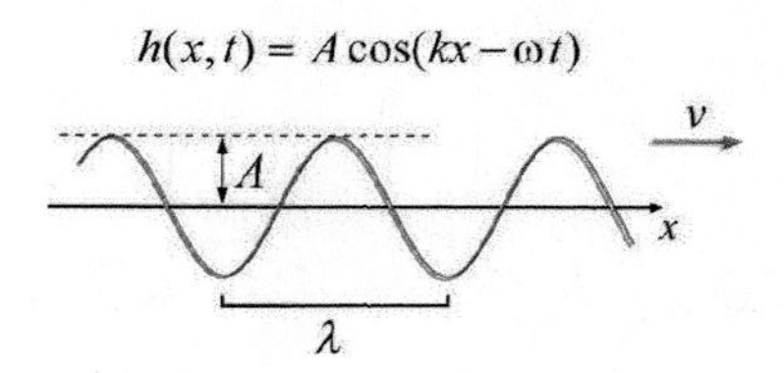

FIGURE 22.4 A harmonic plane wave traveling to the right, having amplitude A, and wavelength λ.

A represents the **amplitude** of the oscillations, the maximum value for disturbance h (see Figure 22.4). The spatial form at any time t is specified by the **wave number** $k = 2\pi/\lambda$, where λ is the **wavelength** of the wave, the distance in x that it takes the form to repeat itself. The time dependence of h at any position x is specified by the **angular frequency** $\omega = 2\pi/T$, where T is the **period** of the wave, the time it takes for the waveform to repeat itself. We also sometimes refer to the frequency of the wave $f = 1/T = \omega/2\pi$. The spatial and time representations are linked by the velocity of the wave, $v = \lambda f = \omega/k$. The velocity of the wave is determined by the **medium** in which the wave is a disturbance.

Armed with this knowledge of waves, we will now demonstrate that Maxwell's equations (specifically the combination of Ampère's law and Faraday's law) have plane wave solutions. We'll start with Ampère's law and Faraday's law in empty space (i.e., a region in which there are no charges or currents).

$$\oint_{loop} \vec{B} \cdot d\vec{l} = \mu_o \varepsilon_o \frac{d\Phi_E}{dt}$$

$$\oint_{loop} \vec{E} \cdot d\vec{l} = -\frac{d\Phi_B}{dt}$$

Here we see that in both cases we have a line integral equal to the time rate of change of flux. If we assume the plane wave is traveling in the z direction (therefore $\vec{E}$ and $\vec{B}$ can only depend on z and t), we can obtain relations between E_x and B_y, for example, by choosing a loop for Faraday's law in the x-z plane and a loop for Ampère's law in the y-z plane.

We start with Faraday's law using the loop in the x-z plane, as shown in Figure 22.5 (left). The magnetic flux through the loop is given by the average value of the y-component of the magnetic field multiplied by the area of the loop ($\Phi_B = B_y \Delta x \Delta z$). The only contributions to the line integral of the electric field will occur along the sides parallel to the electric field. Moving clockwise around the loop, we get a positive contribution going up the left side and a negative contribution coming down the right side, with the net result being equal to the length of the side times the difference in the electric fields at the two z-values. Now this difference can be obtained from the partial derivative of E_x with respect to z ($\Delta E_x = (\partial E_x/\partial z)\Delta z$). Putting these equations together, we see that Faraday's law

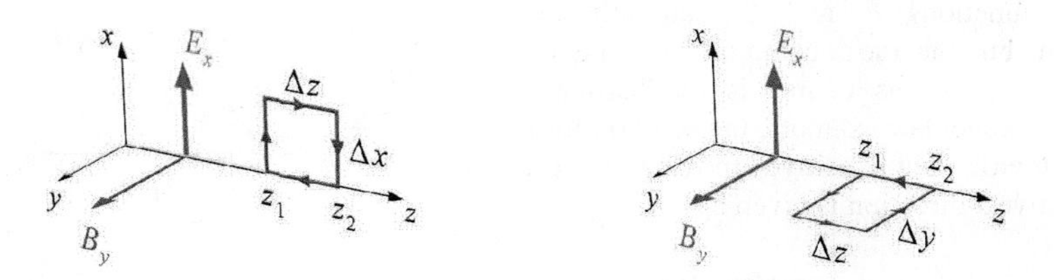

FIGURE 22.5 Applying Faraday's law to the loop in the *x-z* plane (left) and Ampère's law to the loop in the *y-z* plane (right) to obtain the wave equation.

defines a relationship between the partial derivative of the electric field with respect to *z* and the magnetic field with respect to time:

$$\Delta E_x \Delta z \Delta x = -\frac{d}{dt}(B_y \Delta x \Delta z) \qquad \Rightarrow \qquad \frac{\partial E_x}{\partial z} = -\frac{\partial B_y}{\partial t}$$

If we now apply Ampère's law to a loop in the *y-z* plane we get a very similar expression relating the partial derivative of the magnetic field with respect to *z* and the electric field with respect to time.

$$\frac{\partial B_y}{\partial z} = -\mu_o \varepsilon_o \frac{\partial E_x}{\partial t}$$

We can eliminate reference to the magnetic field in these two equations by differentiating both sides of the Ampère's law equation with respect to time and differentiating both sides of the Faraday's law equation with respect *z*. We then end up with a second order differential equation for the electric field.

$$\frac{\partial^2 E_x}{\partial z^2} = \mu_o \varepsilon_o \frac{\partial^2 E_x}{\partial t^2}$$

The only subtlety here was the extra minus sign, which occurs because an increasing E_x leads to a path integral in the counterclockwise sense in the *y-z* plane when viewed from above. Consequently a positive dot product for a positive B_y occurs at z_1 rather than at z_2. We will discuss some features of these electromagnetic waves as revealed in this equation in the next section.

22.6 The Velocity of Electromagnetic Waves

We'll start with a one-dimensional wave equation for E_x that we just derived to discuss some interesting properties of these electromagnetic waves. The first order of business is to calculate the velocity of these waves. We use the standard one-dimensional wave equation to identify the velocity of electromagnetic waves to be

$$v = \frac{1}{\sqrt{\mu_o \varepsilon_o}}$$

Now, μ_o and ε_o are constants that have been determined from electric and magnetic measurements. When we use the measured values for these constants, we determine that the velocity of electromagnetic waves is equal to 3.00×10^8 m/s. This is an amazing number; it happens to be identical to the well-measured value for the speed of light. Consequently, we identify light as an electromagnetic wave!

The velocity of the wave in the wave equation refers to its velocity with respect to the medium. What is the medium for electromagnetic waves? Maxwell named it the aether. The things that are waving in electromagnetic waves are the fields ($\vec{E}$ and $\vec{B}$). Consequently, Maxwell viewed $\vec{E}$ and $\vec{B}$ fields as disturbances in the aether. A major experimental program was launched to determine the speed of the rest frame of the aether with respect to the Earth. All such experiments failed in the sense that they all measured the aether frame to be at rest with respect to the Earth, even though the Earth reverses its velocity every six months.

This puzzle was not resolved until 1905 when Einstein proposed the special theory of relativity. In this theory, the speed of light is a constant, the same in all inertial reference frames. This property of light requires space and time to be quite different from what we have assumed in classical physics. Unfortunately, we don't have time here to develop this theory; we will simply accept that light is an electromagnetic wave that travels, in vacuum, at a constant speed, $c = 3.00 \times 10^8$ m/s, with respect to *all* inertial observers.

22.7 Relationships of *E* and *B* in Electromagnetic Waves

We'll close this unit with a discussion of the relationship between electric and magnetic fields in an electromagnetic wave. We begin by noting that in our derivation in section *E*, we could have just as easily eliminated E_x rather than B_y from our system of equations. Had we done that, we would have obtained a wave equation for B_y. In particular, we would have determined:

$$\frac{\partial^2 B_y}{\partial z^2} = \mu_o \varepsilon_o \frac{\partial^2 B_y}{\partial t^2}$$

We can see from this equation that we have a wave in B_y that moves with the same velocity, *c*, as the wave in E_x. How are these waves related in phase and magnitude?

To make this determination, we'll begin with a specific harmonic solution for E_x. Namely, let's assume that

$$E_x = E_o \cos(kz - \omega t)$$

To determine the B_y wave, we simply go back to the equation we got from applying Faradays' law that links E_x and B_y:

$$\frac{\partial E_x}{\partial z} = -\frac{\partial B_y}{\partial t}$$

Differentiating our form for E_x, we obtain

$$\frac{\partial B_y}{\partial t} = kE_o \sin(kz - \omega t)$$

To find B_y, we just need to integrate this equation over *dt* to obtain

$$B_y = \frac{k}{\omega} E_o \cos(kz - \omega t)$$

Comparing this equation for B_y to the one we assumed for E_x, we see two important features: (1) B_y is *in phase* with E_x (in this case they both oscillate as $\cos(kz - \omega t)$), and (2) the amplitude of the magnetic field oscillations is equal to the amplitude of the electric field oscillations divided by the speed of light:

$$B_o = \frac{E_o}{c}$$

22.8 Summary

In this unit, we introduced the displacement current in order to modify Ampère's law. With this modified version of Ampère's law, we were able to demonstrate the existence of electromagnetic waves.

We began by identifying a lack of symmetry between Faraday's law and Ampère's law in empty space in that Faraday's law contains a term ($d\Phi_B / dt$) that had no counterpart in Ampère's law. Namely, Faraday's law allows that a changing magnetic field can create an electric field, but Ampère's law did not allow that a changing electric field might create a magnetic field.

We then discovered that this lack of symmetry actually causes a problem when we try to apply Ampère's law to a circuit used to charge a capacitor. Namely, Ampère's law states that the integral of $\vec{B} \cdot d\vec{l}$ around a closed loop is proportional to the current through any surface bounded by the closed loop. The problem arises if the bounding surface is chosen to pass between the plates of the capacitor since no real current passes between the plates. This problem was resolved by Maxwell when he observed that the charging current does produce an electric flux in the region between the plates that does change in time. He identified this quantity, $\varepsilon_o \, d\Phi_E / dt$, as the "displacement current" and modified Ampère's law by adding it to the real current on the right-hand side of the equation.

We were then able to use this modified version of Ampère's law in conjunction with Faraday's law to demonstrate that components of the electric and magnetic fields satisfied the wave equation in which the velocity of the wave was identified as

$$v = \frac{1}{\sqrt{\mu_o \varepsilon_o}}$$

The numerical value of this velocity was found to be identical to the speed of light, and thereby supports the identification of light as an electromagnetic wave.

Finally we determined that for an electromagnetic wave traveling in the z direction, the E_x and B_y waves were in phase, and that the ratio of the amplitudes, E_o / B_o, is equal to c, the speed of light.

MAIN POINTS

Displacement Current and Ampère's Law

Maxwell defined the displacement current and added it to the real current in Ampère's law.

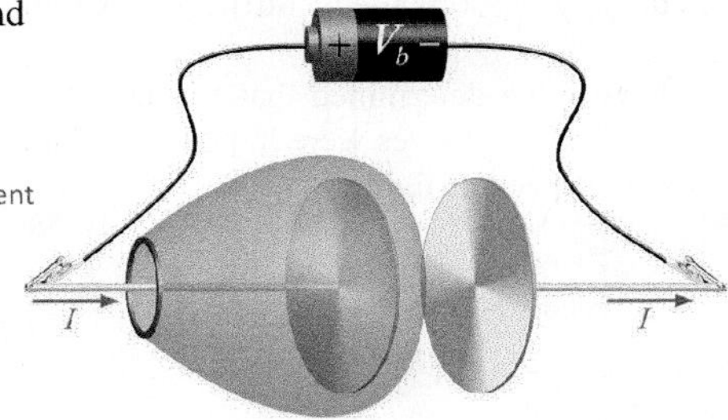

$$I_D \equiv \varepsilon_o \frac{d}{dt}\int \vec{E}\cdot d\vec{A}$$

Maxwell's Displacement Current

$$\oint \vec{B}\cdot d\vec{l} = \mu_o I + \mu_o \varepsilon_o \frac{d}{dt}\int \vec{E}\cdot d\vec{A}$$

Modified Ampère's Law

Electromagnetic Waves

The modified Ampère's law in conjunction with Faraday's law is used to demonstrate that the components of the electric and magnetic fields satisfy the wave equation.

$$\frac{\partial^2 E_x}{\partial z^2} = \mu_o \varepsilon_o \frac{\partial^2 E_x}{\partial t^2}$$

$$\frac{\partial^2 B_y}{\partial z^2} = \mu_o \varepsilon_o \frac{\partial^2 B_y}{\partial t^2}$$

The velocity of electromagnetic waves is equal to the speed of light.

$$v = \frac{1}{\sqrt{\mu_o \varepsilon_o}} = 3.00\times 10^8 \text{ m/s}$$

The electric and magnetic fields in an electromagnetic wave are in phase and the ratio of the amplitudes is equal to the speed of light.

Example: A Harmonic Solution

$$E_x = E_o \cos(kz - \omega t)$$

$$B_y = \frac{k}{\omega} E_o \cos(kz - \omega t)$$

$$c = f\lambda = \frac{\omega}{k} = \frac{E_o}{B_o}$$

PROBLEMS

1. Electromagnetic Waves: The magnetic field in a plane monochromatic electromagnetic wave with wavelength $\lambda = 645$ nm, propagating in a vacuum in the z-direction, is described by

$$\vec{B} = [B_1 \sin(kz - \omega t)](\hat{i} + \hat{j})$$

where $B_1 = 8.5 \times 10^{-6}$ T, and $\hat{i}$ and $\hat{j}$ are the unit vectors in the $+x$ and $+y$ directions, respectively. (a) What is k, the wave number of this wave? (b) What is z_{max}, the distance along the positive z-axis to the position where the magnitude of the magnetic field is a maximum at $t = 0$? (c) What is E_{max}, the amplitude of the electric field oscillations? (d) What is E_y, the y-component of the electric field at ($x = 0$, $y = 0$, $z = z_{max}$) at $t = 0$? (e) Which of the following equations describes the spatial and time dependence of the electric field oscillations?

(i) $\vec{E} = \left[\frac{E_{max}}{\sqrt{2}} \sin(kz - \omega t)\right](\hat{i} + \hat{j})$

(ii) $\vec{E} = \left[\frac{E_{max}}{\sqrt{2}} \sin(kz - \omega t)\right](\hat{i} - \hat{j})$

(iii) $\vec{E} = \left[\frac{E_{max}}{\sqrt{2}} \sin(kz - \omega t)\right](\hat{j} - \hat{i})$

(iv) $\vec{E} = \left[\frac{E_{max}}{\sqrt{2}} \cos(kz - \omega t)\right](\hat{i} + \hat{j})$

(v) $\vec{E} = \left[\frac{E_{max}}{\sqrt{2}} \cos(kz - \omega t)\right](\hat{i} - \hat{j})$

(vi) $\vec{E} = \left[\frac{E_{max}}{\sqrt{2}} \cos(kz - \omega t)\right](\hat{j} - \hat{i})$

(f) What is t_{max}, the first time after $t = 0$, when the magnitude of the electric field at the origin ($x = y = z = 0$) has its maximum value? (g) Compare E_{x1} and E_{x2}, the values of the x-component of the electric field at $t = 0$. E_{x1} is evaluated at ($x = 0$, $y = 0$, $z = z_{max}$), while E_{x2} is evaluated at ($x = \lambda$ nm, $y = \lambda$ nm, $z = z_{max}$).

(i) $E_{x1} < E_{x2}$

(ii) $E_{x1} = E_{x2}$

(iii) $E_{x1} > E_{x2}$

UNIT

23

PROPERTIES OF ELECTROMAGNETIC WAVES

23.1 Overview

In this unit we will look in more detail at some properties of the electromagnetic waves we introduced in the last unit. We will begin by discussing general characteristics of these waves. We will explore the vast range of the electromagnetic spectrum and calculate the Doppler shift, the change in frequency due to relative motion between the wave source and the observer.

We'll then discuss the energy that is carried by electromagnetic waves. We will define the intensity of the wave as the average value of the energy carried by the wave per unit area.

Finally, we will introduce the Poynting vector as a convenient representation of all of this information.

23.2 *E* and *B* in Electromagnetic Waves

We'll start with a brief discussion of the relationship between the electric and magnetic fields in an electromagnetic wave. Our primary example will be a plane harmonic wave since the mathematics is simple, and many waves, far from the source, can be approximated as plane waves.

We will assume the electric field has only an *x*-component given by

$$E_x = E_o \sin(kz - \omega t)$$

Here the oscillations in space are specified by the wave number k and the oscillations in time are specified by the angular frequency ω. These quantities are related by the velocity of the wave:

$$c = \frac{\omega}{k}$$

The velocity of the wave is given by

$$c = \frac{\omega}{k} = \frac{1}{\sqrt{\mu_o \varepsilon_o}} = \frac{E_o}{B_o} = 3\times10^8 \ \frac{\text{m}}{\text{s}}$$

which is numerically equal to the measured speed of light.

If the wave travels in the positive z direction, then the magnetic field will have only a y-component and it will oscillate in phase with E_x:

$$B_y = B_o \sin(kz - \omega t)$$

Since the magnetic field is produced by the changing electric field, and the electric field is produced by the changing magnetic field, their amplitudes are related to the speed of the wave:

$$c = \frac{E_o}{B_o}$$

Such a plane wave is often represented on the printed page by a diagram, such as the one shown in Figure 23.1. Care needs to be taken in interpreting this picture since lots of different kinds of information are being presented here and it is very easy to get confused. The picture represents a snapshot at a given time, with one spatial axis (the direction of propagation, z, in this case) and two field axes (E_x and B_y in this case). It is possible to construct this picture for a plane wave since neither E_x nor B_y have any x or y

dependence. The wave you see in E_x or B_y at a given z-value exists at all x and at all y values.

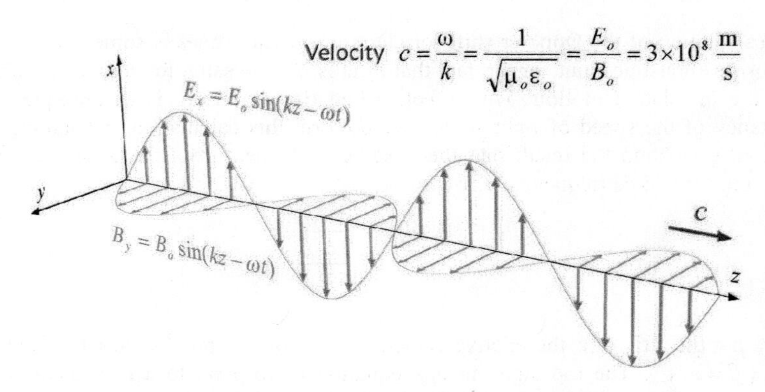

FIGURE 23.1 A (potentially misleading) representation of a plane electromagnetic wave. In particular, the values of $\vec{E}$ and $\vec{B}$ do not vary with x or y at a given z.

23.3 The Electromagnetic Spectrum

All electromagnetic waves travel at the speed of light in empty space. We usually associate "light" with visible light, having wavelengths in the 380–750 nm range, which correspond to frequencies in the range of $8-4\times10^{14}$ Hz. Visible light, though, makes up only a tiny fraction of the electromagnetic spectrum.

To illustrate this statement, let's consider examples of electromagnetic radiation with wavelengths much lower and much higher than visible light. On the large wavelength end, for example, we have radio waves. Wi-Fi operates at a frequency of 2.4 GHz, which corresponds to a wavelength of about 10 cm. On the other end of the spectrum we have gamma rays. At Fermilab, for example, gamma rays can be produced with frequencies as high as 10^{26} Hz. The corresponding wavelength is very tiny ($\sim10^{-18}$ m).

These examples span almost 20 orders of magnitude. Electromagnetic radiation is indeed pervasive, playing an important role in almost everything we experience.

23.4 Doppler Shifts

You are probably familiar with the **Doppler shift** in sound waves. The increase in frequency as a sound source (a race car or a train, for example) approaches you, and the decrease in frequency as it moves away from you, are examples of this Doppler shift. A very similar effect exists for electromagnetic waves.

An observer moving toward a source of electromagnetic waves will encounter these waves at a higher frequency, since the separation between the source and observer is getting smaller.

The calculation of the Doppler shift for electromagnetic waves is somewhat complicated because we must account for the fact that time is not the same for source and observer if they are in relative motion. This effect, called **time dilation**, is a consequence of the constancy of the speed of light. When we perform this calculation, accounting for time dilation, we obtain the result that the observer will measure a frequency f', which is related to the source frequency f by

$$f' = f\sqrt{\frac{1 \pm \beta}{1 \mp \beta}}$$

where β= the ratio of v, the relative velocity of the source and observer, to c, the speed of light ($\beta = v / c$). The top signs in this equation correspond to a decreasing separation between source and observer while the bottom signs correspond to an increasing separation between source and observer. Clearly the frequency will increase if the source and the observer are moving toward each other and will decrease if they are moving away from each other. The magnitude of the shift only depends on their relative velocity v.

For most examples you may be familiar with (e.g., radar guns and such), the relative velocity of source and observer is small compared to the speed of light, so that in the limit ($\beta << 1$), we obtain the result

$$f' \approx f(1 \pm \beta)$$

23.5 Energy in Electromagnetic Waves

Electromagnetic waves carry energy. We know this because electric and magnetic fields contain energy. We already know the expressions for the energy densities (energy per unit volume) for $\vec{E}$ and $\vec{B}$ fields. Namely, $u_E = ½\ \varepsilon_o E^2$ and $u_B = ½\ B^2 / \mu_o$. Consequently, we can calculate these energy densities in an electromagnetic wave at any point at any given time from a knowledge of the electric and magnetic field values at that point at that time.

We'll start with the energy density in the magnetic field. We can use the relationship between the magnitudes of the electric and magnetic fields ($E = cB$) to rewrite this expression in terms of only the electric field:

$$u_B = \frac{1}{2}\frac{B^2}{\mu_o} = \frac{1}{2\mu_o}\left(\frac{E}{c}\right)^2$$

Finally, we can use the relationship between the speed of light and the fundamental constants μ_o and ε_o, $c = 1/\sqrt{\mu_o \varepsilon_o}$ to obtain the result that the energy density in the magnetic field at any point at any time is exactly equal to the energy density in the electric field at that point and time:

$$u_B = \frac{1}{2}\varepsilon_o E^2 = u_E$$

Therefore, the total energy density is just equal to the product of ε_o and the square of the electric field:

$$u = u_B + u_E = \varepsilon_o E^2$$

Since electromagnetic waves exhibit periodic behavior, the more relevant quantity is the *average* energy density. Taking our plane wave example again ($E_x = E_o \sin(kz - \omega t)$), we calculate the average energy density to be

$$\langle u \rangle = \varepsilon_o E_o^2 \langle \sin^2(kz - \omega t) \rangle = \frac{1}{2}\varepsilon_o E_o^2$$

The energy in electromagnetic waves is usually characterized in terms of its **intensity** I defined as the average power per unit area. Indeed, the intensity is the relevant quantity that determines how bright an object appears. We can determine I by first calculating the total energy in a volume defined by a cross-sectional area A and a length equal to the distance the wave moves in time t, which is just equal to ct, to obtain

$$I \equiv \frac{\langle Power \rangle}{Area} = \frac{\langle u \rangle}{t}\frac{A \cdot ct}{A} = c\langle u \rangle = \frac{1}{2}c\varepsilon_o E_o^2$$

In order to formalize these results for the energy transport in an electromagnetic wave, it is usual to define the **Poynting vector** $\vec{S}$ as

$$\vec{S} \equiv \frac{\vec{E} \times \vec{B}}{\mu_o}$$

Note that the *direction* of the Poynting vector is the *direction of the propagation of the wave*, while the *magnitude* of the Poynting vector at any point at any time is just equal to the *instantaneous power per unit area* at that point and time:

$$S = \frac{EB}{\mu_o} = \frac{E^2}{\mu_o c} = c\varepsilon_o E^2$$

Consequently, the average value of the magnitude of the Poynting vector corresponds to the intensity of the wave.

23.6 An Example

To this point, our discussion has been abstract and general. We will now consider a specific ordinary example to get some idea of the magnitude of these quantities.

FIGURE 23.2 The Earth illuminated by the Sun.

Let's look at the energy present in the sunlight that arrives on Earth (Figure 23.2). A typical value for the intensity of sunlight here is $I = 100$ mW/cm^2. Note that this solar energy is non-negligible. If you were able to produce a 100% efficient solar cell with area 1 square meter, it would produce power at the rate of 1 kilowatt!

Finally, let's look at the size of the amplitude of the electric field oscillations in sunlight. We can obtain this value directly from the intensity; in particular, the square of the *rms* value of the electric field is just equal to the product of the intensity and the quantity $\mu_o c$. With a little bit of work, we can see that the units of $\mu_o c$ are ohms. In fact, $\mu_o c$ is numerically equal to 377 Ω; it is usual to call this quantity Z_o, the impedance of free space. Consequently, we obtain the *rms* value for the electric field of sunlight ($E_{rms}^2 = \mu_o cI$) to be equal to 614 V/m.

23.7 Photons

We'll close this unit with an important observation about exactly how this energy is transported in electromagnetic waves. Up until this point, our presentation has been totally classical in the sense that the wave was considered to be a continuous form. In fact, the energy and momentum in an electromagnetic wave are carried by **photons**, the quanta of electromagnetic radiation. Photons are quantum objects that exhibit, at different times, properties of both waves and particles. Individual photons have an energy that is proportional to their frequency,

$$E = hf$$

where h is Planck's constant and equals 1.24×10^{-6} eV-m. Photons also carry momentum:

$$p = \frac{E}{c} = \frac{h}{\lambda}$$

The energy of the individual photons in visible light range from about 1.6 to 3 eV. The energy of Wi-Fi photons are only about 10^{-6} eV, while the highest energy photons at Fermilab are about 1 TeV, or 10^{12} eV!

23.8 Summary

In this unit, we explored some properties of electromagnetic waves. First, we established that the electric and magnetic field oscillations in a plane electromagnetic wave are in phase with each other and that the direction of propagation of these waves can be obtained from the cross product of $\vec{E}$ and $\vec{B}$. We also showed that the ratio of the amplitudes of the oscillations in the electric and magnetic fields is simply equal to the speed of light.

$$c = \frac{E_o}{B_o}$$

We determined that the frequency (and wavelength) of electromagnetic waves vary by almost 20 orders of magnitude from radio waves to photons at the Tevatron at Fermilab.

We then introduced the Doppler shift, the frequency shift due to the relative motion of the source of these waves and the observer. In particular, we presented the result that the observer will measure a frequency f', which is related to the source frequency f by

$$f' = f\sqrt{\frac{1 \pm \beta}{1 \mp \beta}}$$

where β equals the ratio of v, the relative velocity of the source and observer, to c, the speed of light ($\beta = v/c$). The top signs in this equation correspond to a decreasing separation between source and observer, while the bottom signs correspond to an increasing separation between source and observer.

Finally, we used our prior knowledge of the energy densities present in $\vec{E}$ and $\vec{B}$ fields to determine that the energy density in the $\vec{E}$ and $\vec{B}$ fields of an electromagnetic wave are equal, resulting in a total instantaneous energy density of

$$u = \varepsilon_o E^2$$

Averaging over space and time, we obtained the more relevant quantity, the intensity I of the wave, defined as the average power transmitted per unit area, which is given by

$$I = \frac{1}{2}\varepsilon_o E_o^2$$

where E_o is the amplitude of the E field oscillations in the wave. We formalized the energy transport by introducing the Poynting vector $\vec{S}$, defined as the cross product of the electric and magnetic fields, normalized by μ_o:

$$\vec{S} \equiv \frac{\vec{E} \times \vec{B}}{\mu_o}$$

The direction of this vector is the direction of propagation of the wave while its magnitude is equal to the instantaneous power in the wave per unit area. The average over space and time of the magnitude of $\vec{S}$ is just the intensity I of the wave.

MAIN POINTS

Relationships between *E* and *B* in Electromagnetic Waves

The electric and magnetic fields in an electromagnetic wave are in phase, and the ratio of the amplitudes is equal to the speed of light.

Example: A Harmonic Solution

$$E_x = E_o \cos(kz - \omega t)$$

$$B_y = \frac{k}{\omega} E_o \cos(kz - \omega t)$$

$$c = \lambda f = \frac{\omega}{k} = \frac{E_o}{B_o}$$

Doppler Shift

The frequency of an electromagnetic wave is shifted due to the relative motion of the source and observer. The frequency increases if the source and observer are approaching each other and decreases if they are receding from each other.

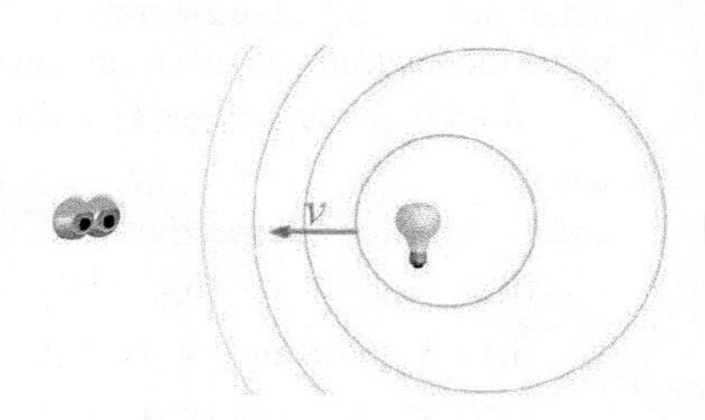

$$f' = f\sqrt{\frac{1 \pm \beta}{1 \mp \beta}}$$ where $$\beta \equiv \frac{v}{c}$$

Energy in Electromagnetic Waves

The energy densities in the $\vec{E}$ and $\vec{B}$ fields in an electromagnetic wave are equal.

$$u_E = u_B = \frac{1}{2}\varepsilon_o E^2$$

$$I \equiv \frac{\langle Power \rangle}{Area}$$

The intensity of the wave is defined as the average power transmitted per unit area.

The Poynting vector $\vec{S}$ is defined such that its direction is the direction of propagation of the wave and its magnitude is the instantaneous power in the wave.

$$\vec{S} \equiv \frac{\vec{E} \times \vec{B}}{\mu_o}$$

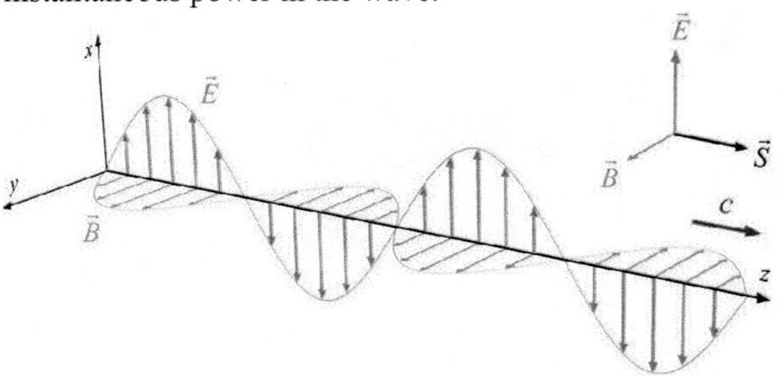

PROBLEMS

1. Electromagnetic Waves 2: A plane monochromatic electromagnetic wave with wavelength $\lambda = 4.3$ cm, propagates through a vacuum. Its magnetic field is described by

$$\vec{B} = (B_x \hat{i} + B_y \hat{j})\cos(kz + \omega t)$$

where $B_x = 3.8\times10^{-6}$ T, $B_y = 3.3\times10^{-6}$ T, and $\hat{i}$ and $\hat{j}$ are the unit vectors in the $+x$ and $+y$ directions, respectively. (a) What is f, the frequency of this wave? (b) What is I, the intensity of this wave? (c) What is S_z, the z-component of the Poynting vector at $(x = 0, y = 0, z = 0)$ at $t = 0$? (d) What is E_x, the x-component of the electric field $(x = 0, y = 0, z = 0)$ at $t = 0$? (e) Compare the sign and magnitude of S_z, the z-component of the Poynting vector at $(x = y = z = t = 0)$ of the wave described above to the sign and magnitude of S_{IIz}, the z-component of the Poynting vector at $(x = y = z = t = 0)$ of another plane monochromatic electromagnetic wave propagating through vacuum described by

$$\vec{B} = (B_{IIx} \hat{i} - B_{IIy} \hat{j})\cos(kz - \omega t)$$

where $B_{IIx} = 3.3\times10^{-6}$ T, $B_{IIy} = 3.8\times10^{-6}$ T, and $\hat{i}$ and $\hat{j}$ are the unit vectors in the $+x$ and $+y$ directions, respectively.

(i) $S_{IIz} < 0$ and $|S_{IIz}| \neq |S_z|$
(ii) $S_{IIz} < 0$ and $|S_{IIz}| = |S_z|$
(iii) $S_{IIz} > 0$ and $|S_{IIz}| \neq |S_z|$
(iv) $S_{IIz} > 0$ and $|S_{IIz}| = |S_z|$

2. EM Wave (INTERACTIVE EXAMPLE): A plane monochromatic radio wave ($\lambda = 0.3$ m) travels in vacuum along the positive x-axis, with a time-averaged intensity $I = 45$ W/m^2. Suppose at time $t = 0$, the electric field at the origin is measured to be directed along the positive y-axis with a magnitude equal to its maximum value. What is B_z, the magnetic field at the origin, at time $t = 1.5$ ns?

UNIT

24

POLARIZATION

24.1 Overview

In this unit we will discuss possible polarizations of electromagnetic waves. We will begin by specifying the requirements for a linearly-polarized wave, and then will move on to discuss how to produce such waves by passing the waves through a polarizer. We will then determine the reduction of intensity that occurs when linear polarizers are used.

We will then discuss other possible polarizations of electromagnetic waves, focusing on circular polarization. We will introduce the property of birefringence and its use in quarter-wave plates to produce circularly-polarized light.

24.2 Linear Polarization

We begin by referring once again to our usual example of a plane harmonic wave. Namely, for a wave traveling in the $+z$ direction, we take $E_x = E_o \sin(kz - \omega t)$ and $B_y = B_o \sin(kz - \omega t)$, where $\omega = kc$, $E_o = cB_o$, and $c = 1/\sqrt{\mu_o \varepsilon_o}$, which is equal to the speed of light.

This wave is a linearly polarized wave in the x direction. We define **polarizations** using just one field, the $\vec{E}$-field, since the $\vec{B}$-field behavior can always be obtained from the $\vec{E}$-field behavior and the direction of propagation of the wave. In fact, in the discussion and diagrams that follow, we will refer only to the components of the $\vec{E}$ field, leaving out the $\vec{B}$ field entirely to make the diagrams easier to understand.

We have used here the example of an $\vec{E}$ field oscillating in the x direction, but for a wave propagating in the z direction, the oscillations of the $\vec{E}$ field could be along any direction in the x-y plane. Consequently, we define the general form for a linearly-polarized wave moving in the $+z$ direction in terms of a unit vector $\hat{e}$ in the x-y plane that defines the polarization direction, and a constant phase angle ϕ that specifies the $\vec{E}$-field strength at $z = t = 0$:

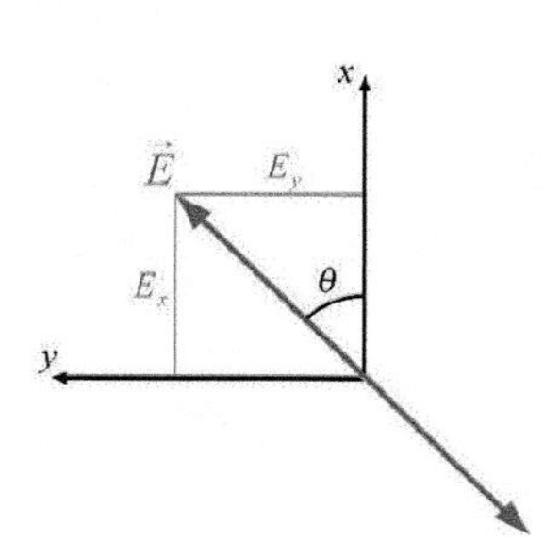

FIGURE 24.1 The definition of the polarization direction of a plane wave propagating in the $+z$ direction.

$$\vec{E} = \hat{e}E_o \sin(kz - \omega t + \phi)$$

We can then obtain the x and y components of the field by expanding $\hat{e}$ in terms of θ, the angle between $\hat{e}$ and the x-axis, as shown in Figure 24.1.

$$E_x = E_o \cos\theta \sin(kz - \omega t + \phi)$$

$$E_y = E_o \sin\theta \sin(kz - \omega t + \phi)$$

24.3 Polarizers

Most natural sources of electromagnetic radiation are, in fact, not polarized. For example, the $\vec{E}$ fields in the light coming from your incandescent light bulb are oscillating in random directions, always perpendicular to the direction of propagation, but not along any special axis. We call this radiation unpolarized. We can, however, create linearly polarized light from this source by passing it through a polarizer.

A **polarizer** is constructed from special materials that pass the $\vec{E}$-field component that is parallel to a transmission axis and totally absorb the $\vec{E}$-field component that is perpendicular to this transmission axis. For example, for visible light, polarizers can be constructed from Polaroid filters, sheets of plastic that have very long organic molecules aligned and embedded in them. $\vec{E}$ fields that are parallel to the molecules are absorbed, forming a polarizer with a transmission axis that is perpendicular to the alignment of the molecules.

Since the polarizer works by absorption, the intensity of the incident wave must be reduced after passing through the polarizer. We can easily calculate this reduction for the ideal polarizer, in which the $\vec{E}$-field component perpendicular to the transmission axis is totally absorbed and the $\vec{E}$-field component parallel to the transmission axis is totally passed.

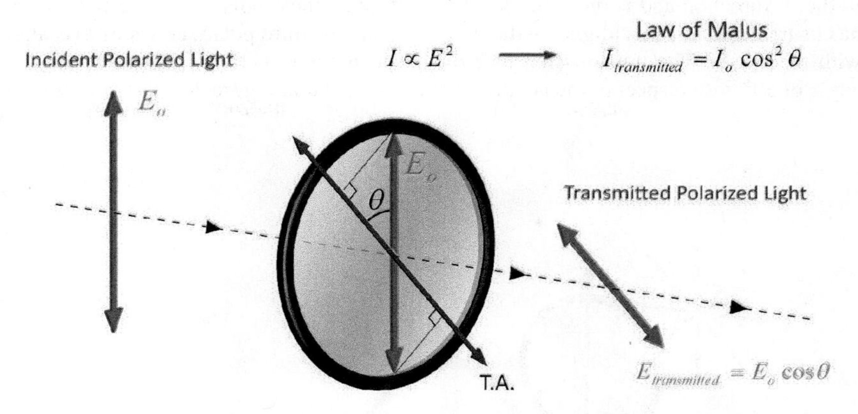

FIGURE 24.2 Vertically polarized light is incident on a linear polarizer whose transmission axis (*T.A.*) makes an angle θ with the incident polarization direction. The transmitted light is linearly polarized along the direction of the transmission axis and its intensity is reduced, being equal to the product of the incident intensity and the square of the angle θ.

The amount of reduction depends on the polarization of the incident wave. For example, if the incident wave is unpolarized, then the sum of the $\vec{E}$-field components parallel to the transmission axis is equal to the sum of the $\vec{E}$-field components perpendicular to the transmission axis, so that ½ of the incident intensity is passed (i.e., $I = ½\ I_o$).

If the incident wave is already linearly polarized, then to find how much of the intensity is passed, we need to find the component of the $\vec{E}$ field that is parallel to the transmission axis. To find this component, we simply take the projection of the $\vec{E}$-field vector along the direction of the transmission axis. If the incident $\vec{E}$ field has amplitude E_o and its direction makes an angle θ with respect to the transmission axis, as shown in Figure 24.2, then the amplitude of the transmitted $\vec{E}$-field oscillations is just given by

$$E_{transmitted} = E_o \cos\theta .$$

Since the intensity is proportional to the square of the amplitude of these $\vec{E}$-field oscillations, we arrive at the **law of Malus**:

$$I = I_o \cos^2\theta$$

where I is the intensity of the wave after the polarizer and I_o is equal to the intensity before the polarizer.

24.4 Example

We will now work through an example to make clear how this reduction in intensity works. Suppose we start with an unpolarized light source of intensity I_o that propagates in the $+z$ direction and is incident on a series of three linear polarizers. The first polarizer has its transmission axis aligned with the y-axis, and the third polarizer has its axis aligned with the x-axis. The transmission axis of the polarizer in between these two makes an angle of 30° with respect to the positive y-axis, as shown in Figure 24.3.

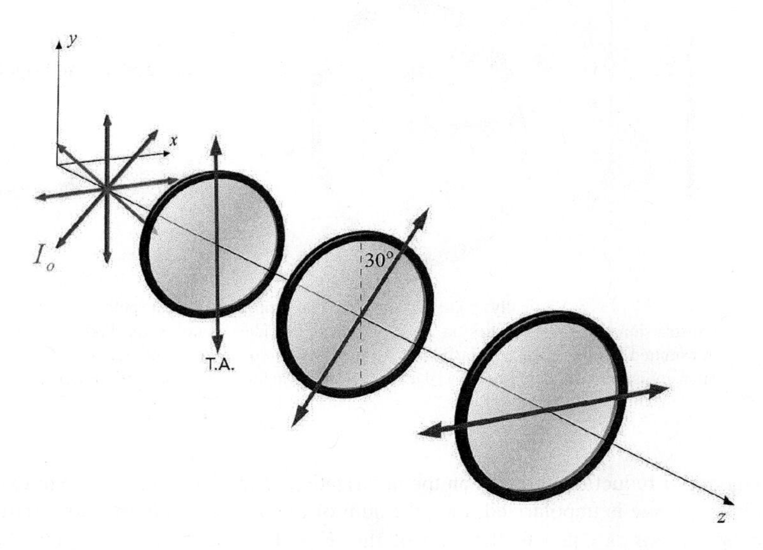

FIGURE 24.3 Unpolarized light is incident on a series of three linear polarizers. The transmission axis of the first polarizer is aligned with the y-axis, while the transmission axis of the third polarizer is aligned with the x-axis. The transmission axis of the polarizer in between these two makes an angle of 30° with respect to the y-axis. The reduction in intensity of the transmitted light is calculated for this example in the text.

In order to calculate the final intensity, we need to work through the polarizers one at a time. The light transmitted beyond the first polarizer is linearly polarized along the y-axis and its intensity I_1 is cut in half since the incident light was unpolarized:

$$I_1 = \frac{1}{2} I_o$$

The light transmitted beyond the second polarizer is linearly polarized along its transmission axis and its intensity is reduced by a factor of (30°). In terms of the initial intensity, we see that

$$I_2 = \left(\frac{\sqrt{3}}{2}\right)^2 I_1 = \frac{3}{8} I_o$$

To determine the final intensity, we note that the polarization direction of the light incident on the third polarizer makes an angle of 60° with respect to its transmission axis.

Consequently, the light transmitted beyond the third polarizer is linearly polarized along the x-axis and its intensity I_3 is given by

$$I_3 = \left(\frac{1}{2}\right)^2 I_2 = \frac{3}{32} I_o$$

Our final result then is that the intensity I_3 is equal to $3/32$ of the initial intensity I_o. We obtained this intensity by calculating the intensity reduction at each polarizer. It's important to realize, however, that we are dealing with a vector wave here and not just an attenuation of intensity due to passing through filters. For example, the addition of more polarizers does not always mean that the final intensity will decrease. Indeed, if you were to remove the second polarizer, the intensity would not increase; in fact, it would go to zero!

24.5 Other Polarization States

To this point, we have dealt pretty exclusively with linearly-polarized electromagnetic waves. It's important to realize, however, that there are many other kinds of polarizations that are possible for electromagnetic waves.

For example, suppose we look at the form for our original linearly polarized wave, but drop the requirement that the constant phase associated with E_x has to be the same as the constant phase associated with E_y. For example, let's assume

$$E_x = E_o \cos\theta \sin(kz - \omega t + \phi_x)$$

$$E_y = E_o \sin\theta \sin(kz - \omega t + \phi_y)$$

These forms will work since we have separate wave equations for E_x and E_y; there is no relation required between ϕ_x and ϕ_y. In fact, we specify different states of polarization by specifying the relationship between ϕ_x and ϕ_y.

For linear polarization, we require the relative phase (defined as $\phi_x - \phi_y$) to be equal to zero. Another important polarization we will discuss in this course is circular polarization. For circular polarization, we require the relative phase ($\phi_x - \phi_y$) to be equal to $\pm\pi/2$ and θ to be equal to $\pi/4$ (i.e., equal x and y amplitudes). For example, if we choose $\phi_x - \phi_y$ to be equal to $+\pi/2$, we see that we can write E_x as being proportional to the cosine, which demands that E_y be proportional to the sine:

$$E_x \propto E_o \cos(kz - \omega t + \phi_y)$$

$$E_y \propto E_o \sin(kz - \omega t + \phi_y)$$

If, on the other hand, we choose *the relative phase* to be equal to $-\pi/2$, we see that we can write E_y as being proportional to the cosine, which demands that E_x be proportional to the sine:

$$E_x \propto E_o \sin(kz - \omega t + \phi_x)$$

$$E_y \propto E_o \cos(kz - \omega t + \phi_x)$$

The effect in either case is that the amplitudes of the E_x and E_y oscillations are equal, but these oscillations are 90° out of phase with each other.

You may not be surprised to learn that these two kinds of circular polarization (corresponding to $\phi_x - \phi_y = +\pi/2$ and $\phi_x - \phi_y = -\pi/2$) are called right-handed and left-handed circular polarization.

If the wave is moving in the $+z$ direction, a relative phase $\phi = \phi_x - \phi_y = +\pi/2$ corresponds to a right-handed circular polarization. To understand this choice of words, let's freeze the wave at an instant of time, as shown in Figure 24.4, and look at the spatial dependence of the wave. E_x oscillates as $+\cos(kz)$, while E_y oscillates as $+\sin(kz)$. Therefore, the $\vec{E}$-field vector starts out pointing in the positive x direction and as z increases, it rotates up to the positive y direction, down to the negative x direction, continues to rotate down to the $-y$ direction, and then finally back up to the $+x$ direction, completing a clockwise rotation. If we look at this movement from behind, i.e., along the direction of propagation, we see that the $\vec{E}$-field vector traces out the spiral movement of an ordinary right-handed screw.

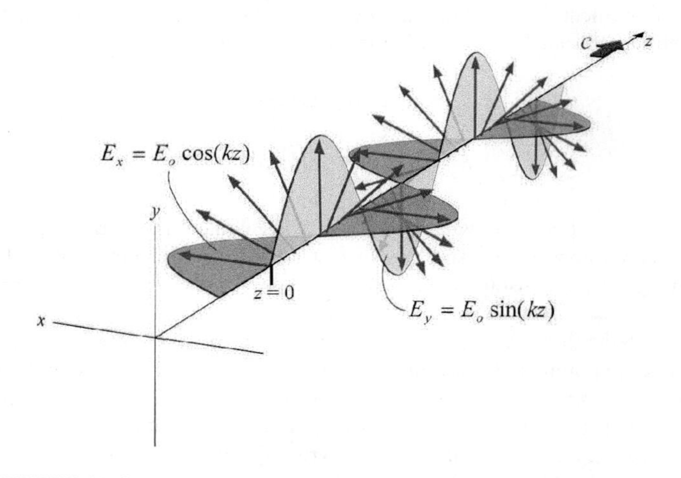

FIGURE 24.4 A snapshot at $t = 0$ of the electric field spatial oscillations of a right-handed circularly polarized electromagnetic wave. Looking at this movement from behind (i.e., along the direction of propagation), we see that the $\vec{E}$-field vector traces out the spiral movement of an ordinary right-handed screw.

Had we chosen the relative phase $\phi = \phi_x - \phi_y = -\pi/2$, we would have E_x oscillating as $\sin(kz)$, while E_y would oscillate as $\cos(kz)$. Here the rotation of the $\vec{E}$-field vector when observed along the direction of propagation would now rotate counterclockwise, corresponding to a left-handed screw.

24.6 Birefringence and Quarter-Wave Plates

Now that we have defined circularly-polarized radiation, how can we actually produce it? Clearly what we need to do is find some way to create a phase difference between two orthogonal transverse components of the $\vec{E}$ -field.

Fortunately, materials exist that do just that. Namely, **birefringent materials**, usually crystals of stressed plastics, have an asymmetric structure in transverse directions, which results in the speed of light being different in each direction! Now, if the speed of light is different in the x direction than it is in the y direction, for example, then after a wave has passed through a fixed thickness of the material, the phase of the E_x oscillation will have advanced by a different amount than that of the phase of the E_y oscillation. Therefore, if the wave entered the material with the x and y oscillations in phase (e.g., a linearly-polarized wave), it will exit the material with these oscillations out of phase and the wave will no longer be linearly polarized.

Let's see how this works. Let's assume the x-axis is the "slow" axis corresponding to a speed of light equaling v_{slow}, and the y-axis is the "fast" axis corresponding to a speed of light equaling v_{fast}. If the thickness of the material is d, then the x-phase advances by an amount ϕ_x equal to the product of the angular frequency of the wave and the time it takes for the wave to move through the material:

$$\phi_x = \omega \frac{d}{v_{slow}}$$

Similarly, the y-phase advances by an amount determined by v_{fast}:

$$\phi_y = \omega \frac{d}{v_{fast}}$$

Consequently, the relative phase at the exit of the material is proportional to the product of the angular frequency, the thickness, and the difference of the inverse speeds:

$$\Delta\phi \equiv \phi_y - \phi_x = \omega d \left(\frac{1}{v_{fast}} - \frac{1}{v_{slow}} \right)$$

For a given birefringent material (i.e., for a certain v_{fast} and v_{slow}), the relative phase change is determined by the thickness d and the frequency ω. Therefore, for a given frequency ω, there exists a definite thickness d_o that will produce a relative phase change between the fast and slow axes of $\pi/2$. Such a piece of birefringent material is called a **quarter-wave plate** and is used to produce circularly-polarized radiation. For example, if radiation incident on a quarter-wave plate is linearly polarized at an angle of 45° to the fast axis, then the radiation leaving the plate will be circularly polarized!

Figure 24.5 shows a linear polarizer aligned at 45° with respect to the x and y axes, producing the linearly polarized light that is incident on a quarter-wave plate whose fast axis is aligned with the y-axis. The light that leaves the quarter-wave plate will have its E_x and E_y oscillations out of phase by 90° and hence be circularly polarized.

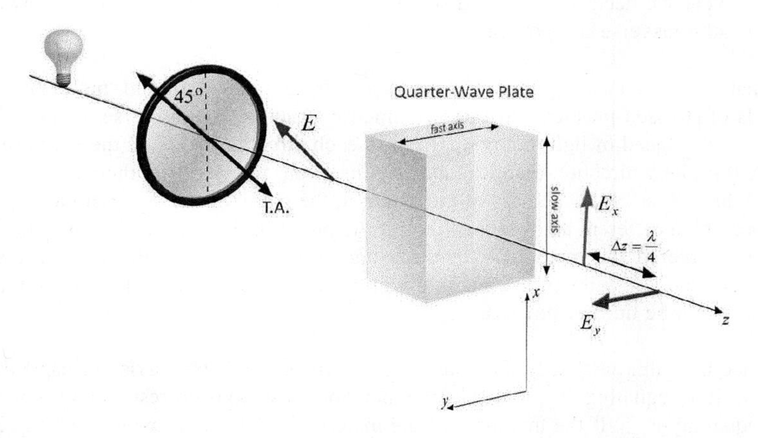

FIGURE 24.5 Light that is polarized at 45° to the x and y axes is incident on a quarter-wave plate of thickness d whose slow axis is aligned with the x-axis and whose fast axis is aligned with the y-axis. The vector representing the peak value of the transmitted $\vec{E}$ -field component along the fast axis is drawn as leading the vector representing the peak value of the transmitted $\vec{E}$ -field component along the slow axis. Curling the fingers of your right hand from the slow component vector toward the fast component vector, your thumb points in the direction of propagation of the light, indicating that the transmitted light is right-circularly polarized.

We can figure out the handedness of the circular polarization using a simple technique. Since E_y is the fast component, put its vector ahead of the slow component, E_x, on the downstream side of the plate. If you curl the fingers of your right hand from the slow component vector toward the fast component vector, your thumb will point in the direction of propagation of the wave if the light is right-circularly polarized and in the opposite direction if the light is left-circularly polarized. Here we see that the light is right-circularly polarized!

You can verify that this is the correct determination using the equations. Namely, before the quarter-wave plate, we have

$$E_x = E_y = E_o \sin(kz - \omega t)$$

After the quarter-wave plate, the phase of the slow component (E_x) must be larger than the phase of the fast component (E_y) by $\pi/2$. Consequently, after the quarter-wave plate, if we write E_y as

$$E_y = E_o \sin(kz - \omega t)$$

then E_x becomes

$$E_x = E_o \cos(kz - \omega t)$$

leading to the same vector representation of E_x and E_y that indicates right-handed circular polarization.

24.7 Summary

In this unit we explored some possible polarizations of electromagnetic waves. We began by defining what we had previously called "plane harmonic waves" as linearly-polarized waves. In particular, we defined the direction of polarization to be the axis of the electric field oscillations of the plane wave, i.e., we defined the general form for linearly polarized waves moving in the $+z$ direction to be

$$\vec{E} = \hat{e}E_o \sin(kz - \omega t + \phi)$$

where $\hat{e}$ is the unit vector in the x-y plane that defines the polarization direction and ϕ is just a constant phase that specifies the E-field strength at $z = t = 0$.

We then discussed how linearly polarized radiation could be produced by using polarizers, materials that totally absorbed the $\vec{E}$-field component perpendicular to its transmission axis, while completely passing the $\vec{E}$-field component parallel to that axis. We determined that half of the intensity of incident unpolarized light would be transmitted through a polarizer. On the other hand, the intensity of linearly-polarized light incident on a polarizer would be reduced by an amount determined by θ, the angle between the initial polarization direction and the transmission axis:

$$I = I_o \cos^2 \theta$$

We then introduced circular polarization, in which the $\vec{E}$-field oscillations in two orthogonal transverse directions were 90° out of phase with each other. We distinguished between right-handed and left-handed circular polarization by observing the sense of rotation of the $\vec{E}$-field oscillations in space at a fixed time.

Finally, we introduced birefringent materials that have an asymmetric structure in transverse directions, which results in the speed of light being different in each direction!

These materials can then be used to construct quarter-wave plates that produce circularly-polarized radiation from radiation that is linearly polarized at an angle of 45° to the "fast" and "slow" axes of the material.

MAIN POINTS

Linearly-Polarized Waves

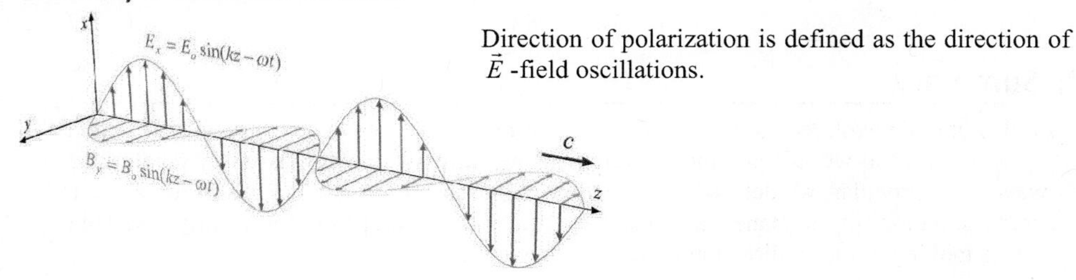

Direction of polarization is defined as the direction of $\vec{E}$-field oscillations.

Linear Polarizers

Polarizers produce polarized light along their transmission axis.

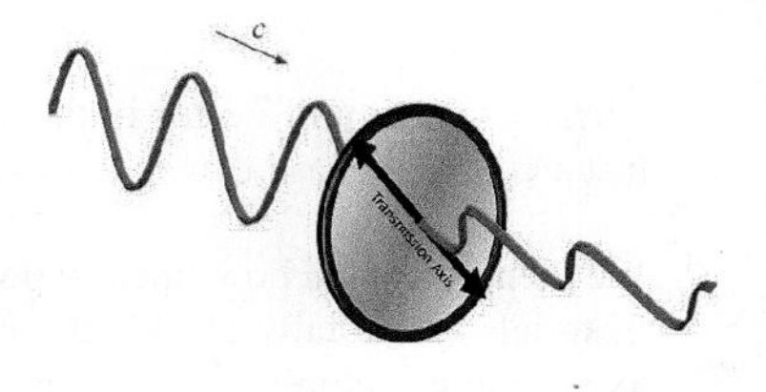

If incident light is polarized, the transmitted intensity is reduced by an amount determined by the angle between the incident polarization and the transmission axis.

$$I_{transmitted} = I_{incident} \cos^2 \theta$$

Law of Malus

Circularly-Polarized Waves

Circularly-polarized light has one $\vec{E}$-field component 90° out of phase with the orthogonal component.

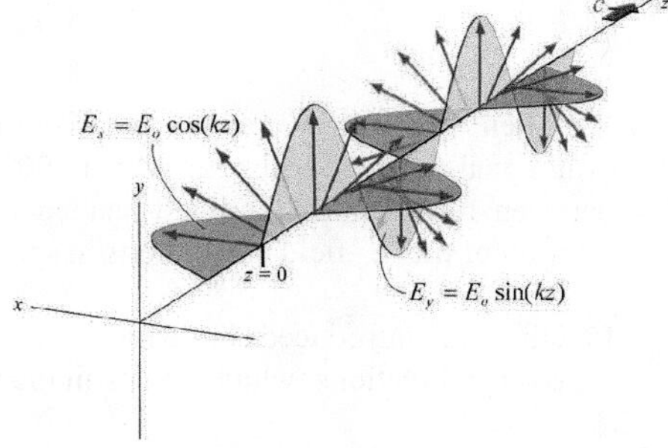

Birefringent materials can be used to produce circularly-polarized light.

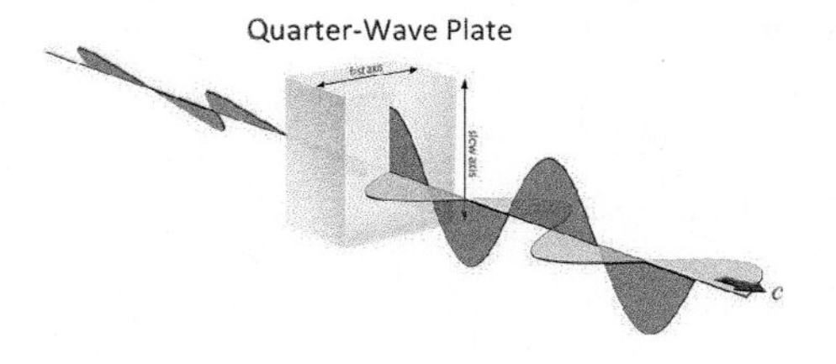

PROBLEMS

1. Polarizers: An unpolarized beam of intensity $I_o = 163\ \text{W/m}^2$ travels in the positive z-direction and is incident on a series of three linear polarizers, as shown in Figure 24.6. The transmission axis of the first polarizer is aligned with the y-axis. The following polarizers make angles of $\theta_2 = 54°$, and $\theta_3 = 27°$ with the positive x-axis.

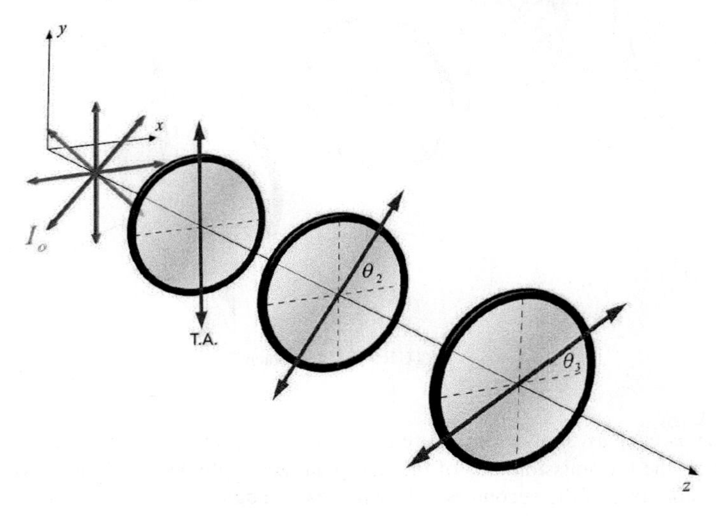

FIGURE 24.6 Problem 1

(a) What is I_2, the intensity of the beam immediately following the second polarizer? (b) What is I_{final}, the intensity of the beam immediately following the third polarizer? (c) The positions of the second and third polarizers are now interchanged. What is $I_{final,new}$, the intensity of the beam after passing through the new arrangement of the three polarizers? (d) Suppose we now move our incident unpolarized beam so that it is incident from the right. That is, the beam is now incident upon the polarizer whose transmission axis makes an angle θ_2 with the x-axis, and exits the last polarizer (with transmission axis aligned with the y-axis) with intensity $I_{final,RL}$. How does $I_{final,RL}$ compare with $I_{final,new}$?

(i) $I_{final,RL} < I_{final,new}$
(ii) $I_{final,RL} = I_{final,new}$
(iii) $I_{final,RL} > I_{final,new}$

(e) Suppose we now rotate the last polarizer by 90° so that its transmission axis becomes aligned with the x-axis. With the beam still incident from the right (on the polarizer whose transmission axis makes an angle θ_2 with the x-axis), what is the $I_{final,RLnew}$, the intensity of the beam immediately following the last polarizer (transmission axis aligned with the x-axis)?

2. Polarizers and a Quarter-Wave Plate: A monochromatic laser beam of intensity $I_o = 704\ \text{W/m}^2$ is polarized in the y-direction and propagates in the positive z-direction. This beam is incident upon a quarter-wave plate whose fast axis makes an angle of 45°

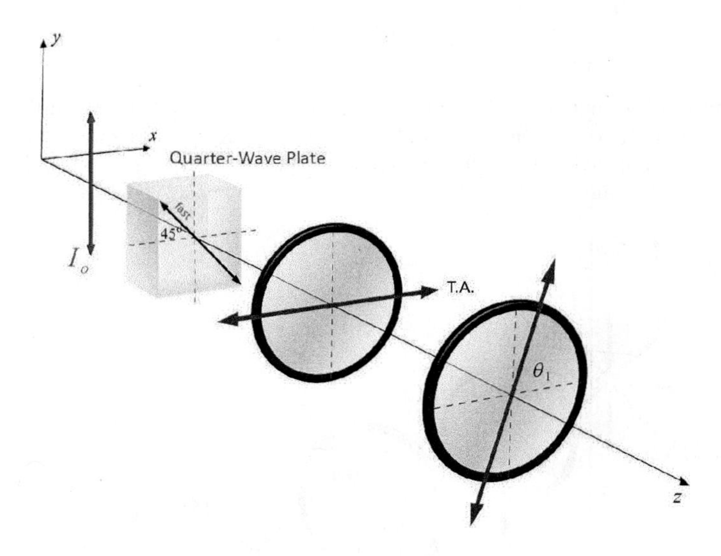

FIGURE 24.7 Problem 2

with the x-axis, as shown in Figure 24.7. Following the quarter-wave plate are two polarizers; the transmission axis of the first polarizer is aligned with the x-axis, while the transmission axis of the second polarizer makes an angle of $\theta_1 = 77°$ with the positive x-axis. (a) What is I_{mid}, the intensity of the beam immediately following the polarizer whose transmission axis is aligned with the x-axis? (b) What is I_{final}, the intensity of the beam immediately following the last polarizer? (c) What is the ratio of $E_{y,final}$, the maximum value of the y-component of the electric field immediately following the last polarizer, to E_o, the amplitude of the electric field oscillations in the incident polarized beam? (d) The positions of the quarter-wave plate and the last polarizer are now interchanged. What is $I_{final,new}$, the intensity of the beam after passing through the new arrangement of the two polarizers and the quarter-wave plate? (e) What is the polarization of the beam immediately following the quarter-wave plate?

(i) Right circularly polarized.
(ii) Left circularly polarized.
(iii) Linearly polarized at 45° to the positive x-axis.
(iv) Linearly polarized at 135° to the positive x-axis.

(f) Suppose you are free to rotate θ_1, the transmission axis of the initial polarizer. How many values of θ_1, $0 \leq \theta_1 < 180°$, will produce a final intensity of zero after the quarter-wave plate: *none*, *one*, or *two*?

3. Polarization (INTERACTIVE EXAMPLE): A circularly-polarized beam of intensity I_o travels in the $+z$ direction and is incident on a stack of three linear polarizers. The transmission axes of the first two polarizers make angles with the x-axis of $\theta_1 = 60°$ and $\theta_2 = 130°$, respectively. The ratio of the intensity of the beam at the exit of the three polarizers to the incident intensity is $I_3 / I_o = 0.02$. What is θ_3, the angle between the x-axis and the transmission axis of the third polarizer?

UNIT 25

REFLECTION AND REFRACTION

25.1 Overview

In this unit we will begin our discussion of geometric optics by discussing reflection and refraction of electromagnetic waves. We will begin by discussing the reflection of electromagnetic waves (usually light) from a surface. We will claim that the angle of incidence is equal to the angle of reflection and will demonstrate this relation for plane waves

We will then discuss the propagation of electromagnetic waves in matter and discover that the speed of the wave in matter is less than in vacuum, with the difference being specified in terms of the index of refraction of the material. We will introduce Snell's law to determine the angle of refraction.

We will then discuss the intensities and polarization of the reflected wave. In particular, we will discuss the requirements on the incident angles to produce total internal reflection and linear polarization of the reflected wave.

25.2 Geometric Optics

To this point, we have been concerned with the propagation and polarization of electromagnetic waves in vacuum. We will now focus on what happens to electromagnetic waves (usually light) in different materials.

To determine electric and magnetic fields in matter, we need to replace the vacuum constants ε_o and μ_o with the constants ε and μ that describe the material of interest. For the remainder of the course we will be studying **geometric optics** in which the wavelength of the radiation (usually light) is small compared to the objects with which it interacts. With this restriction, we can assume that the light travels in straight lines called rays.

Our primary focus will be on the reflection and refraction of these rays at the interface of two materials that are described in terms of their indices of refraction, which are in turn determined by the values of ε for these materials.

25.3 Reflection

We begin by stating the **law of reflection** that the angle of incidence equals the angle of reflection, where both angles are measured with respect to the normal, as shown in Figure 25.1. This law is quite general, and we will offer a derivation for plane waves.

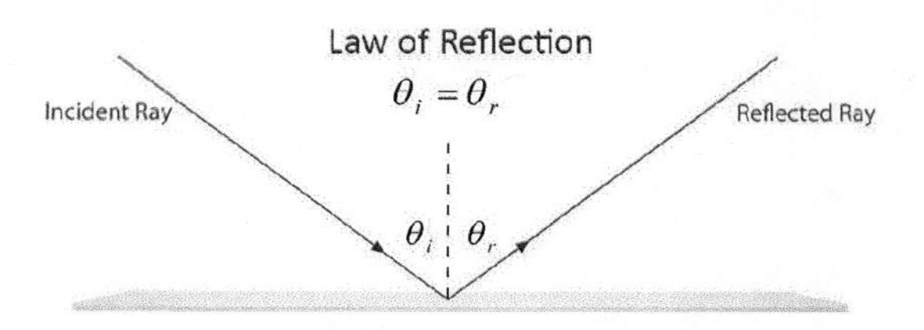

FIGURE 25.1 The law of reflection: The angle of incidence equals the angle of reflection.

Figure 25.2 shows a plane wave, represented by two rays, incident on a reflecting surface, with angle of incidence equal to θ_i. Each of these rays are reflected with an angle of reflection equal to θ_r. We claim that it must take the same amount of time for reflected ray 1 to travel from point *a* to point *d* as it does for incident ray 2 to travel from point *b* to point *c*.

Why? Well we know that points *a* and *b* on the incident rays must be in phase since they define a wavefront perpendicular to the rays. Similarly, points *c* and *d* on the reflected rays must also be in phase. Consequently, the time it takes light to travel from *a* to *d* must

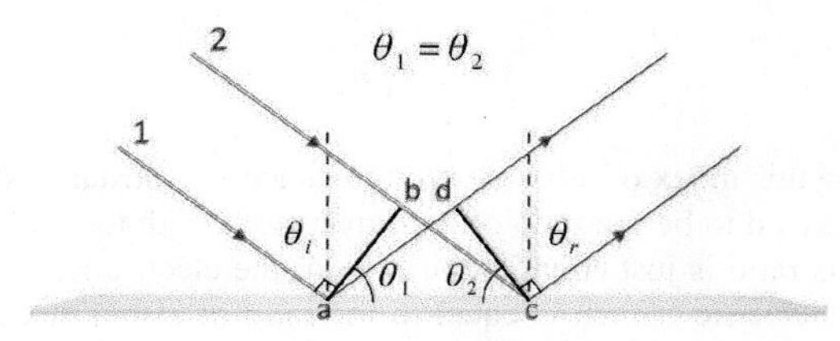

FIGURE 25.2 A proof of the law of reflection for plane waves. The time it takes ray 1 to go from a to d must be equal to the time it takes ray 2 to go from b to c. Therefore, $\theta_2 = \theta_1$. However, $\theta_1 = \theta_i$ and $\theta_2 = \theta_r$. Therefore, $\theta_i = \theta_r$.

be the same as it takes light to travel from b to c. Since the speed of light for both rays is the same, we see that the distance ad must be equal to the distance bc. The distance ad is just equal to $D \sin \theta_2$, while the distance bc is equal to $D \sin \theta_1$. Therefore, we know that θ_2 must be equal to θ_1 .

How are the angles θ_2 and θ_1 related to the angle of incidence and the angle of reflection? Well, a close look at the geometry reveals that θ_1 is equal to the angle of incidence and θ_2 is equal to the angle of reflection. Thus, we have arrived at our result that the angle of incidence equals the angle of reflection.

$$\theta_{incedent} = \theta_{reflected}$$

25.4 Refraction

We will now turn to the general case in which there is also a refracted ray. The first order of business is to consider how Maxwell's equations must be modified to hold in matter, rather than in vacuum. The modification consists simply in replacing the vacuum constants ε_o and μ_o with the constants ε and μ that describe the material of interest.

The main effect of this change is that the speed of electromagnetic waves in matter must be different from what it is in vacuum. Namely, the velocity of the wave in matter is equal to

$$v = \frac{1}{\sqrt{\mu\varepsilon}}$$

which is different from $1/\sqrt{\mu_o \varepsilon_o}$, the speed of the wave in vacuum.

We introduce n, the **index of refraction** of the material, to specify this change in the speed of light. Namely, we define n as the ratio of the speed of light in vacuum to the speed of light in the material. Now, in our study of geometric optics, we will not be interested in magnetic materials, so that we can approximate μ by μ_o and we see that

$$n \approx \sqrt{\frac{\varepsilon}{\varepsilon_o}}$$

We can also relate this index of refraction to the dielectric constant. Namely, the dielectric constant κ was defined to be the ratio of the capacitance with the dielectric to that without the dielectric. This ratio is just equal to the ratio of the electric field with the dielectric to that without the dielectric, which is equal to the ratio of ε to ε_o. Consequently, we can write

$$n \approx \sqrt{\kappa}$$

The index of refraction is frequency dependent. In our study of geometric optics, we will be concerned with visible light. At these frequencies, for example, n ranges from a minimum of 1 for air, is about equal to 1.5 for glass, and reaches a maximum of about 2.4 for diamond.

We now want to determine how the angle of refraction is related to the angle of incidence. This relationship is totally determined by the difference in the speeds of the waves in the materials at the interface.

We can derive this relationship for plane waves much in the same way as we did when we derived the law of reflection. Figure 25.3 shows a plane wave, represented by two rays, incident on a reflecting surface, with angle of incidence θ_1. Each of these rays are refracted with an angle of refraction θ_2. In this case, we claim that it must take the same amount of time for refracted ray 1 to travel from point a to point d as it does for incident ray 2 to travel from point b to point c.

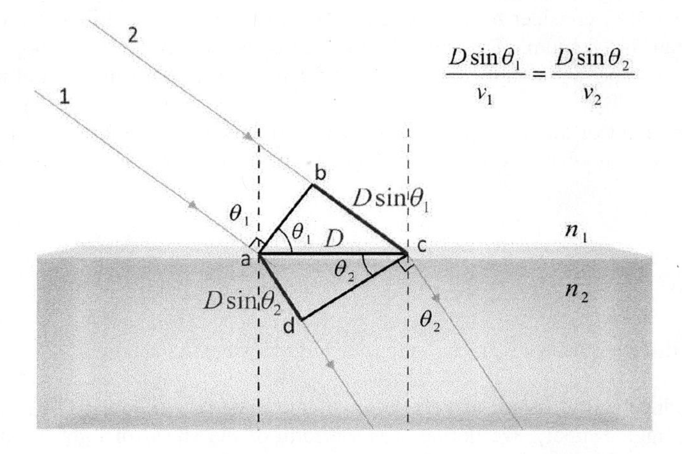

FIGURE 25.3 A proof of Snell's law for plane waves. The time it takes ray 1 to go from a to d must be equal to the time it takes ray 2 to go from b to c. Therefore, $D\sin\theta_1 / v_1 = D\sin\theta_2 / v_2$. Replacing the velocities with the indices of refraction, we arrive at Snell's law: $n_1 \sin\theta_1 = n_2 \sin\theta_2$.

Why? Well, once again we know that points a and b on the incident rays must be in phase since they define a wavefront perpendicular to the rays, and similarly, points c and d on the refracted rays must also be in phase. Consequently, the time it takes light to travel from a to d must be the same as it takes light to travel from b to c. Here the speed of light for the incident wave (v_1) is different from the speed of light for the refracted wave (v_2). The distance from point a to point d is just equal to $D\sin\theta_2$; therefore, the time it takes light to travel from a to d is equal to $D\sin\theta_2 / v_2$. Similarly, the time it takes light to travel from b to c is just equal to $D\sin\theta_1 / v_1$. Setting these times equal, we obtain the relation between the angles and the speeds.

$$\frac{D\sin\theta_1}{v_1} = \frac{D\sin\theta_2}{v_2}$$

Recall that we earlier defined the index of refraction for a material as the ratio of the speed of light in vacuum to the speed of light in the material. We use this definition to obtain the usual form of **Snell's law**:

$$n_1 \sin\theta_1 = n_2 \sin\theta_2$$

Note that while we drew the incident ray in medium 1 and the refracted ray in medium 2, the mathematics of this derivation is unchanged if ray 2 were actually the incident ray and ray 1 were the refracted ray. Snell's law simply states that the product of the index of refraction with $\sin\theta$ in each medium at the interface is the same.

25.5 Intensities

We now know the relationships between all of the angles involved in the reflection and refraction of light. Namely, the angle of incidence equals the angle of reflection and $n_1 \sin\theta_1 = n_2 \sin\theta_2$.

What about the intensities of these waves? How much is reflected and how much is refracted? While the answers to these questions can be obtained from Maxwell's equations, the derivations are beyond the scope of this course. We will be able to determine, however, intensities for a couple of special cases in the following sections.

We simply note here the results for two extreme cases. Namely, for glancing incidence ($\theta_1 \sim 90°$), we have complete reflection ($R \sim 1$), while for normal incidence, it can be shown that the reflected intensity is proportional to the square of the difference in indices of refraction divided by the square of their sum.

$$R = \left(\frac{n_2 - n_1}{n_2 + n_1}\right)^2$$

Note that for an air-glass interface, for example, only 4% of the incident light is reflected at normal incidence.

25.6 Total Internal Reflection

We have just noted that only 4% of the incident light is reflected at an air-glass interface at normal incidence. We will now see that it is possible, however, for some angles of light incident from glass to air to get all of the light reflected.

We start from Snell's law and note that

$$\frac{\sin\theta_2}{\sin\theta_1} = \frac{n_1}{n_2}$$

Now if the light is incident at angle θ_1 from the medium with the larger index of refraction, we see that the light will be refracted at a larger angle θ_2, i.e., $\sin\theta_2 / \sin\theta_1 = n_1 / n_2 > 1$ (see Figure 25.4).

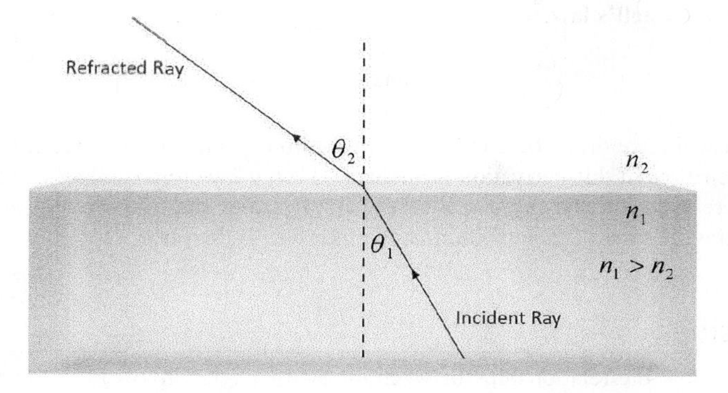

FIGURE 25.4 Light incident from a certain material is refracted into a medium with a smaller index of refraction. This results in an angle of refraction that is larger than the angle of incidence.

Consequently, as θ_1 increases, θ_2 also increases, and is in fact always larger than θ_1. However, there is a limit to this process. Namely, θ_2 can never be larger than 90°. Therefore, for all angles θ_1 such that $\sin\theta_1 > n_2 / n_1$, there will be no refracted ray; all of the incident light will be reflected.

We call this maximum angle of incidence the **critical angle**.

$$\theta_c = \sin^{-1}\left(\frac{n_2}{n_1}\right)$$

For example, light in glass which is incident on a glass-air interface has a critical angle $\theta_c = \sin^{-1}(1.0/1.5) = 41.8°$. Light that is incident at any angle $\theta_1 > \theta_c$ will be totally reflected. This property of **total internal reflection** is the basis for optical fiber communication.

25.7 Polarization

We now know through what angles light rays are reflected and refracted at an interface and something about the relative intensities of the beams. What about the polarization of these beams? Can an unpolarized incident beam produce a polarized reflected beam? The answer to this question is yes, if the unpolarized beam is incident at a particular angle, called **Brewster's angle.** Here's how it works.

Figure 25.5 shows incident and reflected rays each making an angle θ_1 with the normal and a refracted ray making an angle θ_2 with the normal. We know the $\vec{E}$ field for each ray must oscillate in a plane that is perpendicular to that ray. We indicate the components of this oscillation with circles and double-headed arrows. The $\vec{E}$-field component indicated by circles is the component perpendicular to the plane of incidence; this component is parallel to the surface, and is therefore perpendicular to all rays. This component of the polarization will always be present.

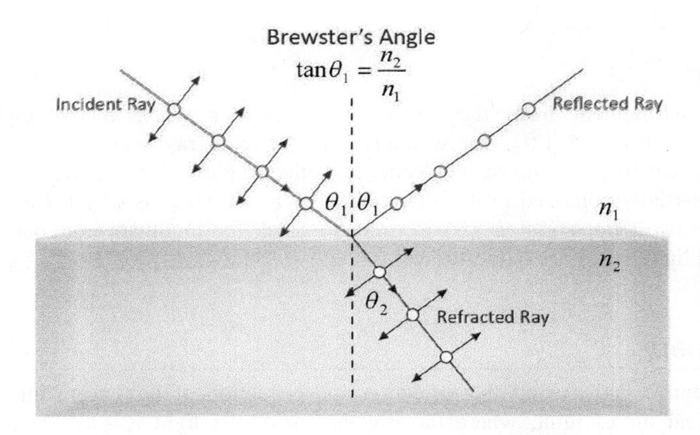

FIGURE 25.5 Unpolarized light incident at a particular angle called Brewster's angle produces a reflected ray that is polarized perpendicular to the plane of incidence.

The $\vec{E}$-field component, indicated by double-headed arrows, is the component in the plane of incidence. It is found empirically that the amplitude of these $\vec{E}$-field oscillations in the reflected ray are not as large as those in the incident ray. In fact, at one angle, called Brewster's angle, in which the angle between the refracted and reflected rays is equal to 90°, the polarization of the reflected ray in the plane of incidence goes to zero!

We can understand this relation by considering how the reflected and refracted rays are generated. These rays are generated by the oscillations of electrons in the surface. These oscillations in the plane of incidence must be perpendicular to the direction of the refracted ray. These oscillations, therefore, will not be perpendicular to the direction of the reflected ray, and only the component of these oscillations that is also perpendicular to the reflected ray can give rise to $\vec{E}$-field oscillation in the plane of incidence for the reflected ray. Now, when the angle between the reflected ray and the refracted ray becomes equal to 90°, we see this component will be zero and there will be no $\vec{E}$-field oscillations in the plane of incidence for the reflected ray. Consequently, when unpolarized light is incident at this angle, the reflected light will be completely polarized perpendicular to the plane of incidence.

We can determine this special angle of incidence by combining Snell's law with our defining condition that the angle between the reflected and refracted rays must be equal to 90°.

$$n_1 \sin\theta_1 = n_2 \sin\theta_2 \qquad \text{(Snell's law)}$$

$$\theta_1 + \theta_2 = 90^\circ \qquad \Rightarrow \qquad \sin\theta_2 = \cos\theta_1$$

We then obtain the final expression for Brewster's angle as the angle whose tangent is equal to the ratio of the indices of refraction.

$$\tan\theta_1 = \frac{n_2}{n_1}$$

If light is incident from the air at an air-glass interface at an angle of 56.3° (i.e., $\tan(56.3^\circ) = 1.5/1.0$), for example, the reflected ray will be linearly polarized perpendicular to the plane of incidence. For other angles of incidence, the reflected ray will be partially polarized in this direction. This result explains why Polaroid sunglasses have a vertical transmission axis in order to block the dominant horizontally-polarized reflected light from horizontal surfaces.

25.8 Summary

In this unit, we discussed the reflection and refraction of light from interfaces in the geometrical optics limit, where the wavelength of the light is small compared to the objects with which it interacts.

We began by introducing the law of reflection, that the angle of incidence equals the angle of reflection. We claimed this law was quite general, but did provide justification for plane waves.

We then introduced the modifications to Maxwell's equations that are necessary to describe electromagnetic fields in matter. By replacing ε_o and μ_o with ε and μ, respectively, we determined that the speed of an electromagnetic wave in matter is less than the speed of light in vacuum. In particular, we obtained the velocity in matter to be equal to *c*, the velocity in vacuum, divided by *n*, the index of refraction for that material. For most materials we will be concerned with in our study of geometric optics, $\mu \approx \mu_o$, so that

$$n \approx \sqrt{\frac{\varepsilon}{\varepsilon_o}} = \sqrt{\kappa}$$

where κ is the dielectric constant of the material.

We then applied our understanding of the speed of electromagnetic waves in matter to determine the relationship between the angle of incidence and the angle of refraction for a plane wave. In particular, we demonstrated that these angles satisfied Snell's law:

$$n_1 \sin\theta_1 = n_2 \sin\theta_2$$

We then discussed the intensities of these rays for a few special cases, most notably deriving an expression for the critical angle

$$\theta_c = \sin^{-1}\left(\frac{n_2}{n_1}\right)$$

For all angles of incidence $\theta_1 > \theta_c$, there is no refracted ray; all of the incident intensity is reflected.

Finally, we determined that the reflected ray is usually partially polarized perpendicular to the plane of incidence, being totally polarized at Brewster's angle, defined as

$$\theta_B = \tan^{-1}\left(\frac{n_2}{n_1}\right)$$

MAIN POINTS

Law of Reflection

The angle of incidence is equal to the angle of reflection.

$$\theta_i = \theta_r$$

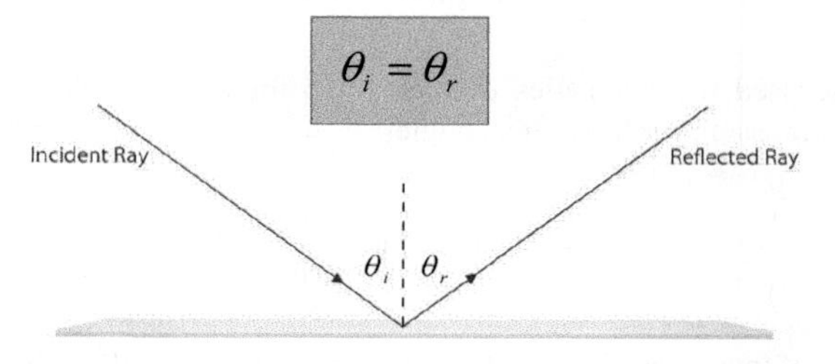

Electromagnetic Fields in Matter

Replacing ε_o and μ_o by ε and μ, we determine the speed of light in matter.

$$v = \frac{c}{n}$$

Law of Refraction (Snell's Law)

The ratio of the sines of the angles of incidence and refraction is equal to the ratio of the indices of refraction.

$$n_1 \sin\theta_1 = n_2 \sin\theta_2$$

Intensities and Polarizations

Light moving from a dense medium to a less dense medium will be totally internally reflected if the angle of incidence is greater than the critical angle.

$$\theta_c = \sin^{-1}\left(\frac{n_2}{n_1}\right)$$

Reflected light is generally partially polarized perpendicular to the plane of incidence and is totally polarized at Brewster's angle.

$$\theta_B = \tan^{-1}\left(\frac{n_2}{n_1}\right)$$

PROBLEMS

1. Monochromatic Light and a 30-60-90 Prism: Red light is incident in air on a 30°-60°-90° prism as shown. The incident beam is directed at an angle of $\phi_1 = 34.3°$ with respect to the horizontal and enters the prism at a height $h = 20$ cm above the base. The beam leaves the prism to the air at a distance $d = 59.4$ m along the base as shown. (a) What is ϕ_2, the angle the beam in the prism makes with the horizontal axis? (b) What is n, the index of refraction of the prism for red light? (c) What is ϕ_3, the angle the transmitted beam makes with the horizontal axis? (d) What is $\phi_{1,\max}$, the maximum value of ϕ_1 for which the incident beam experiences total internal reflection at the horizontal face of the prism? (e) The red beam is now replaced by a violet beam that is incident at the same angle ϕ_1 and same height h. The prism has an index of refraction $n_{violet} = 1.29$ for violet light. Compare d_{violet}, the exit distance for violet light, to d, the exit distance for red light.

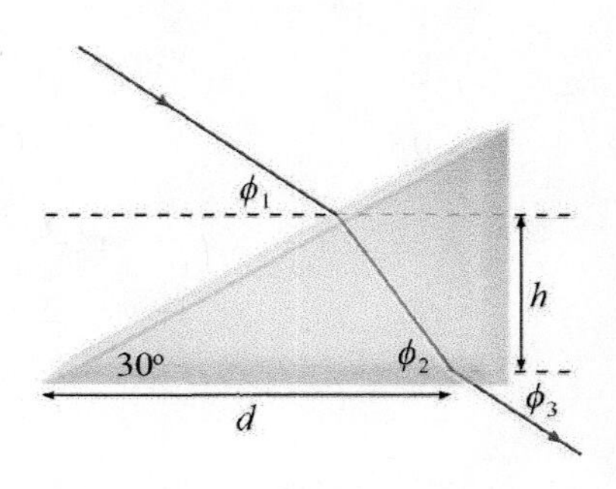

FIGURE 25.6 Problem 1

(i) $d_{violet} < d$
(ii) $d_{violet} = d$
(iii) $d_{violet} > d$

(f) Suppose now that the violet beam is incident at height h, but makes an angle $\phi_{1,y} = 60°$ with the horizontal. What is $\phi_{3,y}$, the angle the transmitted beam makes with the horizontal axis?

2. Refraction (INTERACTIVE EXAMPLE): A transparent object with an isosceles right triangular cross-section has an index of refraction $n_2 = 1.2$. A light beam in air is incident on this object, making an angle $\theta_{in} = 75°$ with respect to the x-axis, as shown. At what angle (with respect to the x-axis), θ_{out}, does the observer see the light beam exit the object?

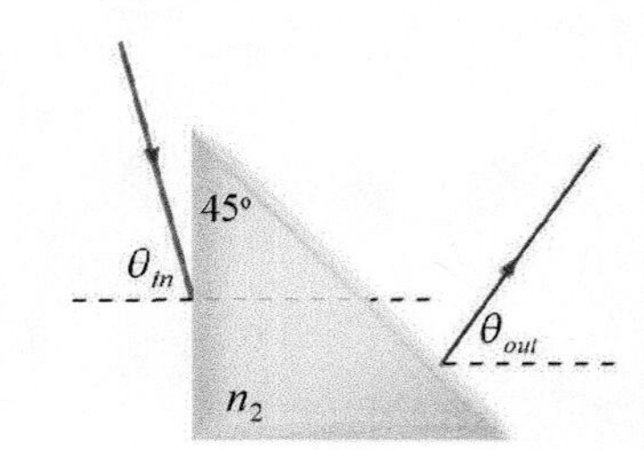

FIGURE 25.7 Problem 2

UNIT

26

LENSES

26.1 Overview

In this unit we will discuss image formation by lenses. We will begin by noting that we already have the fundamental knowledge we need to understand how lenses work. Namely, we will be able to understand the properties of lenses by applying Snell's law of refraction.

We will first discuss the difference between converging and diverging lenses. We will use ray tracing techniques on a converging lens to develop the lens equation. We will then work through both converging and diverging lens examples.

We will conclude by developing the lensmaker's formula for a plano-convex lens, an expression that determines how to produce a lens with a desired focal length. We will extend this formula to the more general case of two curved surfaces and discuss briefly some of the aberrations that are present in simple lenses.

26.2 Formation of Images

For the final three units, we will discuss applications of geometric optics. In particular, we will study lenses, mirrors, and optical instruments. The really good news is that we will need no new physics to do this work. All we will need will be the fundamentals of geometric optics that we presented in the last unit.

In particular, we will consider light to propagate as simple straight line rays, since we will be concerned with situations in which the length scales are much greater than the wavelength of the light. To justify this approach, we first examine what happens to light waves when they hit a **lens**, a transparent object with curved surfaces. Figure 26.1 shows wavefronts emanating from the tip of an arrow. The parts of the wavefront that strike the

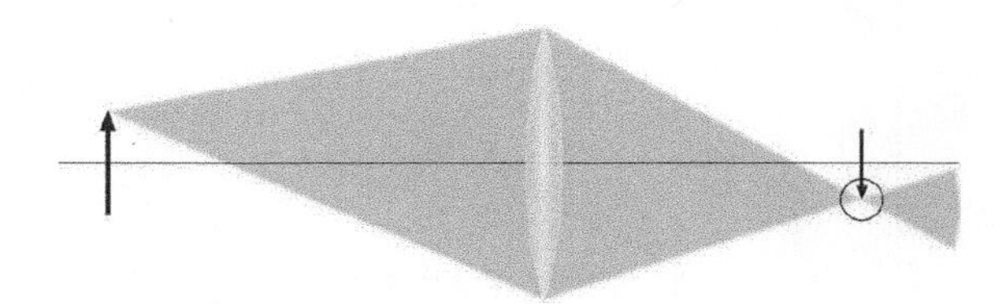

FIGURE 26.1 Wavefronts from the tip of an arrow are refracted through a lens and emerge to create a focus downstream of the lens, as indicated by the image of the tip of the arrow identified by the circle.

lens are refracted, since the index of refraction of the lens is different from that of air. Note that these waves, after they pass through the lens, actually converge to a point forming a focus of the tip of the arrow. Light waves starting from the bottom of the arrow similarly form a focus of the bottom of the arrow at the same distance downstream of the lens but at a point above the tip, forming an inverted image of the arrow. We can simplify our representation by drawing straight lines (the rays) that are perpendicular to the wave front, as shown in Figure 26.2. We can determine the behavior of these rays using Snell's

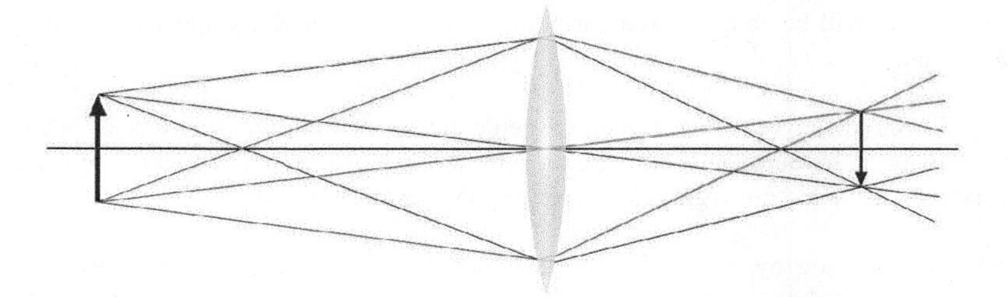

FIGURE 26.2 Rays drawn from the tip and the bottom of an arrow are refracted through a lens, according to Snell's law, and emerge to create an inverted image of the arrow downstream of the lens.

law, that the product of the index of refraction and the sine of the angle with respect to the normal to an interface between two materials is the same on either side of that interface.

Similarly, we can determine the behavior of light rays that are reflected from a mirror using the law of reflection, that the angle of incidence equals the angle of reflection. Consequently, we need only these two laws to understand the properties of lenses (perfect refraction), mirrors (perfect reflection), and optical instruments (combinations of lenses, such as microscopes, telescopes, and eyeglasses).

26.3 Converging and Diverging Lenses

We will now study the properties of lenses, transparent materials that are constructed so that parallel rays are refracted toward a point, the **focus**.

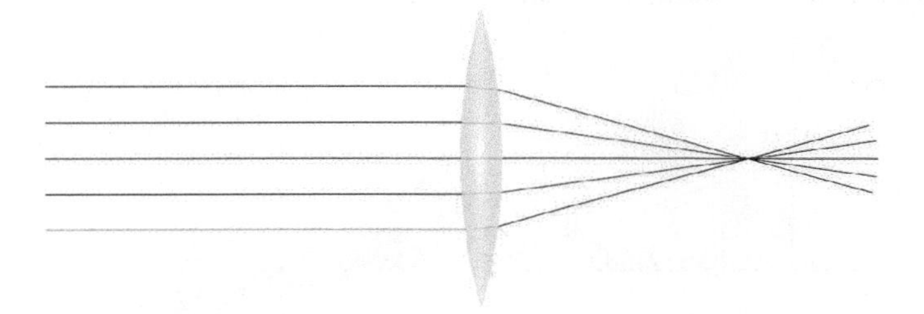

FIGURE 26.3 Parallel rays are incident on a converging lens. These rays are refracted according to Snell's law and emerge to form a real focus on the axis, downstream of the lens.

Figure 26.3 shows a **converging lens**. Parallel rays are first refracted into the glass toward the normal and then refracted back into the air away from the normal. We see these

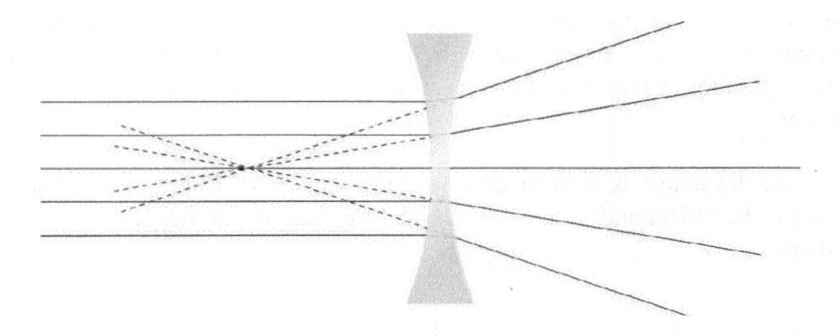

FIGURE 26.4 Parallel rays are incident on a diverging lens. These rays are refracted according to Snell's law and diverge downstream of the lens. The projections of the diverging rays back through the lens (indicated by the dashed lines) form a virtual focus on the axis.

refracted rays are converging (hence the name of the lens) and indeed intersect, forming a **real** focus on the downstream side of the lens.

Figure 26.4 shows a **diverging lens**. Parallel rays are first refracted into the glass toward the normal and then refracted back into the air away from the normal. We see these refracted rays are diverging (hence the name of the lens) and do not intersect. We say these rays form a **virtual** focus on the upstream side of the lens.

26.4 The Lens Equation

We will now derive the lens equation that determines the image distance in terms of the object distance and the focal length. Figure 26.5 shows a converging lens in which the object and image distances are defined (s and s', respectively). Parallel rays are incident from the left on this lens. We will determine the location and size of the image by tracing rays that pass through the tip of the arrow.

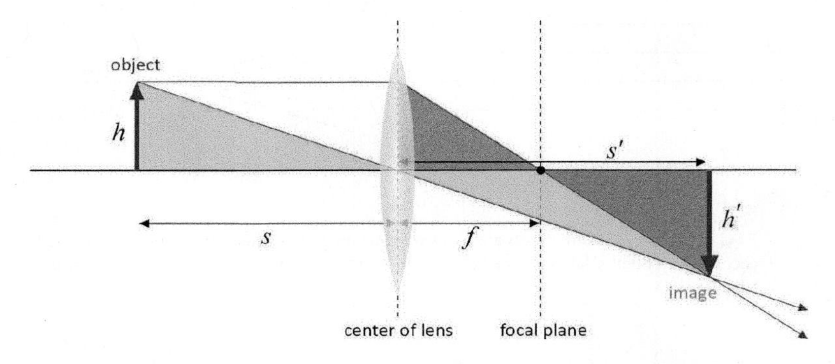

FIGURE 26.5 Illustration of the rays used to derive the lens equation that relates object (s) and image (s') distances for a converging lens with focal length f.

First we consider the ray that passes through the center of the lens. In the thin lens approximation, this ray passes through undeflected. Next, we consider the ray that propagates parallel to the axis of the lens; this ray is refracted and passes the axis at the **focal point**.

Here we see the image is formed downstream of the lens. Using the highlighted similar triangles for the object and image, we see that the ratio of the heights is equal to the ratio of the distances.

$$\frac{|h'|}{h} = \frac{s'}{s}$$

Using the smaller similar triangles to the right of the lens, we obtain an expression for the ratio of the heights in terms of the focal length and image distance.

$$\frac{|h'|}{s'-f} = \frac{h}{f}$$

We can combine these two equations to eliminate the heights, and then rewrite the expression to obtain the **lens equation**:

$$\frac{1}{s} + \frac{1}{s'} = \frac{1}{f}$$

We can see that the image size is magnified (that is, it is larger than the object size) if the image distance is bigger than the object distance. Formally, we define the **magnification** to be equal to minus the ratio of the image size to the object size, which we see is equal to minus the ratio of the image distance to the object distance.

$$M \equiv \frac{h'}{h} = -\frac{s'}{s}$$

You may be wondering why we included the minus sign in this definition. Well, the answer is that the sign of the magnification then tells us whether the image is inverted or upright. In this case, taking both s and s' to be *positive*, we see that we obtain a *negative magnification*, indicating that the image is inverted. We will see in the next slide that it is possible to generate an image upstream of the lens (s' negative), which will then give a positive magnification, indicating an upright image.

26.5 Examples

We will now draw rays and apply the lens equation for both converging and diverging lens examples. We start with a converging lens shown in Figure 26.6. Our object is the arrow to the left of the lens. We draw the two principal rays through the tip of the arrow. The first ray passes through the center of the lens undeflected, while the second ray enters the lens parallel to the axis and is bent and passes through the focal point as shown. The real inverted image is formed downstream of the lens.

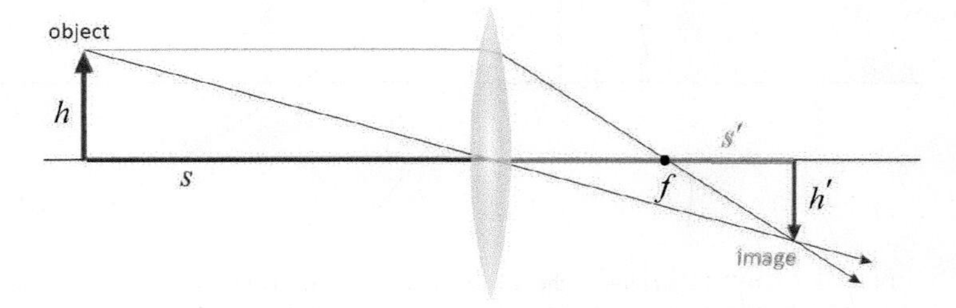

FIGURE 26.6 Illustration of the rays used to determine the real image (inverted arrow) formed of the object (upright arrow) by a converging lens.

We apply the lens equation to determine that the object distance is equal to the product of the image distance and the focal length divided by the difference of these distances.

$$s' = \frac{f\,s}{s - f}$$

If we substitute this expression for the object distance into the magnification equation, we obtain the result that the magnification is equal to minus the focal length divided by the difference of the object distance and the focal length.

$$M = \frac{-f}{s - f}$$

Note what happens if the object distance is less than the focal length. In this case we see that the image distance becomes negative and the magnification becomes positive. The positive magnification indicates an upright image, but what does the negative image distance indicate? Let's move the object now so that its distance to the lens is less than the focal length, as shown in Figure 26.7, and redraw the rays to find out. Once the object is inside the focal length, the refracted rays diverge. In fact, these rays appear to diverge from a point that we identify as the image point, which is indeed located upstream of the lens. We call this kind of an image "virtual" to distinguish it from the "real" image formed when the object distance is larger than the focal length. In a **real image**, the refracted rays intersect; if you place a screen at the image distance, you will see the image on the screen. In a **virtual image**, the refracted rays do not intersect; they appear to come from a point upstream of the lens. If you were to place a screen at the position of the virtual image, you would NOT see an image there since there is no real light emanating from that point. Note that the virtual image is indeed *upright* in agreement with the prediction from the magnification equation.

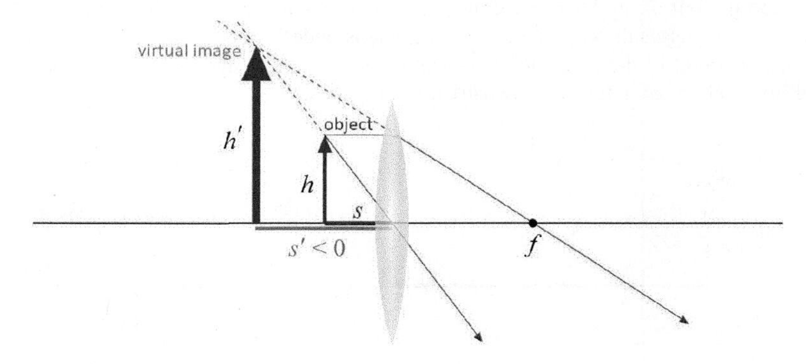

FIGURE 26.7 Illustration of the rays used to determine the virtual image formed of the object by a converging lens when the object is inside the focal length ($s - f < 0$).

We now will look at a diverging lens example. Our object is the arrow to the left of the lens, as shown in Figure 26.8. We draw the two principal rays through the tip of the arrow. The first ray passes through the center of the lens undeflected, while the second ray enters the lens parallel to the axis and is refracted upward as shown. These two refracted rays do not intersect. The extension of the second ray passes through the focal point located on the upstream side of the lens as shown. The result is a virtual upright image formed upstream of the lens.

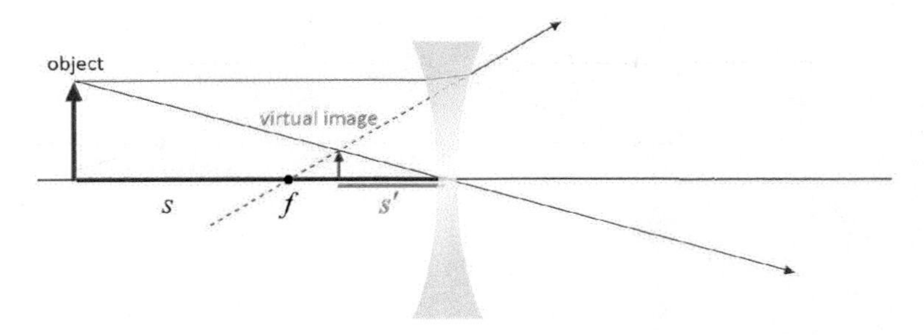

FIGURE 26.8 Illustration of the rays used to determine the virtual image formed of the object by a diverging lens.

Let's look at the lens equation and see if this result is correctly predicted. The expressions for the image location and the magnification are, of course, identical to those we obtained in the preceding example. What does change when we replace a converging lens by a diverging lens is the *sign* of the focal length of the lens. Our convention has been that the focal length of a converging lens is positive. Consequently, we will define the focal length of a diverging lens to be negative. Taking f to be negative, we see that the image distance becomes negative and the magnification becomes positive, indicating a virtual upright image, as expected.

$$s' = \frac{fs}{s-f} < 0$$

$$M = \frac{-f}{s-f} > 0$$

Consequently, we see that the lens equations work for *both* converging and diverging lenses, as long as we remember that a converging lens has a positive focal length, a diverging lens has a negative focal length, and that a negative image distance indicates a virtual image, one that is formed upstream of the lens.

26.6 The Lensmaker's Formula

So far, we have determined the images formed by lenses by specifying the focal length of the lens. We now want to determine exactly how you would go about making a lens to have a specified focal length. Since all that is happening in a lens is refraction at surfaces, we should be able to just use Snell's law to predict the focal length produced by a lens with a certain geometry.

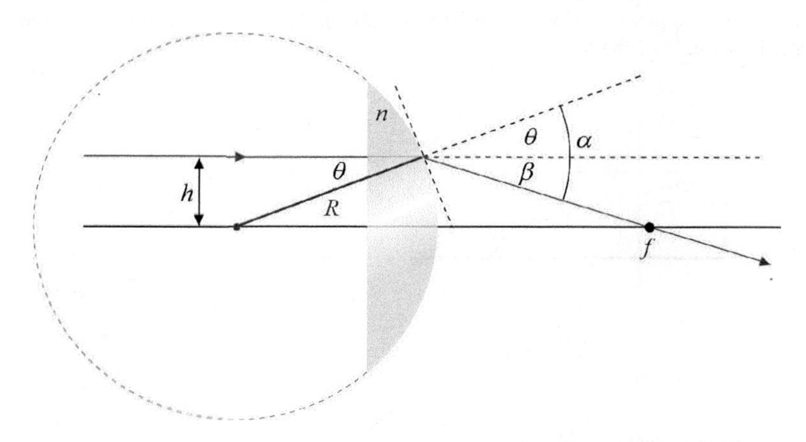

FIGURE 26.9 Illustration of the rays used to derive the Lensmaker's formula for a plano-convex lens. The dotted circle represents the radius of curvature of the lens.

We will start with a plano-convex lens, as shown in Figure 26.9. If a light ray parallel to the axis is incident from the left on the plane surface, it will simply be transmitted without a change of angle since the angle of incidence is zero. This ray will be refracted, however, at the convex-air interface. We apply Snell's law at this interface

$$n \sin \theta = \sin \alpha$$

to obtain an expression for α. In the small angle approximation we see the refracted angle is just equal to the product of the incident angle and the index of refraction of the lens.

$$\alpha \approx n\theta$$

Next we define the bend angle β as the difference between the refracted ray and the incoming ray. Substituting our result for the refracted angle α, we get an expression for the bend angle in terms of the incident angle and the index of refraction.

$$\beta = \alpha - \theta \approx (n-1)\theta$$

This bend angle also defines the focal length, since the incident ray was parallel to the axis. Using the right triangle involving β, we can obtain an expression for the bend angle as a function of h, the distance of the incoming ray from the principle axis, and the focal length.

$$\tan \beta \approx \beta = \frac{h}{f}$$

Once again, in the small angle approximation, θ can be written in terms of the radius of curvature of the lens and the height of the ray above the principle axis.

$$h = R\theta$$

Putting the above three equations together, we can ultimately eliminate h and β to obtain an expression for the focal length as a function of the index of refraction of the material and the radius of curvature of the lens.

$$\frac{1}{f} = (n-1)\frac{1}{R}$$

Note that since *the radius of curvature* is positive in this example, *the focal length* will also be positive, indicating that this lens will be a converging lens.

26.7 Real Lenses

We have just derived the form for the lensmaker's formula for a plano-convex lens. We can follow exactly the same procedure to determine the focal length for a lens constructed of two curved surfaces. If these surfaces have radii of curvature R_1 and R_2, as shown in Figure 26.10, then we can show that the focal length can be determined from the more general expression

$$\frac{1}{f} = (n-1)\left(\frac{1}{R_1} - \frac{1}{R_2}\right)$$

Here the convention is that $R > 0$ if the surface is convex when a light ray hits it.

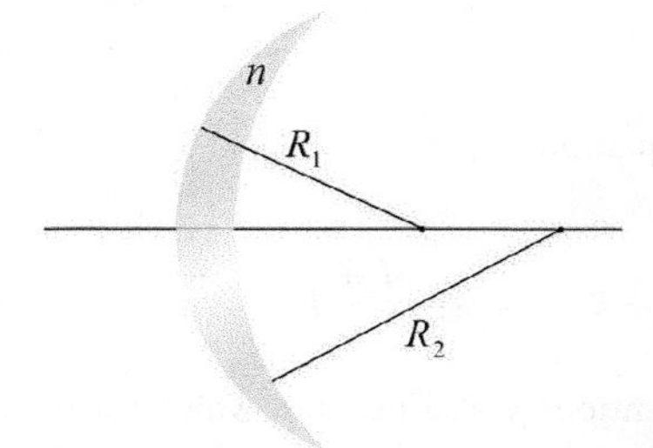

FIGURE 26.10 A lens having surfaces with different radii of curvature.

Of course, in the real world, things are a bit more complicated. You may know that really good lenses are quite expensive. The main reason for this added expense is that simple spherical lenses do have **aberrations** that make the focus fuzzy. For example, we made use of the paraxial approximation in all of our derivations. In the real world, not all angles are small and the focal points that we've been calculating are only approximate focal points. To accommodate larger angles, parabolic lenses can be used. These lenses are much more expensive to grind and polish, however. Another source for the disruption of the focus is chromatic aberration. Namely, since the index of refraction of most materials does in fact depend on frequency, the focal length can be different for different colors. Another source of problems comes from astigmatism, when the curvature of the lens is not symmetric in the transverse directions. One technique to limit these aberrations is to abandon the single lens and construct a system of lenses that can address the combination of aberrations. We will not be concerned in this course with these real aberrations, but we will address, in the last unit, the systems of lenses that are used to create optical instruments.

26.8 Summary

In this unit we developed an understanding of the images formed by lenses in the geometrical optics limit, where the wavelength of the light is small compared to the objects with which it interacts.

We began by considering a converging lens. We found the image formed by this lens by tracing the principal rays, one passing through undeflected through the center of the lens, and one incident parallel to the axis that is refracted, passing through the focal point on the axis. We determined that the nature of the image depended on whether the object distance was greater than or less than the focal length of the lens. In particular, we found that if the object distance is greater than the focal length, then the image is real and inverted, whereas if the object distance is less than the focal length, then the image is virtual and upright.

We used ray tracing to determine algebraic equations that describe the image location and magnification. In particular, we derived the lens equation

$$\frac{1}{s}+\frac{1}{s'}=\frac{1}{f}$$

and the magnification equation

$$M \equiv \frac{h'}{h} = -\frac{s'}{s}$$

We used the sign conventions that a negative value for s' indicated an image location upstream of the lens (i.e., virtual) and that $f > 0$ for a converging lens and $f < 0$ for a diverging lens. With these conventions, a positive magnification indicated an upright image while a negative magnification indicated an inverted image.

Finally, we used Snell's law to derive the lensmaker's formula that relates the focal length of the lens to the index of refraction and the radii of curvature of the surfaces. We used the plano-convex lens to develop the formula, which we generalized to the case of two curved surfaces.

Main Points

The Lens Equation

We derived the lens equation, which is valid in the paraxial approximation, by drawing the principal rays.

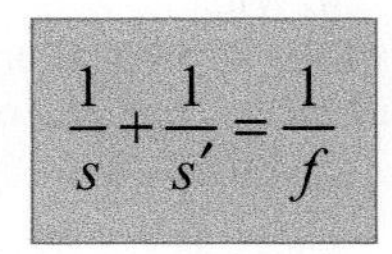

Sign conventions are indicated in the drawing.

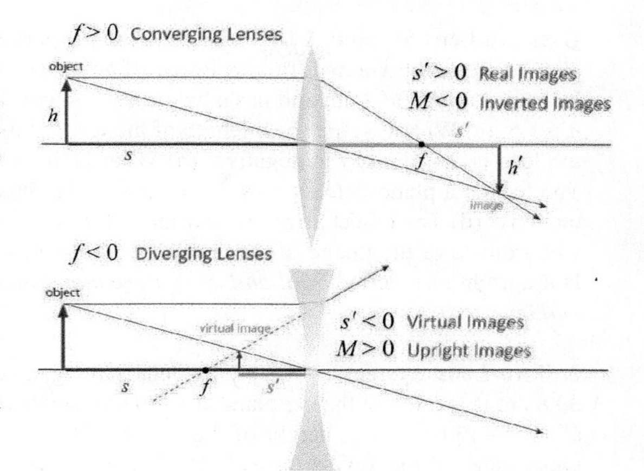

The Magnification Equation

Magnification is defined as the ratio of the image size to the object size.

$$M \equiv \frac{h'}{h} = -\frac{s'}{s}$$

Lensmaker's Formula

We used Snell's law to obtain the relation between the focal length and the radii of curvature of a spherical lens.

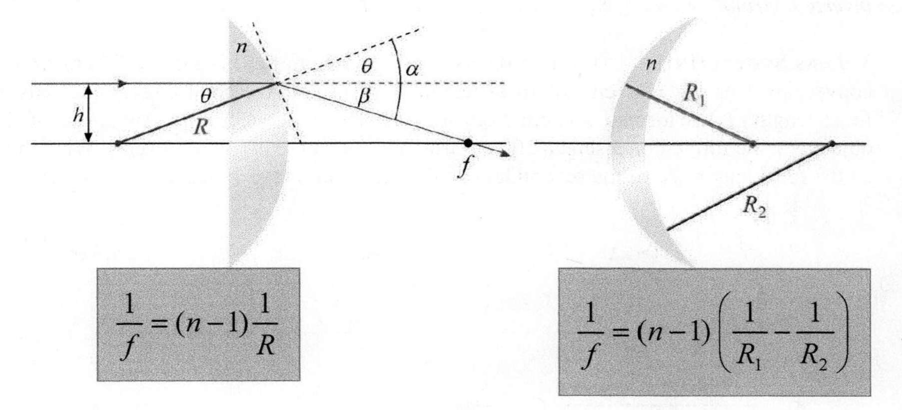

PROBLEMS

1. Single-Lens System: A lens located in the *y*-*z* plane at $x = 0$ and whose principle axis is aligned along the *x*-axis forms an image of an arrow at $x_2 = 51.4$ cm. The object arrow is located at $x_1 = -30.2$ cm and has a height of 4.64 cm. The index of refraction of the lens is $n = 1.5$. (a) What is *f*, the focal length of the lens? If the lens is converging, *f* is positive. If the lens is diverging, *f* is negative. (b) What is the height of the image of the arrow? (c) The lens is a plano-convex lens. What is *R*, the radius of curvature of the convex side of the lens? (d) The object arrow is now moved to $x_{1,new} = -11.8$ cm. What is $x_{2,new}$, the new *x* coordinate of the image of the arrow? (e) Is the new image of the arrow real or virtual? Is it upright or inverted: *real and upright*, *real and inverted*, *virtual and upright*, or *virtual and inverted*?

2. Two-Lens System: A diverging lens having a focal length of magnitude equal to 30.8 cm is located in the *y*-*z* plane at $x = 0$. An arrow to the left of the lens forms an image at $x_2 = -13.6$ cm. The height of the image of the arrow is 3.6 cm. (a) What is x_1, the *x* coordinate of the object arrow? (b) What is the height of the object arrow? (c) A converging lens of focal length $f_{converging} = 6.75$ cm is now inserted at $x_3 = -15.1$ cm. In the absence of the diverging lens, at what *x* coordinate, x_4, would the image of the arrow form? (d) To determine the image of the arrow from the combined converging plus diverging lens system, we take the image from the converging lens (in the absence of the diverging lens) to be the object for the diverging lens. If this image is downstream of the diverging lens, the object for the diverging lens is virtual. All this means is that the rays entering the diverging lens are converging toward a location downstream of the diverging lens. The final image can be calculated using this virtual object distance and the focal length of the diverging lens. If this description seems unclear to you, you can check out the more detailed description of multiple lens systems given in the Optical Instruments unit. What is x_5, the *x* coordinate of the final image of the combined system? (e) Is the final image of the arrow real or virtual? Is it upright or inverted: *real and upright*, *real and inverted*, *virtual and upright*, or *virtual and inverted*?

3. Lens System (INTERACTIVE EXAMPLE): An object 5 cm high is located 75 cm from a converging lens ($x_1 = 75$ cm) of focal length $f_1 = 50$ cm. A second converging lens of focal length f_2 is located 200 cm from the first lens ($x_2 = 200$ cm). An image of the object is to be formed on a screen 200 cm from the second lens ($x_3 = 200$ cm). What must be the focal length f_2 of the second lens so that the final image appears on the screen?

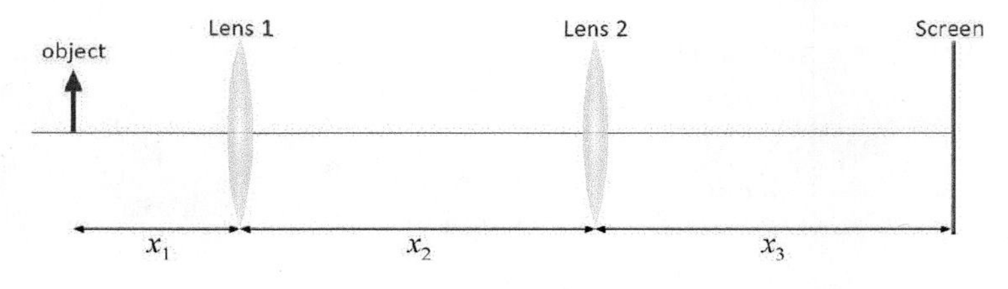

FIGURE 26.11 Problem 3

UNIT 27

MIRRORS

27.1 Overview

In this unit we will discuss image formation by mirrors. We will use the same approach for mirrors that we followed for lenses. Indeed, we will arrive at a mirror equation that relates object and image locations, which is absolutely identical to the lens equation. We will also obtain a magnification equation that is identical to what we obtained for lenses.

We will discuss concave and convex mirrors and the necessary sign conventions. We will also discuss plane mirrors and develop, using a small angle approximation, an expression for the focal point of a spherical mirror.

27.2 Concave and Convex Spherical Mirrors

We'll start by simply drawing rays to determine the focal properties of concave and convex mirrors.

Figure 27.1 shows the reflection of incident parallel rays from a **concave** mirror. These rays are reflected from the surface and then intersect in a point on the axis, the focal point of the mirror. We define the distance between this point and the mirror to be the focal length of the mirror. Our convention is to define this focal length to be positive for concave mirrors.

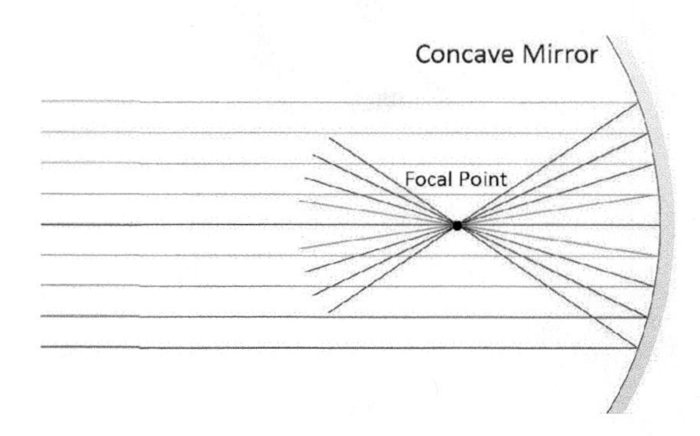

FIGURE 27.1 Parallel rays are incident on a concave mirror and are reflected back, forming a real focus on the upstream side of the mirror. The focal length of a concave mirror is positive.

Figure 27.2 shows the reflection of incident parallel rays from a **convex** mirror. These rays are reflected from the surface and then diverge. These reflected rays do *not* intersect, but their continuations behind the mirror do intersect at a point on the axis. We define the distance between this point and the mirror to be the focal length of the mirror. Our convention is to define this focal length to be negative for convex mirrors.

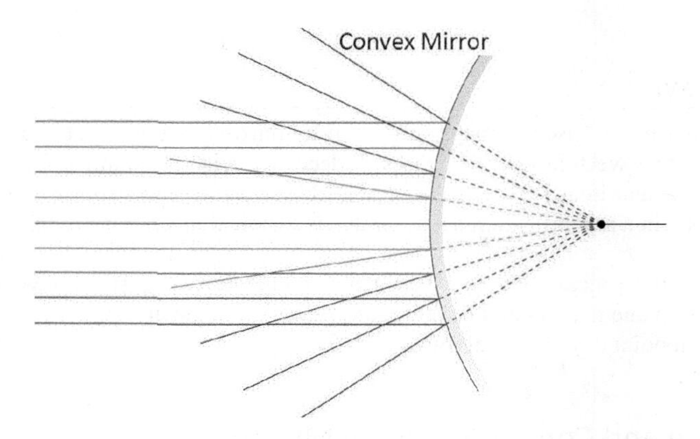

FIGURE 27.2 Parallel rays are incident on a convex mirror and are reflected back and diverge. The projections of these rays behind the mirror appear to come from a point we call the focus. The focal length of a convex mirror is negative.

27.3 The Mirror Equation and Magnification

We will now derive the mirror equation that determines the image distance in terms of the object distance and the focal length. Figure 27.3 shows rays incident on a concave mirror. We will determine the location and size of the image by tracing rays that pass through the tip of the arrow located to the left of the mirror.

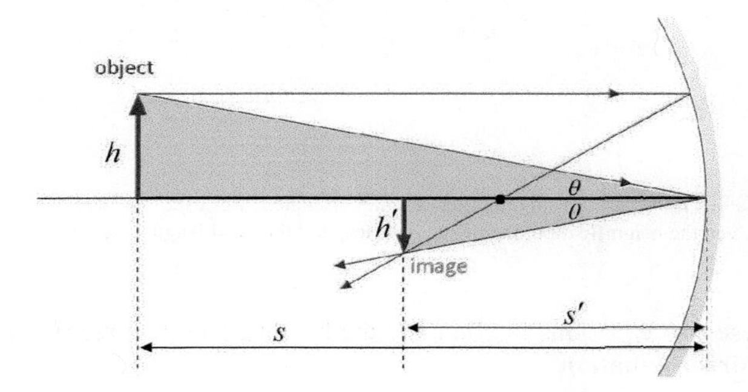

FIGURE 27.3 Two rays are drawn from the tip of the arrow and are reflected by the concave mirror surface, forming a real image in front of the mirror. The similar triangles are used to obtain the relation between the object and image distances (s, s') and sizes (h, h') as shown.

First we consider the ray that propagates parallel to the axis of the mirror; this ray is reflected and passes back through the focal point. Next we consider the ray that strikes the mirror on the axis, making an angle of incidence equal to θ, as shown. This ray is reflected back at an equal angle θ and intersects the first ray at the image point of the tip of the arrow, forming an image in front of the mirror.

Using the highlighted similar triangles for the object and image, we see that the ratio of the heights is equal to the ratio of the distances.

$$\frac{|h'|}{h} = \frac{s'}{s}$$

This result is exactly the same as it was for the lens in the preceding unit! Using the smaller similar triangles shown in Figure 27.4, we obtain an expression for the ratio of the heights in terms of the focal length and image distance.

$$\frac{|h'|}{h} = \frac{s' - f}{f}$$

Note that this relation is approximate since the mirror is curved so that the relevant side of the triangle is not exactly f, but a little smaller than f. The approximation is good in the limit that the angles are small or the radius of curvature of the mirror is large. We can

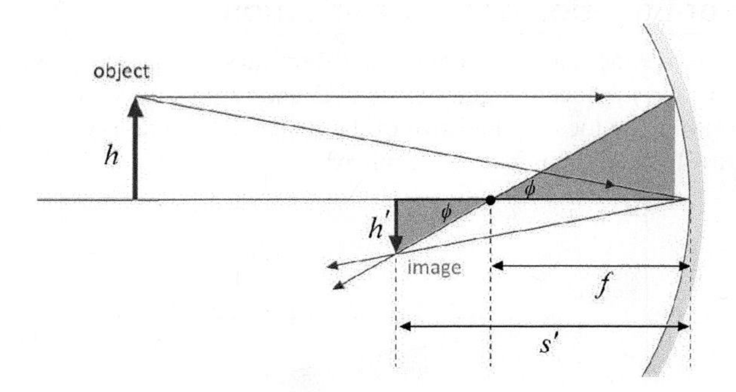

FIGURE 27.4 The highlighted similar triangles are used to obtain the relation between the magnification, the image distance, and the focal length as shown.

combine these two equations to eliminate the heights, and then rewrite the expression to obtain the **mirror equation**.

$$\frac{1}{s} + \frac{1}{s'} = \frac{1}{f}$$

Amazing! We have recovered the lens equation if we identify s' as being *positive* with a real image upstream from the mirror and identify f as *positive* for a concave mirror.

As we noted earlier, we also see that the magnification equation for lenses also holds here. Since we have a positive image distance in this case, the magnification becomes negative, indicating an inverted image, in agreement with our result obtained from ray tracing.

$$M \equiv \frac{h'}{h} = -\frac{s'}{s}$$

Since the equations are identical to those we obtained with lenses, we expect that moving the object location to a point inside the focal point should produce a virtual upright image. We do just that in Figure 27.5 and observe that once inside the focal point, the reflected rays do diverge, giving rise to a virtual upright image.

How do we understand this situation in terms of the mirror equation? If we assume the mirror equation also holds for this case, we see that the image distance s' is now negative ($1/s' = 1/f - 1/s < 0$). How do we interpret this negative image distance? The natural interpretation is that s' being negative means that it is located *behind* the mirror, as it indeed is in this case! If we assume the magnification equation also holds for this case, we see that a negative s' leads to a positive magnification, i.e., we would expect an upright image, as it indeed is in this case!

Consequently, we see that for a concave spherical mirror, the nature of the image depends on whether the object distance is less than or greater than the focal length. If the object distance is greater than the focal length, we have a real inverted image, whereas if the

FIGURE 27.5 Rays drawn from the tip of the arrow located inside the focal length of a concave mirror are reflected back and diverge. The projections of these reflected rays behind the mirror intersect, forming an upright virtual image.

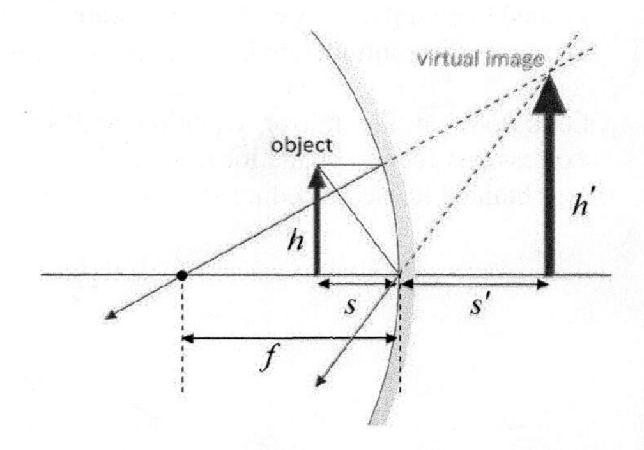

object distance is less than the focal length, we have a virtual upright image. All of these features are represented in the mirror equation and the magnification equation, if we take a negative image distance to indicate a virtual image (i.e., behind the mirror).

27.4 Convex Mirrors

We will now look at a convex mirror example. Our object is the arrow to the left of the mirror in Figure 27.6. Two principal rays are drawn through the tip of the arrow. The first ray is directed so that it strikes the mirror at normal incidence. Consequently, it is reflected directly back along the incident direction. The second ray is drawn parallel to the axis, making an angle of θ with the normal. This ray is reflected with an equal angle of reflection. Since this incident ray was parallel to the axis, we know, from the definition of the focus, that the continuation of the reflected ray must pass through the focal point

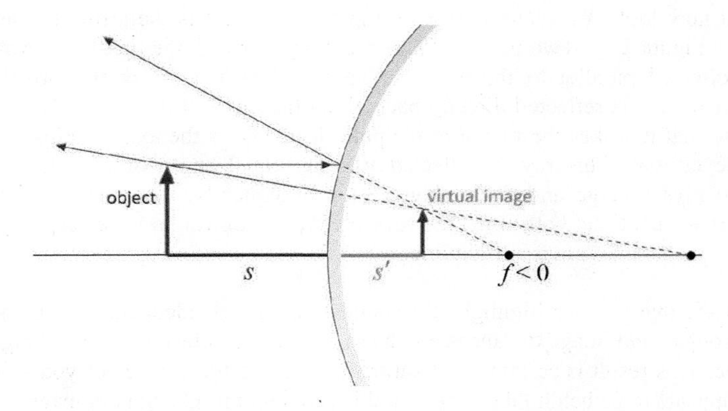

FIGURE 27.6 Two rays are drawn from the tip of the arrow and are reflected from a convex mirror. The reflected rays diverge, appearing to come from a point behind the mirror. The intersection of these reflected rays behind the mirror locates the position of the virtual image of the arrow. This location can also be determined from the mirror equation if we adopt the convention that the focal length of the mirror is negative.

located behind the mirror. If we now continue the first ray behind the mirror, we see that it intersects the continuation of the second ray, forming a virtual upright image.

Let's look at the mirror equation and see if this result is correctly predicted. The expressions for the image location and the magnification are, of course, identical to those we obtained in the preceding section.

$$\frac{1}{s'} = \frac{1}{f} - \frac{1}{s}$$

$$M \equiv \frac{h'}{h} = -\frac{s'}{s}$$

What does change when we replace a concave mirror by a convex mirror is the *sign* of the focal length of the mirror. Recall that our convention is that the focal length of a *concave mirror* is *positive*, while the focal length of a *convex mirror* is *negative*.

Taking f to be negative, we see that the image distance (s') becomes negative and the magnification becomes positive (since s' is negative), indicating a virtual upright image, as expected.

Consequently, we see that the mirror equations work for both concave and convex mirrors, as long as we remember that a concave mirror has a positive focal length, a convex mirror has a negative focal length, and that a negative image distance indicates a virtual image, one that is formed behind the mirror.

27.5 Plane Mirrors

We will now look at a plane mirror example. Our object is the arrow to the left of the mirror in Figure 27.7. Two principal rays are drawn through the tip of the arrow. The first ray is directed parallel to the axis so that it strikes the mirror at normal incidence. Consequently, it is reflected directly back along the incident direction. The second ray is drawn so that it strikes the mirror at the point it intersects the axis, making an angle of θ with the normal. This ray is reflected with an equal angle of reflection. These two reflected rays diverge and do not intersect. Consequently, we expect to find a virtual image. If we continue both reflected rays behind the mirror, we see that they do indeed intersect, forming a virtual upright image.

We can see that the two highlighted triangles are, in fact, identical, leading to the result that the object and image distances are the same and the object and image heights are also the same. This result is certainly not surprising. The image you see of yourself in a plane mirror appears to be behind the mirror, and is certainly upright and not magnified.

How does this result fit with the mirror equation? Well, the image is formed behind the mirror, which indicates that the image distance (s') is negative. We expect the magnification ($M = -s'/s$) to be equal to +1, in agreement with our ray tracing results. What can we say about the focal length? If the mirror equation holds for plane mirrors, then we would predict that the focal length f would have to be infinity!

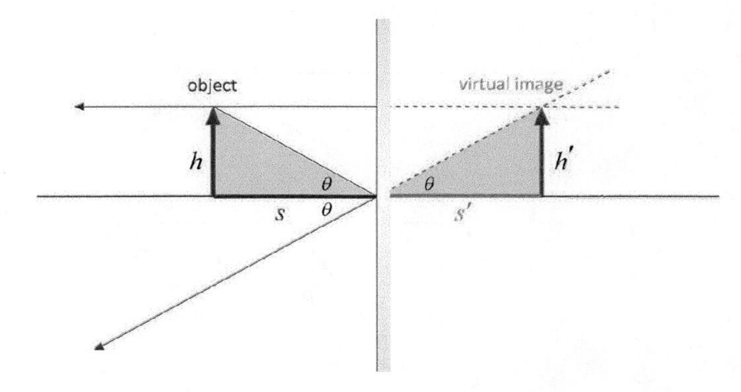

FIGURE 27.7 Two rays are drawn from the tip of the arrow and are reflected from a plane mirror. The reflected rays diverge, appearing to come from a point behind the mirror. The intersection of these reflected rays behind the mirror locates the position of the virtual image of the arrow. This location can also be determined from the mirror equation if we adopt the convention that the focal length of the plane mirror is infinity.

$$\frac{1}{f} = \frac{1}{s} + \frac{1}{s'}$$

This prediction is of course correct. Parallel rays incident along the axis all are reflected directly backward along the incident direction. The reflected rays are then all parallel to the axis. Neither the reflect rays nor their continuation behind the mirror intersect! Therefore, we say that the focal point is "at infinity."

27.6 The Focal Length of a Spherical Mirror

We now want to look at exactly how to construct a mirror to have a desired focal length. Since the behavior of all rays is just determined by the law of reflection, this project reduces to an exercise in geometry. The geometries can get complicated, however. We will restrict ourselves to the case of a spherical concave mirror.

To determine the focal length of a spherical concave mirror defined by a radius of curvature R, we need to find the point that defines the intersection of reflected rays that have been produced from incident rays that are parallel to the axis with the axis itself.

Figure 27.8 shows such an incident ray, originating a distance h from the axis. This ray reflects from the mirror at point b with an angle of reflection equal to θ, the angle of incidence. The ray intersects the axis at point c. We want to find the distance f, the focal length of the mirror.

We start by constructing the line from point c that is perpendicular to the radial line from point a to point b. This line creates two identical triangles. Consequently, the distances d_{ac} and d_{cb} are equal. If we can relate the distance d_{cb} to f, we will be done, since f plus d_{ac} must be equal to R. For an arbitrary incident angle θ, this task is a bit involved.

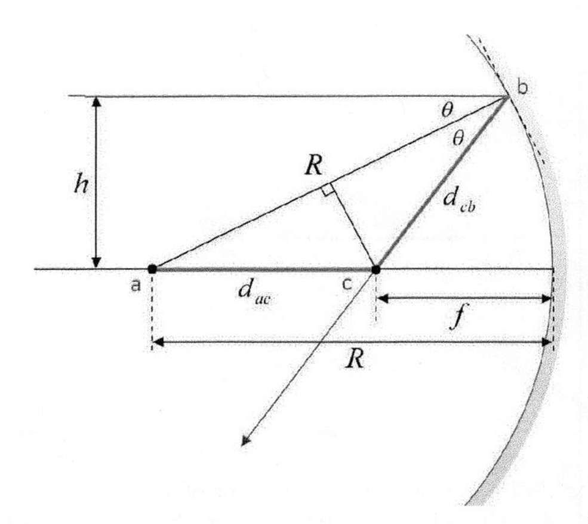

FIGURE 27.8 An incident ray, parallel to the axis of a spherical concave mirror, is reflected and crosses the axis at a distance f from the mirror surface, defining the focal length of the mirror. In the small angle (θ) approximation, this focal length is equal to one half of the radius of curvature of the mirror.

However, in the limit that the angle θ becomes small, the distance d_{cb} must approach the distance f. Therefore, in the small angle approximation, we obtain our result that the focal length of a spherical concave mirror is just equal to one half of the radius of curvature.

$$f = \frac{R}{2} \qquad \text{(small angle approximation)}$$

27.7 Summary

In this unit we developed an understanding of the images formed by mirrors in the geometrical optics limit, where the wavelength of the light is small compared to the objects with which it interacts.

We began by tracing principal rays to develop the mirror equation, the equation that relates the image distance to the object distance for a single mirror, and the magnification equation that relates the image size to object size.

$$\frac{1}{s} + \frac{1}{s'} = \frac{1}{f}$$

$$M \equiv \frac{h'}{h} = -\frac{s'}{s}$$

We found that these equations are identical to the corresponding equations for lenses, provided we adopt the following conventions. Concave mirrors have positive focal lengths while convex mirrors have negative focal lengths. The focal length of a plane mirror is infinity. Image distances are positive if the image is real, i.e., when the image is located in front of the mirror. Image distances are negative if the image is virtual, that is, when the image is located behind the mirror.

Finally, we determined that, in the small angle approximation, the focal length of a concave spherical mirror is one half the radius of curvature of the mirror.

MAIN POINTS

The Mirror and Magnification Equations

We traced principal rays to develop equations to relate the image and object distances to the focal length (mirror equation) and to relate the image size to the object size (magnification equation).

Mirror Equation

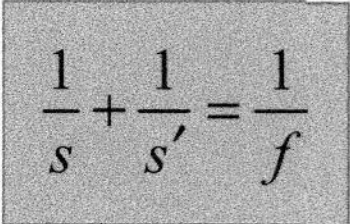

$$\frac{1}{s}+\frac{1}{s'}=\frac{1}{f}$$

Magnification

$$M \equiv \frac{h'}{h} = -\frac{s'}{s}$$

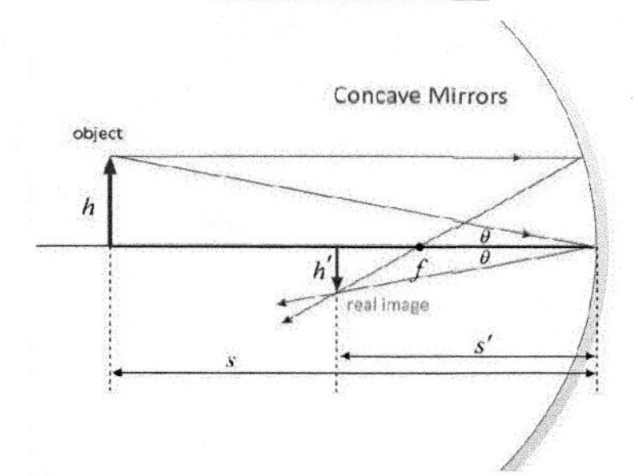

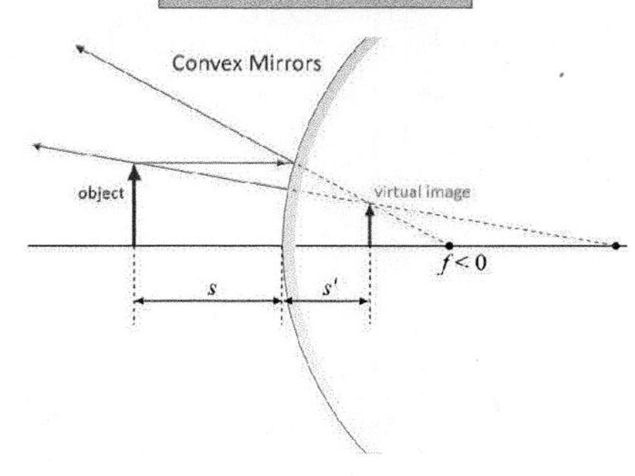

Sign Conventions

Focal Lengths

– Concave mirrors: $f > 0$

– Convex mirrors: $f < 0$

– Plane mirrors: $f = \infty$

Image Distances

–Real image (in front of mirror): $s' > 0$

– Virtual image (behind mirror): $s' < 0$

Spherical Mirror Focal Length (small θ)

$$f = \frac{R}{2}$$

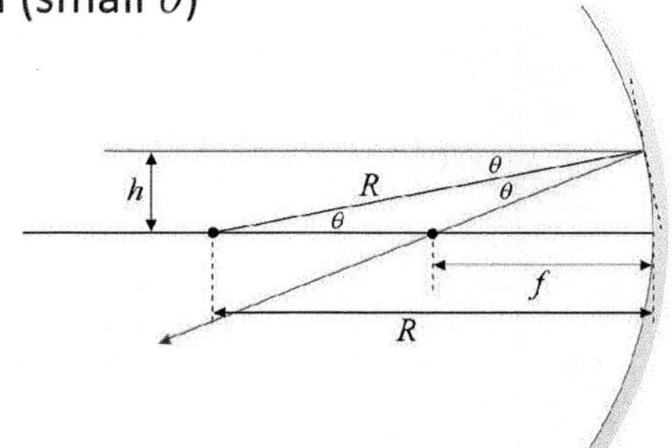

PROBLEMS

1. Single Mirror System: A spherical mirror forms an image of an arrow that is upstream from the mirror at a distance of 73.3 cm. The object is located a distance of 88 cm upstream from the mirror and has a height of 5.67 cm. (a) What is f, the focal length of the mirror? If the mirror is concave, f is positive. If the mirror is convex, f is negative. (b) What is the height of the image of the arrow? (c) The object arrow is now moved such that image distance doubles, i.e., the image is now 146.6 cm upstream from the mirror. What is the new height of the image of the arrow? (d) The object arrow is now moved to a distance of 18.8 cm upstream from the mirror. What is the new height of the image of the arrow?

2. Convex Mirror: The convex mirror shown in Figure 27.9 forms a virtual image of an arrow at $x_2 = 16.1$ cm. The image of the tip of the arrow is located at $y_2 = 6.6$ cm. The magnitude of the focal length of the convex mirror is 31 cm. (a) What is x_1, the x coordinate of the object arrow? (b) What is y_1, the y coordinate of the tip of the object arrow? (c) The object arrow is now moved such that image distance is halved, i.e., $x_{image,new} = 8.05$ cm. What is $x_{1,new}$, the new x coordinate of the object arrow? (d) What is $y_{2,new}$, the y coordinate of the image of the tip of the arrow when the x coordinate of the object arrow is equal to $x_{1,new}$? (e) Which of the following statements concerning a single convex mirror is false?

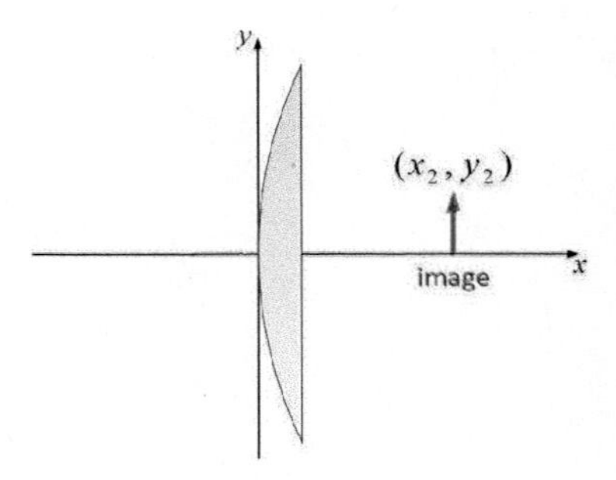

FIGURE 27.9 Problem 2

- (i) The image formed from any real object is always upright.
- (ii) The image formed from any real object is always smaller than the object.
- (iii) The image formed from any real object is always virtual and the magnitude of the image distance can be larger than the focal length of the mirror.
- (iv) The image formed from a virtual object that is located inside the focal length of the mirror is real.

3. Spherical Mirror (INTERACTIVE EXAMPLE): An object (blue arrow) is located in front of a convex spherical mirror of radius $R = 45$ cm, as shown in Figure 27.10. The tip of the arrow is located at $(x, y) = (18 \text{ cm}, -7 \text{ cm})$ from the mirror. What is y_{image}, the y coordinate of the image of the tip of the arrow?

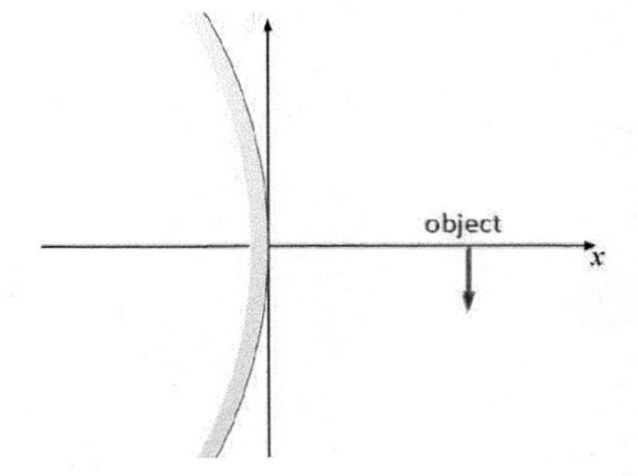

FIGURE 27.10 Problem 3

UNIT

28

OPTICAL INSTRUMENTS

28.1 Overview

In this unit we will complete our study of geometric optics by discussing optical instruments and the human eye. We'll begin by analyzing systems of lenses in which the image of one lens is treated as the object for the following lens. We will then move on to model the human eye as a converging lens with a somewhat variable focal length. We will discuss the lenses that are necessary to correct near-sightedness and far-sightedness.

We will close by discussing simple telescopes and microscopes, which are best described in terms of an angular magnification.

28.2 Multiple Lenses

We begin by determining the image formed by a combination of lenses. In order to make this determination, we simply move through the combination one lens at a time, considering the object for a given lens to be the image from the preceding lens.

Figure 28.1 shows a system of two lenses, a converging lens with a focal length of 1 m and a diverging lens with a focal length of −4 m; the separation between the lenses is 4 m. The object is the upright arrow at a distance of 1.5 m in front of the first lens.

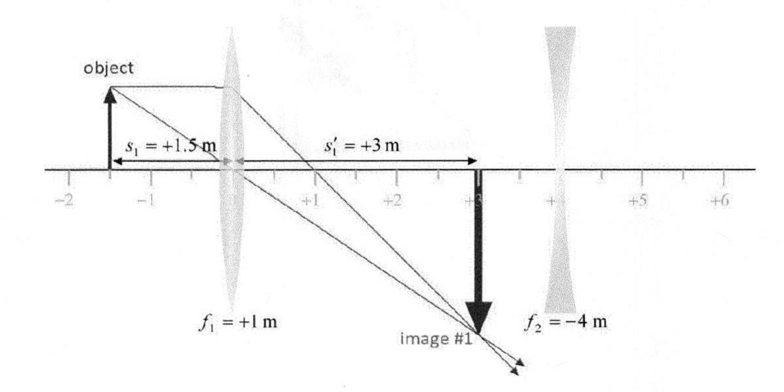

FIGURE 28.1 Two rays from the tip of the object arrow are incident on a converging lens and are refracted to form a real inverted image. This image from the converging lens will be used as the object for the diverging lens to obtain the final image.

The object is beyond the focal point of the first lens so we expect a real inverted image to be formed. Shown in the figure are two principal rays from the object intersecting at a distance of about 3 m downstream of the first lens, giving the expected real inverted image.

We now take this image as the object for the second lens. We expect this image to be virtual and upright. Figure 28.2 shows the two principal rays diverging downstream of the second lens, which indicates a virtual image. The intersection of the projection of these rays is located about $3/4$ m upstream of the second lens, giving the expected virtual upright image. Here, upright just means that the object and image are not inverted. Of course, the object was the inverted image of the initial arrow so that the final image is still inverted.

We will now go through the algebra to get the numbers. We start with the first lens. We apply the lens equation

$$\frac{1}{s} + \frac{1}{s'} = \frac{1}{f}$$

with the object distance $s_1 = +1.5$ m and focal length $f_1 = +1$ m to determine the image distance s_1' to be equal to +3 m, in agreement with our drawing. We apply the magnification equation

$$M_1 \equiv \frac{h_1'}{h_1} = -\frac{s_1'}{s_1}$$

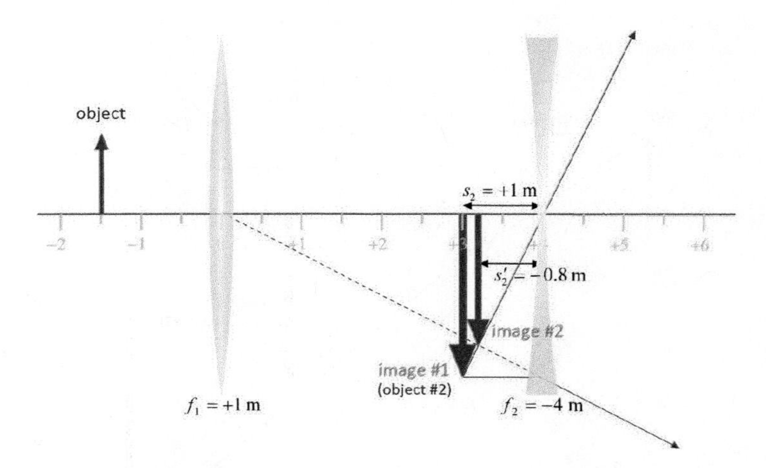

FIGURE 28.2 The real image formed by the converging lens is used as the object for the diverging lens. Two rays from the tip of the inverted arrow (image #1 and object #2) are incident on the diverging lens and are refracted and diverge, appearing to come from a point upstream of the diverging lens (image #2). This final image is virtual.

to determine that $M_1 = -2$.

We now take this image to be the object for the second lens. Therefore, we will apply the lens equation once more, this time with the object distance $s_2 = +1$ m and focal length $f_2 = -4$ m to determine the image distance s_2' to be equal to −0.8 m, in agreement with our drawing. The magnification, M_2, is given by $-s_2'/s_2$, which is equal to $+4/5$.

Combining the results we see that the overall magnification M, which is equal to the product of M_1 and M_2, is equal to

$$M = M_1 M_2 = (-2)\cdot\left(\frac{4}{5}\right) = -\frac{8}{5}$$

The image from the system of lenses is virtual and inverted.

The example of a system of lenses we have just completed illustrated our procedure for determining the ultimate image. It was chosen because it needed only a straightforward application of the procedure. It's certainly possible to imagine less straightforward examples, namely, when the image of the first lens is produced downstream of the second lens. In this case, we need to consider the object for the second lens to be virtual!

If we reduce the separation of the lenses in the previous case from 4m to 1m, we obtain such an example, as shown in Figure 28.3. The calculation of the image of the object from the first lens is as before. It is located 3 m downstream of the first lens and is real and inverted. However, this image never really forms since it is downstream of the second lens. The rays that would have created the real image diverge at the second lens.

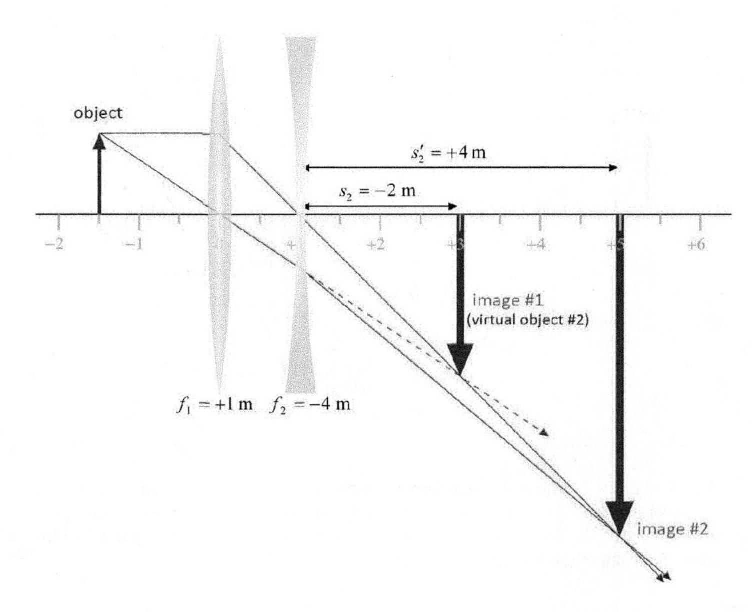

FIGURE 28.3 Illustration of rays passing through a system of two lenses placed sufficiently close together so that the image from the first lens alone would be formed downstream of the second lens, i.e., the object for the second lens is *virtual*!

How do we figure out what will happen? The prescription is as follows: Since the image from the first lens would have been downstream of the second lens, we will call that object distance negative, i.e., a virtual object. If, therefore, we substitute $s_2 = -2$ m and $f_2 = -4$ m in the lens equation, we obtain an image distance $s_2' = +4$ m. Hence, we expect a real image located 4 m downstream of the second lens. The magnification M_2 from the second lens alone is equal to +2, indicating an upright image. Here again, however, the object for the second lens (which is virtual) is inverted. Consequently, the final image is inverted and has a total magnification, the product of M_1 and M_2, equal to −4.

28.3 The Normal Eye

We have just learned how to determine the image from a system of lenses. In some sense, we must always deal with a system of lenses, if a human observer is involved (i.e., we only see something when light passes through our cornea and produces a real image on our retina, which in turn can trigger electrical impulses that travel ultimately to the brain). The human eye can be modeled as a variable focal length converging lens.

The human eye is usually described in terms of two points: the **far point** (the maximum distance that a relaxed eye can focus onto the retina) and the **near point** (the closest distance that can be focused onto the retina). In the so-called normal eye, the far point is essentially at infinity, while the near point is around 25 cm.

Before going any further, we should say something about this "at infinity" phrase. We certainly do not mean that the normal eye can see ordinary objects that are infinitely far away. What we mean is simply that a normal eye can focus parallel light rays to a point on its retina. In other words, a relaxed normal eye has a focal length approximately equal to the lens-retina distance, or about 2.5 cm. In this case, "infinity" refers to object distances that are large compared to the focal length. For example, objects that are just a few meters away are over a hundred focal lengths away. The difference in the image distance for an object "at infinity" and one that is 100 focal lengths away is only 1% of the focal length, which would be about a quarter of a millimeter in this case.

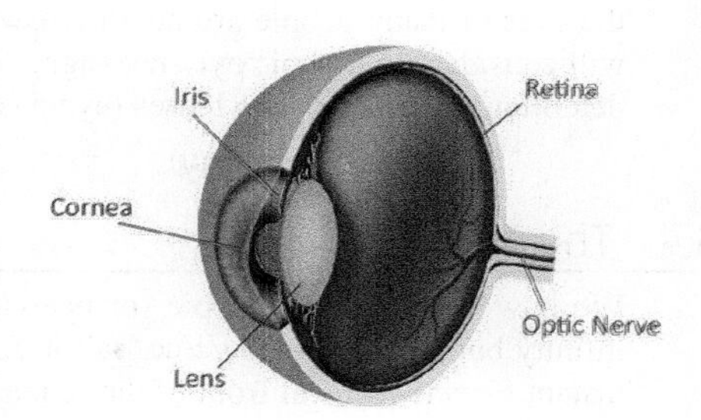

FIGURE 28.4 A cross-section of the human eye.

In order to form a focus on the retina for close objects, however, the focal length of the eye must be reduced. For example, for an object 25 cm away to focus at an image distance of 2.5 cm would require a lens with a focal length that equals 25(2.5) / (25 + 2.5), which is equal to about 2.3 cm. This variability that is needed is illustrated in Figure 28.5.

Consequently, we say that a normal eye is a converging lens with a focal length that can vary over a range of about 10% (e.g., from about 2.3 to 2.5 cm). We know, however, that

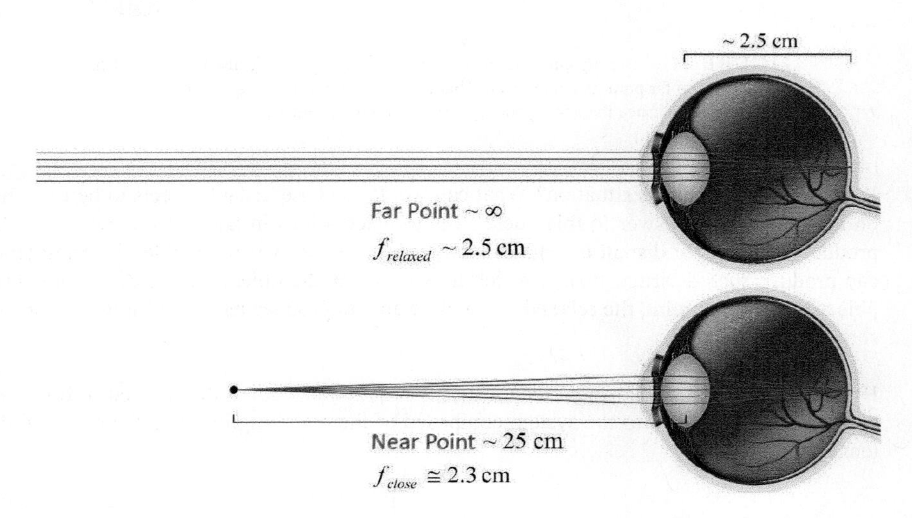

FIGURE 28.5 In order to focus on objects both far away and near (~25 cm), the normal eye must have a focal length that can vary over a range of about 10% (e.g., from 2.3 cm for near objects to 2.5 cm for far objects). Note that distances have not been drawn to scale.

the eyes of many people are not "normal," by this definition. In the next two sections we will consider the myopic eye (nearsightedness) and the hyperopic eye (farsightedness) and determine what corrective lenses (eyeglasses) are needed to produce normal eyesight.

28.4 The Myopic Eye

We now consider the myopic (or near-sighted) eye. The far point for this eye is not at infinity but at some finite value, call it L_{fp}. Our model for this eye is that it is elongated; distant objects focus in front of the retina, i.e., the maximum focal length of the eye is less than the lens-retina distance, as shown in Figure 28.6.

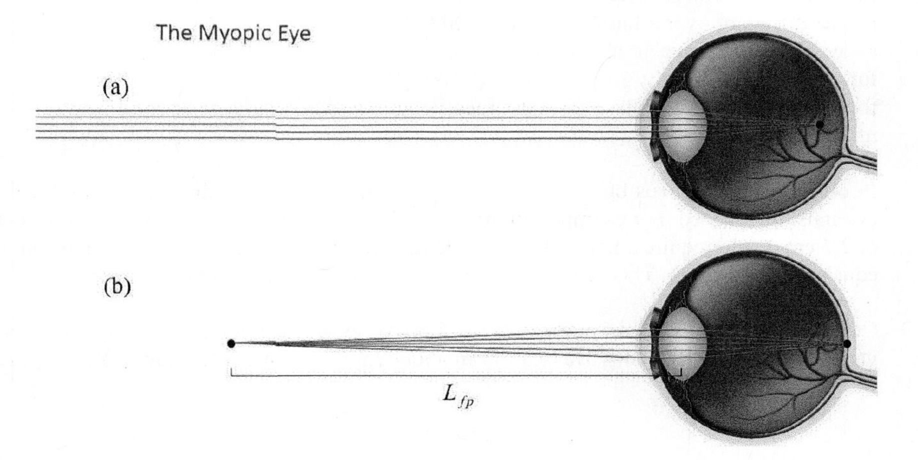

FIGURE 28.6 The myopic eye: (a) Incident parallel rays are focused in front of the retina. (b) The far point is not at infinity but at some finite distance L_{fp} (i.e., L_{fp} is the largest object distance that the myopic eye can focus on the retina).

How can we correct this situation? What can we do to cause distant objects to be focused on the retina? The answer to this question is to place a lens in front of the eye that will produce an image of distant objects at the far point of the eye. A suitable diverging lens can produce such a virtual image, which now becomes the object for the eye. Since this object is at the far point, the relaxed eye will be able to produce an image on the retina, as shown in Figure 28.7.

How do we determine the focal length that is required? Clearly, all we need to do is to apply the lens equation to the system composed of the diverging lens and the converging lens of the eye.

$$\frac{1}{s}+\frac{1}{s'}=\frac{1}{f}$$

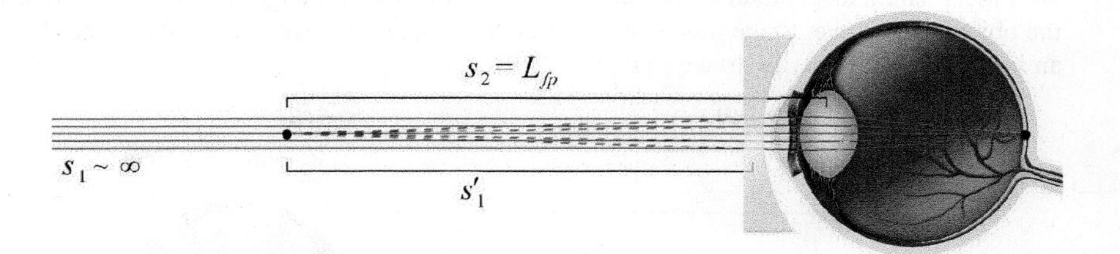

FIGURE 28.7 A diverging lens is placed in front of the myopic eye so that incident parallel rays ($s_1 \sim \infty$) focus at the far point of the eye ($s_1' = -L_{fp}$). This image becomes the object for the second lens (the cornea). Since this object is at the far point ($s_2 = L_{fp}$), it can be focused on the retina.

To simplify the calculation, let's assume the distance between the diverging lens and the eye is small compared to the far point so that the image distance of the diverging lens is essentially equal to the object distance for the eye (i.e., $s_1' = -s_2 = -L_{fp}$). In this case, we determine the focal length of the diverging lens must be equal to minus the far point of the eye.

$$\frac{1}{f} = \frac{1}{s_1} + \frac{1}{s_1'} = 0 - \frac{1}{L_{fp}} \qquad \Rightarrow \qquad f = -L_{fp}$$

28.5 The Hyperopic Eye

We now consider the hyperopic (or far-sighted) eye. The near point for this eye is not at 25 cm but at some larger value, call it L_{np} . Our model for this eye is that it is shortened; close objects focus behind the retina, as shown in Figure 28.8. The minimum focal length of the eye is greater than the lens-retina distance.

How can we correct this situation? What can we do to cause close objects to be focused on the retina? The answer to this question is to place a lens in front of the eye that will produce an image of an object at 25 cm at the near point of the eye. A converging lens

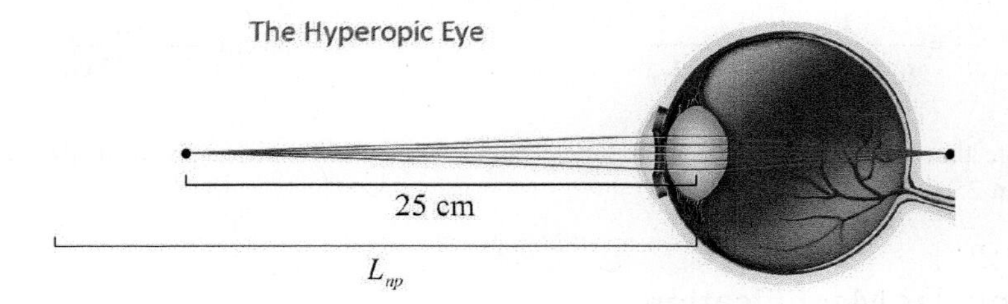

FIGURE 28.8 The hyperopic eye: Incident rays originating from a distance of 25 cm in front of the eye are focused behind the retina. The near point is not at 25 cm but at some distance L_{np} (L_{np} is the smallest object distance that the hyperopic eye can focus on the retina), which is larger than 25 cm.

with focal length larger than 25 cm can produce such a virtual image, which now becomes the object for the eye. Since this object is at the near point, the eye will be able to produce an image on the retina, as shown in Figure 28.9.

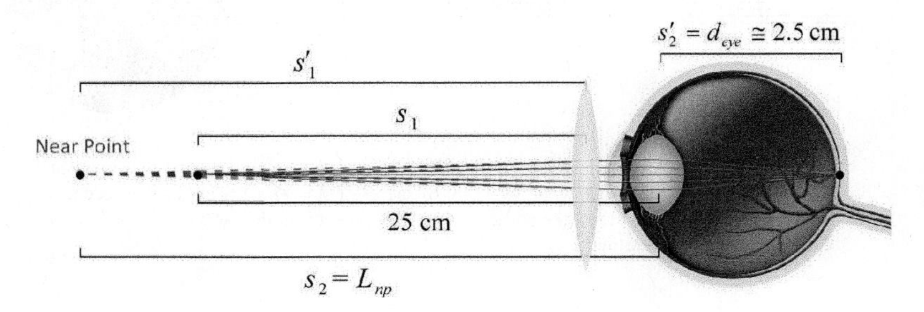

FIGURE 28.9 A converging lens is placed in front of the hyperopic eye so that incident rays originating 25 cm in front of he eye ($s_1 = 25\,\text{cm}$) focus at the near point of the eye ($s_1' = -L_{np}$). This image becomes the object for the second lens (the cornea). Since this object is at the near point ($s_2 = L_{np}$), it can be focused on the retina.

How do we determine the focal length that is required? Clearly, all we need to do is to apply the lens equation to the system composed of the converging lens and the converging lens of the eye.

$$\frac{1}{s} + \frac{1}{s'} = \frac{1}{f}$$

To simplify the calculation, let's once again assume the distance between the converging lens and the eye is small compared to the near point so that the image distance of the converging lens is essentially equal to the object distance for the eye (i.e., $s_1' = -s_2 = -L_{np}$). Applying the lens equation, we obtain an expression for $1/f$, sometimes called the power of the lens. Inverting this equation, we arrive at the formula for the desired focal length of the converging lens in terms of the near points of the normal eye and the hyperopic eye.

$$\frac{1}{f} = \frac{1}{s_1} + \frac{1}{s_1'} = \frac{1}{25\text{ cm}} - \frac{1}{L_{fp}} \qquad \Rightarrow \qquad f = 25\text{ cm}\left(\frac{L_{np}}{L_{np} - 25\text{ cm}}\right)$$

Note that since the near point is larger than 25 cm, the focal length must also be larger than 25 cm, as expected.

28.6 Angular Magnification

We would now like to pull together our knowledge of multiple lens systems and the human eye to discuss simple telescopes and microscopes. Clearly, the purpose of both of these instruments is to produce magnified images. The magnification that is relevant in these instruments, however, is the angular magnification.

Our sense of the size of an object is determined by the size of the image on the retina. Consequently, the magnification factor of a system of lenses that is most relevant for direct human observation is just the ratio of the angular size with the lens to the angular size without the lens, as shown in Figure 28.10.

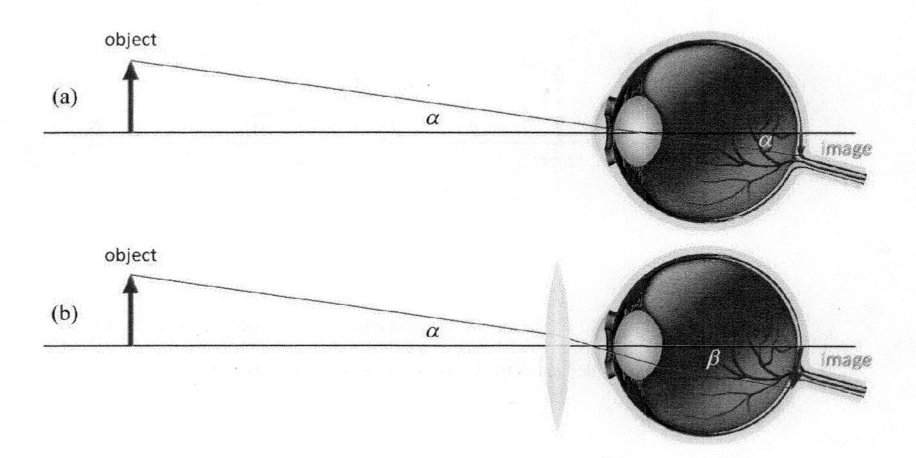

FIGURE 28.10 Our sense of the size of an object is determined by the size of the image on the retina. The addition of a lens in front of the eye, as shown above, produces a larger image on the retina ($\beta > \alpha$). This angular magnification ($M \equiv \beta / \alpha$) is the magnification relevant to the description of the functioning of telescopes and microscopes.

We'll illustrate this concept by considering an object placed inside the focal length of a converging lens, as shown in Figure 28.11. An enlarged virtual image is created and observed by a human eye. The image subtends a larger angle (β) at the eye than does the object (α). The angular magnification is defined as the ratio of β to α.

$$M \equiv \frac{\beta}{\alpha}$$

As the eye moves closer to the lens, the angular magnification decreases, approaching a limiting value of 1 when the eye is next to the lens, as shown in Figure 28.12.

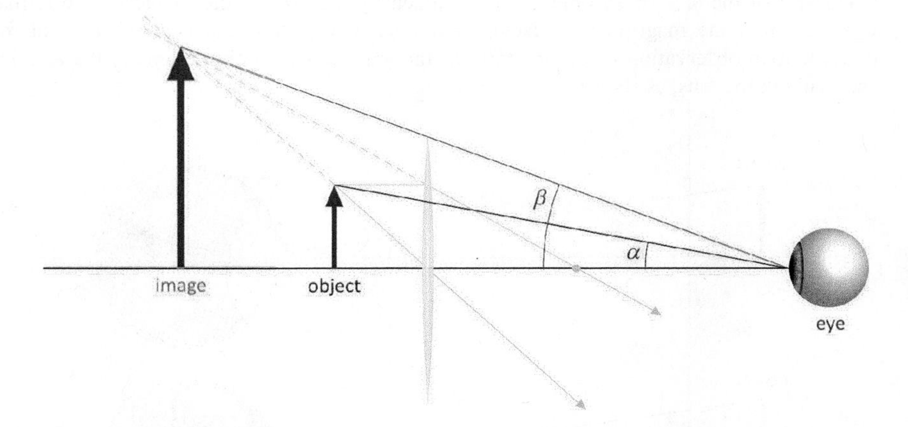

FIGURE 28.11 Rays from the tip of the object arrow placed inside the focal length of a converging lens produce a virtual image that is larger than the original object. The angular magnification is defined as the ratio of the angles β and α ($M \equiv \beta / \alpha$).

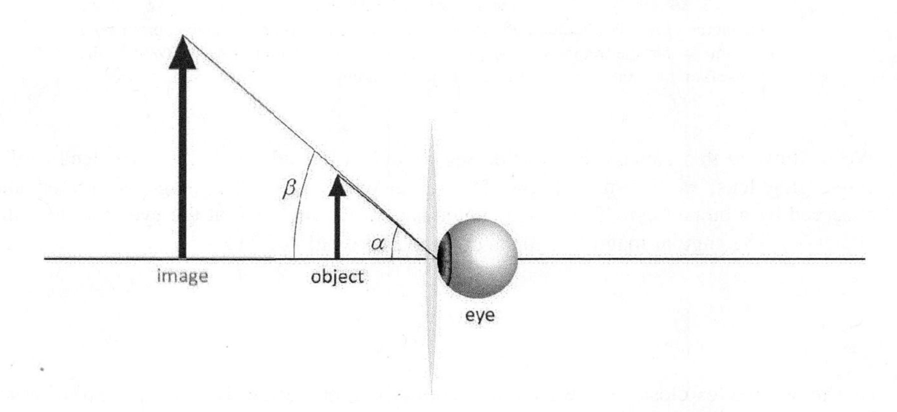

FIGURE 28.12 A limiting case of Figure 28.11: The distance between the eye and the lens goes to zero, eliminating the angular magnification from the lens ($M \equiv \beta / \alpha = 1$).

You probably know that if you hold a magnifying glass close to your eye, though, you definitely see an enlarged image of an object. How does this work? Well, the key here is that the short focal length of the magnifier can produce an image at your near point of an object placed well inside your near point. Consequently, the magnification you see is the ratio of the angle subtended by the image to the angle subtended by the object if it were placed at the near point, as shown in Figure 28.13.

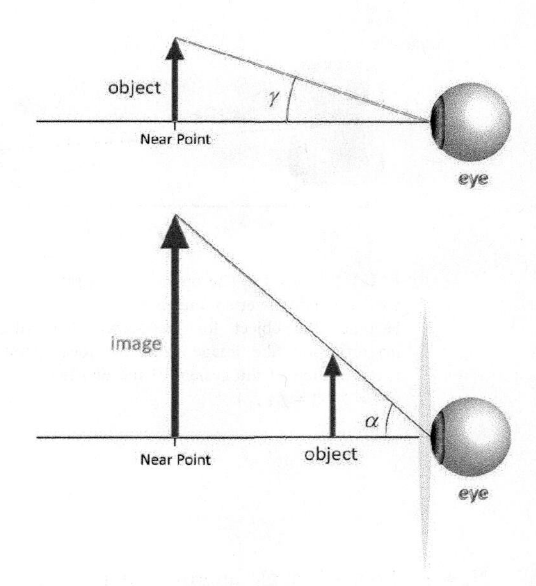

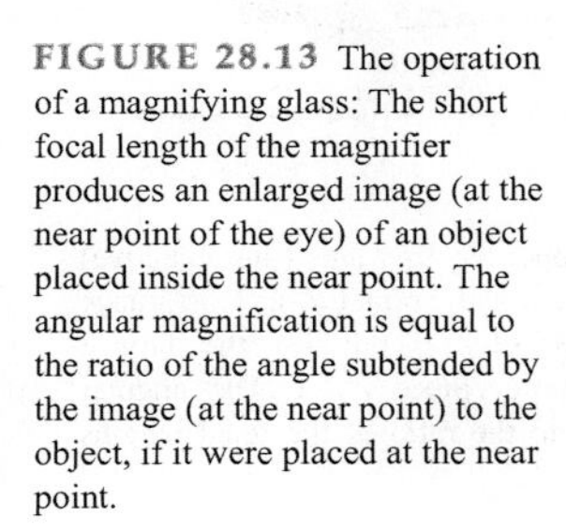
FIGURE 28.13 The operation of a magnifying glass: The short focal length of the magnifier produces an enlarged image (at the near point of the eye) of an object placed inside the near point. The angular magnification is equal to the ratio of the angle subtended by the image (at the near point) to the object, if it were placed at the near point.

28.7 Telescopes

The purpose of a **telescope** is to gather light from distant objects and produce a magnified image. There are two kinds of telescopes, reflecting and refracting. Most astronomical telescopes are reflectors, since the most important feature for these telescopes is the light-gathering ability, and it is much easier to make a large mirror than it is to make a large lens. The mirror then produces a real image at the focal point, which can be directly detected and recorded or viewed through the **eyepiece**.

FIGURE 28.14

Most terrestrial telescopes (as well as Galileo's original telescope used to discover the first four moons of Jupiter) are refracting telescopes (Figure 28.14). Figure 28.15 shows the two lenses that make up a simple refracting telescope. The first lens, the **objective**, creates a real inverted image of a distant object at its focal point. This image of size h', then, becomes the object for the eyepiece. To produce the largest magnification, we want this image to be at the focal point of the eyepiece.

To calculate the angular magnification, we first need to determine the angular size of the object without the telescope. This angular size is approximately equal to θ_1 since the object is assumed to be far away, i.e., the distance to the object is much greater than the length of the telescope. Since the eye is close to the eyepiece, the angular size of the image is approximately equal to θ_2. The angular magnification is simply the ratio of these two angles.

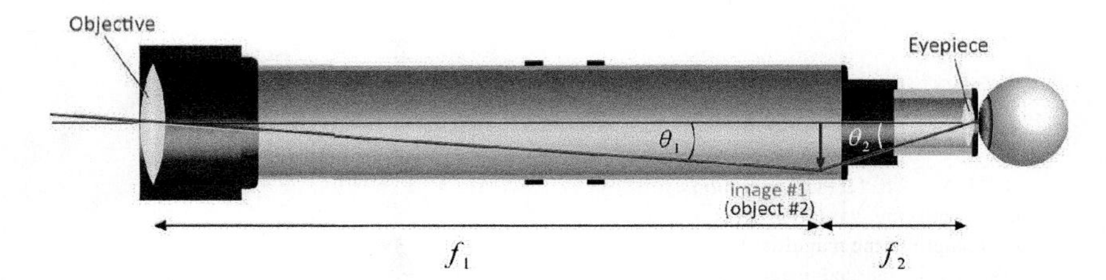

FIGURE 28.15 The operation of a refracting telescope: the first lens (the objective) creates a real inverted image of a distant object at its focal point (f_1). This image becomes the object for the second lens (the eyepiece). To produce the largest magnification, the image is at the focal point of the eyepiece (f_2). The angular magnification of the system of the two lenses is equal to the ratio of the focal lengths ($M = \theta_2 / \theta_1 = f_1 / f_2$).

$$M \approx \frac{\theta_2}{\theta_1}$$

Now in the small angle approximation, θ_1 is equal to the image height divided by the focal length of the objective and θ_2 is equal to the image height divided by the focal length of eyepiece. Therefore, the angular magnification of this telescope is just equal to the ratio of the focal lengths of the objective and the eyepiece.

$$M \approx \frac{f_1}{f_2}$$

Consequently, to produce a large magnification, we need a long telescope (f_1 large) with a strong eyepiece (f_2 small).

28.8 Microscopes

The purpose of a **microscope** is to produce a magnified image of something small and near. Figure 28.16 shows a simple compound microscope composed of two lenses, an objective and an eyepiece separated by a distance L. The arrangement of these lenses is different from what it was in the telescope, however.

The object is placed just beyond the focal point of the objective lens. A real inverted image of size h' is produced as shown. Once again, we want this image to be just inside the focal point of the eyepiece. The angular size observed by the eye (θ_2) is just equal to the ratio of the size of the image produced by the objective to the focal length of the eyepiece.

$$\theta_2 \approx \frac{h'}{f_e}$$

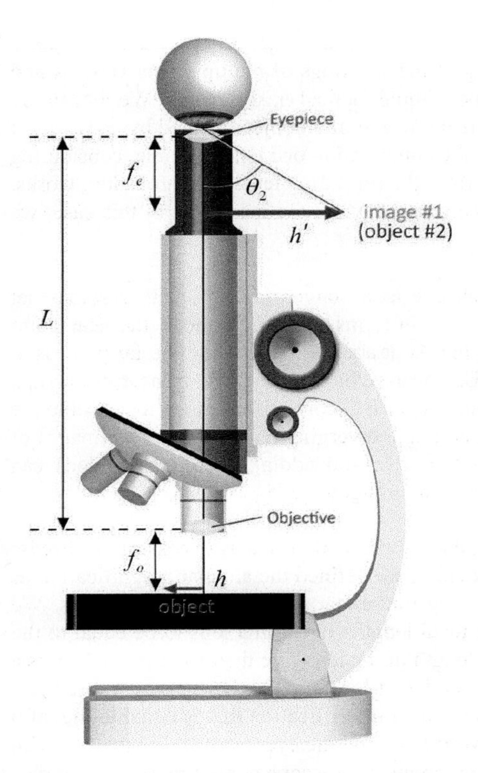

FIGURE 28.16 The operation of a simple compound microscope: The object of size h is placed just beyond the focal point (f_o) of the first lens (the objective), producing a real inverted image of size h' , which becomes the object for the second lens (the eyepiece). This original image is located just inside the focal point of the eyepiece (f_e). The angular magnification of the system of the two lenses is equal to the ratio of the angle of the original image (θ_2) to the angle of the original image had it been placed at the near point of the eye (γ).

We can determine this image size (h') from the lens equation as a function of the initial object size, the length of the telescope and the focal lengths.

$$M_1 \equiv \frac{h'}{h} = -\frac{s'}{s} = -\frac{L-f_e}{f_o} \qquad \Rightarrow \qquad h' = -h\frac{L-f_e}{f_o}$$

The angular size of the object without the microscope is just equal to the initial object size divided by the near point distance.

$$\gamma \approx \frac{h}{L_{np}}$$

The angular magnification, then, is just obtained by taking the ratio of these two angles (θ_2 and γ) to give our final result.

$$M = \frac{\theta_2}{\gamma} = -L_{np}\frac{L-f_e}{f_o f_e}$$

Therefore, to produce a large magnification, we need both the objective and the eyepiece to have small focal lengths and the distance between the objective and the eyepiece to be large.

28.9 Summary

In this unit we developed an understanding of the workings of multiple lens systems and the human eye in order to discuss some simple optical instruments. We began by introducing the procedure used to determine the image of an object formed by a system of lenses. Namely, we simply moved through the combination one lens at a time, considering the object for a given lens to be the image from the preceding lens. This procedure works, even if the image produced by one lens is located beyond the next lens. In this case, we treat the object as virtual and proceed.

We then introduced a model of the human eye as a converging lens with a somewhat variable focal length. We characterized an eye in terms of two distances, the near point and far point. For a normal eye, the near point is at about 25 cm, while the far point is at infinity. For eyes with a finite far point (i.e., near-sighted) and a distant near point (i.e., far-sighted), we determined the focal lengths of the corrective lenses that would be required. In particular, we determined that adding a diverging lens can produce images of distant objects on the retina of a near-sighted eye and adding a converging lens can produce images of near objects on the retina of a far-sighted eye.

Finally, we introduced the concept of angular magnification that is necessary to discuss simple microscopes and telescopes. In particular, we defined the angular magnification as the ratio of the angular sizes of the images on the retina with and without the lens. We found the angular magnification of a short focal length converging lens to be equal to the ratio of the near point distance to the focal length of the lens. We then went on to discuss a simple telescope in which the real image produced by the objective lens is located just inside the focal length of the eyepiece, producing a magnification that is equal to the ratio of the focal length of the objective to the focal length of the eyepiece. We concluded with a determination of the magnification of a compound microscope consisting of an objective and an eyepiece in a different arrangement, in which the object is positioned close to the focal point of the objective, producing a real image that can be focused by the eyepiece. The resulting magnification was inversely proportional to the product of the focal points of the objective and the eyepiece.

MAIN POINTS

Combinations of Lenses

To determine the final image produced by a combination of lenses, simply proceed through the combination one lens at a time, considering the object for a given lens to be the image from the preceding lens.

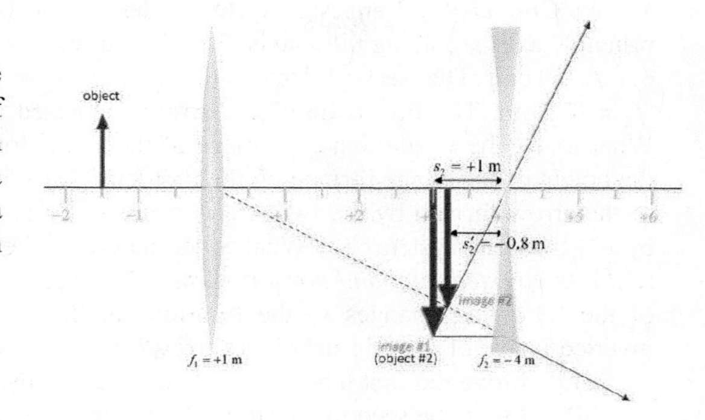

The Human Eye

The human eye is a convergent lens with a variable focal length.

The eye is specified by a near point and far point. The addition of corrective lenses can make the effective far point at infinity and the effective near point at 25 cm.

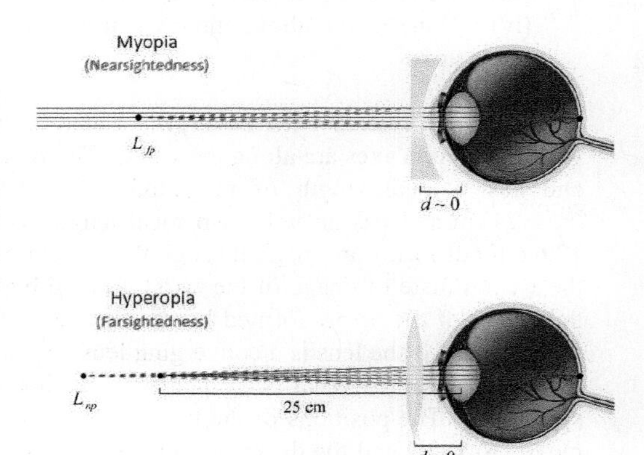

Microscopes and Telescopes

Angular magnification (the ratio of angular image size on the retina with a lens combination to that without any lenses) is used to describe microscopes and telescopes.

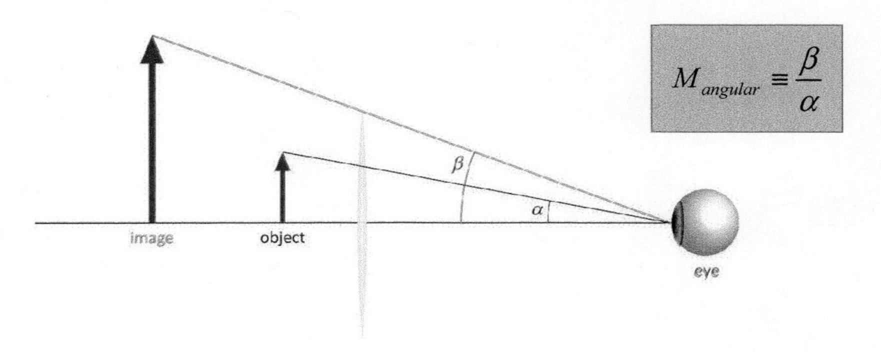

$$M_{angular} \equiv \frac{\beta}{\alpha}$$

PROBLEMS

1. Two Converging Lenses: A system of lenses consists of two converging lenses whose principle axes are along the x-axis. The first lens is located at $x = 0$ and has a focal length of $f_1 = 8$ cm. The second lens is located at $x_2 = 49$ cm and has a focal length of $f_2 = 17.3$ cm. The tip of the object arrow is located at $(x_o, y_o) = (-12$ cm, 6.7 cm). (a) What is x_1, the x coordinate of image of the arrow formed by the first lens? (b) What is the height of the image formed by the first lens? (c) What is x_3, the x coordinate of image of the arrow formed by the two-lens system? (d) What is the height of the image formed by the two-lens system? (e) What is the nature of the final image relative to the object: *real and inverted, real and upright, virtual and inverted,* or *virtual and upright*? (f) Which of the following changes to the locations of the lenses would result in a virtual and inverted image of the original object arrow?

(i) Move the first lens to $x = -8$ cm, keeping the second lens at $x = 49$ cm.
(ii) Move the second lens to $x = 57.65$ cm, keeping the first lens at $x = 0$.
(iii) Move the second lens to $x = 32.65$ cm, keeping the first lens at $x = 0$.
(iv) None of the above moves will produce a virtual inverted image of the object arrow.

2. Two-Lens System with Diverging Lens: A system of lenses consists of two lenses whose principle axes are along the x-axis. The first lens is a diverging lens located at $x = 0$ and has a focal length of magnitude $f_1 = 16.8$ cm. The second lens is located at $x_2 = 24$ cm and has an unknown focal length. An arrow located at $x_o = -34$ cm that is 17.6 cm tall forms an image through the system of lenses at $x_3 = 52$ cm. (a) What is x_1, the x coordinate of image of the arrow formed by the first lens? (b) What is the height of the image of the arrow formed by the first lens? (c) What is f_2, the focal length of the second lens? If the lens is a converging lens, f_2 is positive. If the lens is a diverging lens, f_2 is negative. (d) What is the height of the image of the arrow formed by the two-lens system? (e) The positions of the two lenses are now interchanged, i.e., the second lens is moved to $x = 0$ and the diverging lens is moved to $x_2 = 24$ cm. What is the nature of the final image in this new system: *real and inverted, real and upright, virtual and inverted,* or *virtual and upright*?

APPENDIX A: NUMERICAL DATA

Some Fundamental Physical Constants*

Avogadro's number	N_A	$6.0221415(10)\times10^{23}$ particles/mol
Coulomb constant	k	$8.987551788...\times10^{9}\ \mathrm{N\cdot m^2/C^2}$
Electron rest mass	m_e	$9.10938215(45)\times10^{-31}$ kg
Elementary charge	e	$1.602176487(40)\times10^{-19}$ C
Gravitational constant	G	$6.67428(67)\times10^{-11}\ \mathrm{N\cdot m^2/kg^2}$
Neutron rest mass	m_n	$1.674927211(84)\times10^{-27}$ kg
Permeability constant	μ_o	$4\pi\times10^{-7}\ \mathrm{N/A^2}$
Permittivity constant	ε_o	$8.85418781...\times10^{-12}\ \mathrm{C^2/(N\cdot m^2)}$
Proton rest mass	m_p	$1.672621637(83)\times10^{-27}$ kg
Speed of light in a vacuum	c	299,792,458 m/s

*The values for these constants may be found on the Internet at http://physics.nist.gov/cuu/Constants/index.html. The numbers in the parenthesis represent the uncertainties in the last two digits. For example, the number 6.67428(67) equals 6.67428 ± 0.00067 . Values without parenthesis represent exact values without uncertainties.

Astronomical Data*

Earth	
Mass	5.97×10^{24} kg
Radius	6.37×10^{6} m
Distance from Sun†	1.496×10^{11} m
Moon	
Mass	7.35×10^{22} kg
Radius	1.737×10^{6} m
Period	27.32 days
Distance from Earth†	3.844×10^{8} m
Acceleration of gravity at surface	$1.62\ \mathrm{m/s^2}$
Sun	
Mass	1.99×10^{30} kg
Radius	6.96×10^{8} m

*Data for our solar-system can be found on the Internet at http://nssdc.gsfc.nasa.gov/planetary/planetfact.html.
†Center to center.

APPENDIX B: SI UNITS

Base Units*

Meter (m) The *meter* is the length of the path travelled by light in vacuum during a time interval of $1/299{,}792{,}458$ of a second.

Kilogram (kg) The *kilogram* is the unit of mass; it is equal to the mass of the international prototype of the kilogram.

Second (s) The *second* is the duration of 9,192,631,770 periods of the radiation corresponding to the transition between the two hyperfine levels of the ground state of the cesium 133 atom.

Ampere (A) The *ampere* is that constant current that, if maintained in two straight parallel conductors of infinite length, of negligible circular cross-section, and placed 1 meter apart in vacuum, would produce between these conductors a force equal to 2×10^{-7} newton per meter of length.

Kelvin (K) The *kelvin* is the fraction $1/273.16$ of the thermodynamic temperature of the triple point of water.

Candela (cd) The *candela* is the luminous intensity, in a given direction, of a source that emits monochromatic radiation of frequency 540×10^{12} hertz and that has a radiant intensity in that direction of $1/683$ watt per steradian.

Mole (mol) The *mole* is the amount of substance of a system that contains as many elementary entities as there are atoms in 0.012 kilogram of carbon 12.

*These definitions are found on the Internet at http://physics.nist.gov/cuu/Units/current.html

Derived Units

Quantity	Unit	Definition
Force	newton (N)	$1\ \text{N} = 1\ \text{kg}\cdot\text{m/s}^2$
Work (Energy)	joule (J)	$1\ \text{J} = 1\ \text{N}\cdot\text{m}$
Power	watt (W)	$1\ \text{W} = 1\ \text{J/s}$
Frequency	hertz (Hz)	$1\ \text{Hz} = 1\ \text{cycle/s}$
Pressure	pascal (Pa)	$1\ \text{Pa} = 1\ \text{N/m}^2$
Charge	coulomb (C)	$1\ \text{C} = 1\ \text{A}\cdot\text{s}$
Potential	volt (V)	$1\ \text{V} = 1\ \text{J/C}$
Capacitance	farad (F)	$1\ \text{F} = 1\ \text{C/V}$
Current	ampere (A)	$1\ \text{A} = 1\ \text{C/s}$
Resistance	ohm (Ω)	$1\ \Omega = 1\ \text{V/A}$
Magnetic field	tesla (T)	$1\ \text{T} = 1\ \text{N/(A}\cdot\text{m)}$
Magnetic flux	weber (Wb)	$1\ \text{Wb} = 1\ \text{T}\cdot\text{m}^2$
Inductance	henry (H)	$1\ \text{H} = 1\ \text{J/A}^2$

APPENDIX C: CONVERSION FACTORS

Length
- 1 m = 3.281 ft = 39.37 in
- * 1 in = 2.54 cm
- * 1 ft = 12 in = 30.48 cm
- 1 mi = 5280 ft = 1.609 km
- 1 km = 0.6214 mi

Time
- * 1 h = 60 min = 3600 s
- * 1 day = 24 h = 1440 min = 86,400 s
- 1 yr = 365.25 days = 3.156×10^{7} s

Angular Measurement
- * π rad = 180° = ½ rev
- 1 rad = 57.30° = 0.1592 rev
- 1° = 0.01745 rad

Speed
- 1 km/h = 0.2778 m/s = 0.6215 mi/h
- 1 mi/h = 0.4470 m/s = 1.609 km/h

Acceleration
- 9.8 m/s^2 = 32.2 ft/s^2

Force
- 1 N = 0.2248 lb
- * 1 lb = 4.448222 N
- A 1-kg mass weighs 2.203 lb, where g = 9.80 m/s^s

Pressure
- * 1 Pa = 1 N/m^2
- * 1 atm = 1.01325×10^{5} Pa
- 1 atm = 760 mmHg = 29.9 inHg
- 1 bar = 10^5 Pa
- 1 lb/in^2 = 6.895×10^{3} Pa

Energy
- 1 J = 0.7376 ft · lb
- * 1 cal = 4.184 J
- 1 kW · h = 3.6×10^{6} J
- 1 eV = 1.602×10^{-19} J

Power
- * 1 W = 1 J/s
- 1 W = 0.7376 ft · lb/s
- 1 W = 1.341×10^{-3} hp
- 1 hp = 550 ft · lb/s = 745.7 W

*Exact values.

INDEX